Compendium of Hydrogen Energy

Related titles

Solid-State Hydrogen Storage
(ISBN 978-1-84569-270-4)

Solid Oxide Fuel Cell Technology
(ISBN 978-1-84569-628-3)

Polymer Electrolyte Membrane and Direct Methanol Fuel Cell Technology Volume 1: Fundamentals and Performance of Low Temperature Fuel Cells
(ISBN 978-1-74569-773-0)

Woodhead Publishing Series in Energy:
Number 86

Compendium of Hydrogen Energy

Volume 4
Hydrogen Use, Safety and the Hydrogen Economy

Edited by

Michael Ball, Angelo Basile,
T. Nejat Veziroğlu

ELSEVIER

AMSTERDAM • BOSTON • CAMBRIDGE • HEIDELBERG
LONDON • NEW YORK • OXFORD • PARIS • SAN DIEGO
SAN FRANCISCO • SINGAPORE • SYDNEY • TOKYO
Woodhead Publishing is an imprint of Elsevier

WP
WOODHEAD
PUBLISHING

Woodhead Publishing is an imprint of Elsevier
80 High Street, Sawston, Cambridge, CB22 3HJ, UK
225 Wyman Street, Waltham, MA 02451, USA
Langford Lane, Kidlington, OX5 1GB, UK

ISBN: 978-1-78242-364-5 (print)
ISBN: 9781782423867 (online)

British Library Cataloguing-in-Publication Data
A catalogue record for this book is available from the British Library

Library of Congress Control Number: 2015937280

For information on all Woodhead Publishing publications visit our website at http://store.elsevier.com/

Transferred to Digital Printing in 2016

Contents

List of contributors

J.R. Anstrom The Pennsylvania State University, University Park, PA, USA

M. Ball Shell Global Solutions International, HW Amsterdam, The Netherlands

U. Bünger Ludwig-Bölkow-Systemtechnik GmbH (LBST), Ottobrunn, Germany

F. Crotogino KBB Underground Technologies GmbH (KBB), Hannover, Germany

K. Dutta University of Calcutta, Kolkata, India

A. Elgowainy Argonne National Laboratory, Argonne, IL, USA

N. Frischauf CERN, Geneva, Switzerland

A. Godula-Jopek Airbus Group Innovations, Munich, Germany, and Institute of Chemical Engineering, Polish Academy of Sciences, Gliwice, Poland

M. Harada INPEX Corporation, Minato-ku, Tokyo, Japan

T. Ichikawa Hiroshima University, Higashi-Hiroshima, Japan

R. Judd On Behalf of GERG, European Gas Research Group, Brussels, Belgium

M. Kaltschmitt Hamburg University of Technology (TUHH), Hamburg, Germany

O. Kruck KBB Underground Technologies GmbH (KBB), Hannover, Germany

P.P. Kundu University of Calcutta, Kolkata, India

G. Manjavacas Ortiz Centro Nacional del Hidrógeno, Puertollano, Spain

J. Michalski Ludwig-Bölkow-Systemtechnik GmbH (LBST), Ottobrunn, Germany

M. Mintz Argonne National Laboratory, Argonne, IL, USA

B. Nieto Calderón Centro Nacional del Hidrógeno, Puertollano, Spain

D. Pinchbeck On Behalf of GERG, European Gas Research Group, Brussels, Belgium

K. Reddi Argonne National Laboratory, Argonne, IL, USA

G. Sattler Independent Consultant, Lübeck, Germany

M. Steen European Commission, Petten, the Netherlands

E. Sutherland Fuel Cell Technologies Office, Washington, DC, USA

H. Takagi National Institute of Advanced Industrial Science and Technology, Tsukuba, Ibaraki, Japan

H. Uchida Tokai University, Hiratsuka-City, Kanagawa, Japan

F. Vogler Independent Consultant, Bad Schwartau, Germany

M. Weeda Energy Research Centre of the Netherlands (ECN), Petten, The Netherlands

A. Westenberger Airbus Operations GmbH, Hamburg, Germany

J. Wind Daimler AG, Kirchheim/Teck-Nabern, Germany

C. Wulf Hamburg University of Technology (TUHH), Hamburg, Germany

R. Wurster Ludwig-Bölkow-Systemtechnik GmbH, Ottobrunn, Germany

Woodhead Publishing Series in Energy

1 **Generating power at high efficiency: Combined cycle technology for sustainable energy production**
Eric Jeffs

2 **Advanced separation techniques for nuclear fuel reprocessing and radioactive waste treatment**
Edited by Kenneth L. Nash and Gregg J. Lumetta

3 **Bioalcohol production: Biochemical conversion of lignocellulosic biomass**
Edited by Keith W. Waldron

4 **Understanding and mitigating ageing in nuclear power plants: Materials and operational aspects of plant life management (PLiM)**
Edited by Philip G. Tipping

5 **Advanced power plant materials, design and technology**
Edited by Dermot Roddy

6 **Stand-alone and hybrid wind energy systems: Technology, energy storage and applications**
Edited by John K. Kaldellis

7 **Biodiesel science and technology: From soil to oil**
Jan C. J. Bart, Natale Palmeri and Stefano Cavallaro

8 **Developments and innovation in carbon dioxide (CO_2) capture and storage technology Volume 1: Carbon dioxide (CO_2) capture, transport and industrial applications**
Edited by M. Mercedes Maroto-Valer

9 **Geological repository systems for safe disposal of spent nuclear fuels and radioactive waste**
Edited by Joonhong Ahn and Michael J. Apted

10 **Wind energy systems: Optimising design and construction for safe and reliable operation**
Edited by John D. Sørensen and Jens N. Sørensen

11 **Solid oxide fuel cell technology: Principles, performance and operations**
Kevin Huang and John Bannister Goodenough

12 **Handbook of advanced radioactive waste conditioning technologies**
Edited by Michael I. Ojovan

13 **Membranes for clean and renewable power applications**
Edited by Annarosa Gugliuzza and Angelo Basile

14 **Materials for energy efficiency and thermal comfort in buildings**
Edited by Matthew R. Hall

15 **Handbook of biofuels production: Processes and technologies**
Edited by Rafael Luque, Juan Campelo and James Clark

16 **Developments and innovation in carbon dioxide (CO_2) capture and storage technology Volume 2: Carbon dioxide (CO_2) storage and utilisation**
Edited by M. Mercedes Maroto-Valer
17 **Oxy-fuel combustion for power generation and carbon dioxide (CO_2) capture**
Edited by Ligang Zheng
18 **Small and micro combined heat and power (CHP) systems: Advanced design, performance, materials and applications**
Edited by Robert Beith
19 **Advances in clean hydrocarbon fuel processing: Science and technology**
Edited by M. Rashid Khan
20 **Modern gas turbine systems: High efficiency, low emission, fuel flexible power generation**
Edited by Peter Jansohn
21 **Concentrating solar power technology: Principles, developments and applications**
Edited by Keith Lovegrove and Wes Stein
22 **Nuclear corrosion science and engineering**
Edited by Damien Féron
23 **Power plant life management and performance improvement**
Edited by John E. Oakey
24 **Electrical drives for direct drive renewable energy systems**
Edited by Markus Mueller and Henk Polinder
25 **Advanced membrane science and technology for sustainable energy and environmental applications**
Edited by Angelo Basile and Suzana Pereira Nunes
26 **Irradiation embrittlement of reactor pressure vessels (RPVs) in nuclear power plants**
Edited by Naoki Soneda
27 **High temperature superconductors (HTS) for energy applications**
Edited by Ziad Melhem
28 **Infrastructure and methodologies for the justification of nuclear power programmes**
Edited by Agustín Alonso
29 **Waste to energy conversion technology**
Edited by Naomi B. Klinghoffer and Marco J. Castaldi
30 **Polymer electrolyte membrane and direct methanol fuel cell technology Volume 1: Fundamentals and performance of low temperature fuel cells**
Edited by Christoph Hartnig and Christina Roth
31 **Polymer electrolyte membrane and direct methanol fuel cell technology Volume 2: *In situ* characterization techniques for low temperature fuel cells**
Edited by Christoph Hartnig and Christina Roth
32 **Combined cycle systems for near-zero emission power generation**
Edited by Ashok D. Rao
33 **Modern earth buildings: Materials, engineering, construction and applications**
Edited by Matthew R. Hall, Rick Lindsay and Meror Krayenhoff
34 **Metropolitan sustainability: Understanding and improving the urban environment**
Edited by Frank Zeman
35 **Functional materials for sustainable energy applications**
Edited by John A. Kilner, Stephen J. Skinner, Stuart J. C. Irvine and Peter P. Edwards
36 **Nuclear decommissioning: Planning, execution and international experience**
Edited by Michele Laraia
37 **Nuclear fuel cycle science and engineering**
Edited by Ian Crossland

38 **Electricity transmission, distribution and storage systems**
Edited by Ziad Melhem
39 **Advances in biodiesel production: Processes and technologies**
Edited by Rafael Luque and Juan A. Melero
40 **Biomass combustion science, technology and engineering**
Edited by Lasse Rosendahl
41 **Ultra-supercritical coal power plants: Materials, technologies and optimisation**
Edited by Dongke Zhang
42 **Radionuclide behaviour in the natural environment: Science, implications and lessons for the nuclear industry**
Edited by Christophe Poinssot and Horst Geckeis
43 **Calcium and chemical looping technology for power generation and carbon dioxide (CO_2) capture: Solid oxygen- and CO_2-carriers**
Paul Fennell and E. J. Anthony
44 **Materials' ageing and degradation in light water reactors: Mechanisms, and management**
Edited by K. L. Murty
45 **Structural alloys for power plants: Operational challenges and high-temperature materials**
Edited by Amir Shirzadi and Susan Jackson
46 **Biolubricants: Science and technology**
Jan C. J. Bart, Emanuele Gucciardi and Stefano Cavallaro
47 **Advances in wind turbine blade design and materials**
Edited by Povl Brøndsted and Rogier P. L. Nijssen
48 **Radioactive waste management and contaminated site clean-up: Processes, technologies and international experience**
Edited by William E. Lee, Michael I. Ojovan, Carol M. Jantzen
49 **Probabilistic safety assessment for optimum nuclear power plant life management (PLiM): Theory and application of reliability analysis methods for major power plant components**
Gennadij V. Arkadov, Alexander F. Getman and Andrei N. Rodionov
50 **The coal handbook: Towards cleaner production Volume 1: Coal production**
Edited by Dave Osborne
51 **The coal handbook: Towards cleaner production Volume 2: Coal utilisation**
Edited by Dave Osborne
52 **The biogas handbook: Science, production and applications**
Edited by Arthur Wellinger, Jerry Murphy and David Baxter
53 **Advances in biorefineries: Biomass and waste supply chain exploitation**
Edited by Keith Waldron
54 **Geological storage of carbon dioxide (CO_2): Geoscience, technologies, environmental aspects and legal frameworks**
Edited by Jon Gluyas and Simon Mathias
55 **Handbook of membrane reactors Volume 1: Fundamental materials science, design and optimisation**
Edited by Angelo Basile
56 **Handbook of membrane reactors Volume 2: Reactor types and industrial applications**
Edited by Angelo Basile
57 **Alternative fuels and advanced vehicle technologies for improved environmental performance: Towards zero carbon transportation**
Edited by Richard Folkson

58 **Handbook of microalgal bioprocess engineering**
Christopher Lan and Bei Wang
59 **Fluidized bed technologies for near-zero emission combustion and gasification**
Edited by Fabrizio Scala
60 **Managing nuclear projects: A comprehensive management resource**
Edited by Jas Devgun
61 **Handbook of Process Integration (PI): Minimisation of energy and water use, waste and emissions**
Edited by Jiří J. Klemeš
62 **Coal power plant materials and life assessment**
Edited by Ahmed Shibli
63 **Advances in hydrogen production, storage and distribution**
Edited by Ahmed Basile and Adolfo Iulianelli
64 **Handbook of small modular nuclear reactors**
Edited by Mario D. Carelli and Dan T. Ingersoll
65 **Superconductors in the power grid: Materials and applications**
Edited by Christopher Rey
66 **Advances in thermal energy storage systems: Methods and applications**
Edited by Luisa F. Cabeza
67 **Advances in batteries for medium and large-scale energy storage**
Edited by Chris Menictas, Maria Skyllas-Kazacos and Tuti Mariana Lim
68 **Palladium membrane technology for hydrogen production, carbon capture and other applications**
Edited by Aggelos Doukelis, Kyriakos Panopoulos, Antonios Koumanakos and Emmanouil Kakaras
69 **Gasification for synthetic fuel production: Fundamentals, processes and applications**
Edited by Rafael Luque and James G. Speight
70 **Renewable heating and cooling: Technologies and applications**
Edited by Gerhard Stryi-Hipp
71 **Environmental remediation and restoration of contaminated nuclear and NORM sites**
Edited by Leo van Velzen
72 **Eco-friendly innovation in electricity networks**
Edited by Jean-Luc Bessede
73 **The 2011 Fukushima nuclear power plant accident: How and why it happened**
Yotaro Hatamura, Seiji Abe, Masao Fuchigami and Naoto Kasahara. Translated by Kenji Iino
74 **Lignocellulose biorefinery engineering: Principles and applications**
Hongzhang Chen
75 **Advances in membrane technologies for water treatment: Materials, processes and applications**
Edited by Angelo Basile, Alfredo Cassano and Navin Rastogi
76 **Membrane reactors for energy applications and basic chemical production**
Edited by Angelo Basile, Luisa Di Paola, Faisal Hai and Vincenzo Piemonte
77 **Pervaporation, vapour permeation and membrane distillation: Principles and applications**
Edited by Angelo Basile, Alberto Figoli and Mohamed Khayet
78 **Safe and secure transport and storage of radioactive materials**
Edited by Ken Sorenson

79 **Reprocessing and recycling of spent nuclear fuel**
Edited by Robin Taylor
80 **Advances in battery technologies for electric vehicles**
Edited by Bruno Scrosati, Juergen Garche and Werner Tillmetz
81 **Rechargeable lithium batteries: From fundamentals to applications**
Edited by Alejandro A. Franco
82 **Calcium and chemical looping technology for power generation and carbon dioxide (CO_2) capture**
Edited by Paul Fennell and Ben Anthony
83 **Compendium of Hydrogen Energy Volume 1: Hydrogen Production and Purificiation**
Edited by Velu Subramani, Angelo Basile and T. Nejat Veziroğlu
84 **Compendium of Hydrogen Energy Volume 2: Hydrogen Storage, Distribution and Infrastructure**
Edited by Ram Gupta, Angelo Basile and T. Nejat Veziroğlu
85 **Compendium of Hydrogen Energy Volume 3: Hydrogen Energy Conversion**
Edited by Frano Barbir, Angelo Basile and T. Nejat Veziroğlu
86 **Compendium of Hydrogen Energy Volume 4: Hydrogen Use, Safety and the Hydrogen Economy**
Edited by Michael Ball, Angelo Basile and T. Nejat Veziroğlu

Part One

Hydrogen applications in transport and industry

Hydrogen-fueled road automobiles – Passenger cars and buses

1

J. Wind
Daimler AG, Kirchheim/Teck-Nabern, Germany

1.1 Introduction

In this chapter, the use of hydrogen in road vehicle propulsion is described. Today, all of the main original equipment manufacturers (OEMs) are developing fuel cell electric vehicles (FCEVs) directly fueled with hydrogen. The goal of the most advanced OEMs is to enter the market between 2015 and 2018 (Toyota Motor Corporation, 2015a; Honda AG, 2015a; Daimler AG, 2015a; Thomas Reuters Deutschland GmbH, 2015). Hydrogen internal combustion engine vehicles (H_2-ICE) are no longer considered to be a reasonable solution for the future of road mobility (Die Welt, n.d.). Thus, hydrogen-fueled (fuel cell electric) vehicles are pure electrical vehicles (EVs).

Due to the very high efficiency of all drive-train components used in FCEVs, the vehicle's overall efficiency is about twice the efficiency of a gasoline- or diesel-fueled vehicle with an internal combustion engine (ICE) (Europäische Kommission, 2013). Further, hydrogen-fueled FCEVs emit no polluting components nor any greenhouse gas (GHG). FCEVs, when fueled with hydrogen, are true zero-emission vehicles (ZEVs) and thus have a very high potential to contribute significantly to the reduction of GHGs and energy (fuel) consumption.

Unlike battery electric vehicles (BEVs), which are also pure EVs and ZEVs, FCEVs are suitable for large driving ranges and can be fueled in a few minutes. However, the overall efficiency of BEVs is significantly higher. Vehicles combining ICEs and electric drive trains, such as hybrid electric vehicles (HEV), plug-in hybrid electric vehicles and range extender electric vehicles, also contribute to a reduction of GHG emissions and energy consumption due to increasing efficiency with increase of the electrification rate. However, those drive trains all need fossil fuels for the ICE and emit harmful air polluting components as well as GHG emissions. This chapter focuses on FCEVs for the reasons explained here.

1.2 Comparison of different hydrogen-fueled drive systems

Hydrogen as a fuel for propulsion of road vehicles was considered in the early nineteenth century, when Francois Isaac de Rivaz built the world's first hydrogen-fueled vehicle in 1807 (Ginlex and Cahen, 2011). It took more than a century for OEMs to

Compendium of Hydrogen Energy. http://dx.doi.org/10.1016/B978-1-78242-364-5.00001-4

take up the idea of using hydrogen as a fuel and develop several prototypes of passenger cars and buses with H_2-ICEs or fuel cells as the drive train. From the 1970s on, development of alternative drive trains and use of alternative fuels have gained increasing interest. The goal was (and still is) to reduce GHG emissions, energy consumption and dependency on fossil fuels (Wallentowitz et al., 2003).

ICEs using hydrogen as a fuel are very similar to ICEs using other fuels. The H_2-ICE drive train consists of the modified ICE itself, gear box, transmission and the hydrogen storage system. Besides the modification of the ICE, the main difference is the hydrogen storage system, which is much larger, heavier and more complex than a gasoline or diesel tank. The technological effort needed to develop a H_2-ICE vehicle is much lower than that to develop a complete FCEV. Thus, it was obvious for most OEMs to start with H_2-ICEs.

Most car manufacturers stopped their H_2-ICE development in the 1990s, although others continued their work until about 2010 (Beissmann, n.d.). The main reason for terminating the work on H_2-ICEs was the high well-to-wheel (WTW) energy consumption, when the production of hydrogen is taken into account in addition to the energy consumption of driving. A H_2-ICE has about the same efficiency as an ICE fueled with diesel, which is in the range of 24% in the New European Driving Cycle (NEDC) (Rosbach, 2012). As the production of hydrogen and delivery to the filling station causes significant energy losses (approximately 50%) (Joint Research Centre, 2014), the overall energy consumption of a vehicle using a H_2-ICE calculated from WTW is much higher than the WTW energy consumption of an ICE vehicle using diesel or gasoline as a fuel. Further, it is also necessary to store large amounts of hydrogen on board the vehicle when reasonable driving ranges are expected. This is not possible with currently available hydrogen storage systems, especially in passenger cars, due to the large space needs of any hydrogen storage system. Thus, the range of vehicles with a H_2-ICE is limited and does not meet customer expectations.

FCEVs offer many advantages compared to ICE vehicles. However, the drive train of an FCEV is much more complex and the effort to develop it is greater than the effort for a H_2-ICE. Figure 1.1 shows the main components of an FCEV. The vehicle is driven by an electric motor (power range typically around 100 kW peak power), which is supplied with electricity from the fuel cell system (power range slightly below 100 kW) and possibly a small battery (power range up to 40 kW) (Braess and Seiffert, 2013). The electric motors used in modern EVs are mainly alternating current (AC) motors. Thus, an inverter, converting the direct current (DC) produced by the fuel cell system into AC, is needed.

The main power source of an FCEV is the fuel cell system, which is fueled with hydrogen. To store the necessary amount of hydrogen on board the vehicle, a hydrogen storage system is needed. Currently, the solution chosen by almost all OEMs is a composite (carbon fibre wrapped metal or plastic cylinders) storage system of type IV for gaseous hydrogen with a storage pressure of 700 bars to provide for a sufficient driving range. This system consists of the pressure cylinders, valves, sensors and piping.

Finally, all current FCEVs have a small battery on board, which is used to store the recuperated braking energy and to allow for an optimized operation of the fuel cell system. Additionally, a number of sensors and controllers is needed.

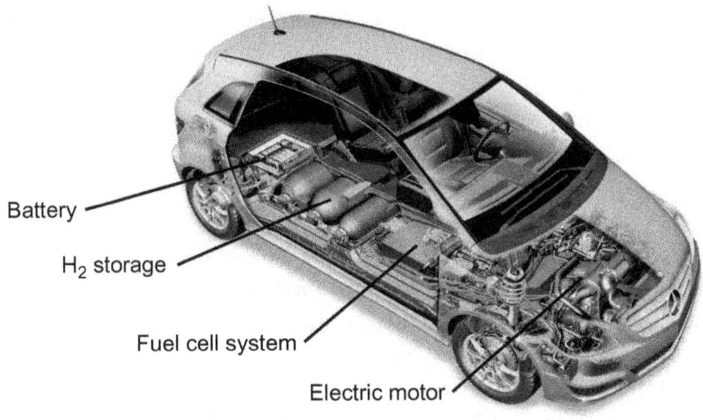

Figure 1.1 Main components of an FCEV.
Source: Daimler (2008).

Due to the very high efficiency of all components of an FCEV, the overall tank-to-wheel efficiency is twice as high as that of an ICE. Thus, the fuel consumption (on energy content basis) is about half the consumption of ICEs (Joint Research Centre, 2011a). As a result, the WTW energy consumption of the whole chain (including fuel production) is still lower than that of a vehicle with an ICE (Joint Research Centre, 2011b). Further, an FCEV does not emit anything except water vapour, provided it is fueled with hydrogen. Due to the low hydrogen consumption, the amount of hydrogen needed to be stored on board the vehicle is much less than in a H_2-ICE, leading to reasonable driving ranges of more than 500 km.

The costs of FCEVs are still significantly higher than those for the current vehicles with ICEs that dominate the market. As the costs of the hydrogen storage system are also still too high, vehicles with a H_2-ICE would also not yet be able to achieve the cost targets needed for economic success (McKinsey, 2010).

1.3 Technical solutions for FCEVs

The design and development of FCEVs is rather complex due to the variety of available components on the one hand and the possibility of combining a fuel cell system with a battery on the other hand. Different concepts for FCEVs have been developed and tested or are still being tested, either inside private companies or in publicly funded research demonstration projects. Roughly, the concepts can be classified into four different categories:

- Lean FCEV without any battery (no longer considered).
- Hybridized FCEVs with a small battery (currently the concept preferred by most OEMs), now called simply FCEV, since pure FCEVs are no longer being considered.
- Plug-in FCEVs combining a slightly smaller FC system and a larger, externally rechargeable battery.
- Fuel cell range extenders, which are BEVs with a slightly smaller chargeable battery and a small FC system to recharge the battery in order to extend the range of the BEV.

Certainly, some FCEV concepts will not fit completely into one of these categories, as they might have properties of more than one of the concepts. In the following sections all four concepts are described, followed by a description of the different options for the main components of FC drive trains.

1.3.1 Lean FCEV

Until the early 1990s only a few prototypes of FCEVs (mainly passenger cars and buses) had been developed and built. Most of those did not consider the use of a battery in the vehicle at all. Thus, the FC drive train of these lean FCEVs consisted of the FC system, hydrogen storage system, inverter and electric motor. One example of such a vehicle is Daimler's NECAR 1 from 1994 (Figure 1.2).

This prototype was the first FCEV built by an OEM that was certified to run on public roads (Wallentowitz and Reif, 2011). It is one of the most important milestones in FCEV development, as its successful use on normal roads and the announcement by Daimler in 1997 that it would aim at a market introduction of FCEVs motivated many other OEMs to start development of FCEVs with the goal of market introduction. Thus, the NECAR 1 can be considered the start of a number of large development programs for FCEVs. The OEMs soon realized the advantages of combining the FC system with a small battery, which are described in the next section. The main components of such a puristic FCEV are already listed in Section 1.2.

1.3.2 Hybridized FCEV

Electric vehicles with a high power FC system and a small battery are the mainstream of FCEV development today. The battery type in this concept is a power battery, which is designed to provide high power in a short time. Figure 1.3 shows a selection of such FCEVs of the latest generation from different car manufacturers. Table 1.1 lists the main properties of these vehicles. As can be seen, the concepts are very similar in the dimensioning of the main components.

Figure 1.2 Daimler's NECAR 1 from 1994.
Source: Daimler AG (2015b).

Figure 1.3 Daimler B-Class F-CELL, General Motors HydroGen 4, Honda FCX Clarity, Hyundai ix35 Fuel Cell and Toyota Mirai.
Sources: Opel AG (2015), Honda AG (2015b), Daimler AG (2009), Hyundai Motor Company (2015a) and Toyota Motor Corporation (2015b).

As the current–voltage characteristics of the FC system and the battery are different, it is necessary to integrate at least one DC/DC converter in the drive system in order to adjust the voltage of one of the electricity sources to the other. Usually the additional DC/DC converter is applied to the battery and the battery voltage adjusted to that supplied by the FC system. In some concepts, two DC/DC converters are applied, which leads to higher costs and weight of the drive train.

Compared to the lean FCEVs in the early phase of FCEV development, the addition of a battery to the FC drive train offers a number of advantages for a small additional effort. In principle, the advantages are very similar to those which are used to reduce fuel consumption in HEVs. All EVs can recuperate braking energy in principle by generating electricity during braking, using the electric motor as a generator. However, an FC system cannot store the generated electricity. The electricity generated during braking can be stored in the small battery. Typically these batteries have a storage capacity of about 1 kWh of electricity.

Another advantage is the possibility of using both electricity sources for propulsion, allowing the FC system to be operated in its optimal operation regions. Usually the battery provides additional electricity to the FC system or can even be used as the only electricity source in the low power region, where the FC system has a rather low efficiency. If the electricity is provided by the battery, the FC system can be shut off, avoiding the low efficiency area.

Another possibility is to operate the fuel cell system in a higher power area where the efficiency is high and store the excess electricity that is not needed for propulsion in the battery. Further, the battery can be used during acceleration and generally in the very higher power area, where the efficiency of the FC system is again lower. Figure 1.4 shows a typical efficiency curve of an FC system, indicating the described effects. All these operation modes lead to an increase of the total efficiency of the FCEV and less demanding dynamics for the FC system. In addition, cold start ability is also better.

1.3.3 Plug-in FCEV

A plug-in FCEV is a hybridized FCEV with a significantly larger (larger than 1 kWh) battery (high energy density type), which can be charged from the electricity grid with an on-board charger. In this type of FCEV the FC system is still the main power source, but it is also possible to drive certain distances only using the battery. Due to the high efficiency of the battery, this leads to a significant reduction of energy

Table 1.1 Main properties of a selection of FCEVs

Vehicle		Daimler B-Class F-CELL	General Motors HydroGen 4	Honda FCX Clarity	Hyundai ix35 Fuel Cell	Toyota Mirai
Release date	a	2009	2008	2008	2012	2015
Fuel cell stack	kW	90	93	100	100	114
Battery	kW	24	35	40	24	N/A
Electric motor	kW	70	73	100	100	113
H_2 storage	kg H_2	3.7	4.2	4	5.64	~5
H_2 consumption	kg H_2/100 km	0.97	1.3125	0.87	0.95	~0.71[a]
Driving range NEDC[b]	km	385	320	460	594	700 (JC08[a])
Acceleration 0–100 km	s	11.4	12	9	12.5	9.6
Maximum Speed	km/h	170	160	160	160	178

[a] Japanese driving cycle.
[b] NEDC=New European driving cycle.
Sources: Hyundai Motor Company (2015a,b), Stolten and Grube (2010), Honda AG (2015c,d), Daimler AG (2015c,d) and Toyota Motor Corporation (2015c).

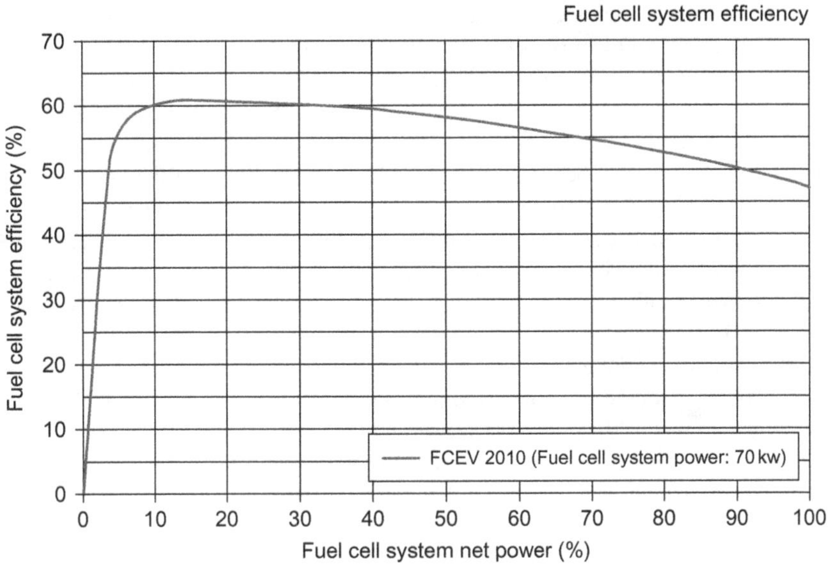

Figure 1.4 FC system efficiency curve.
Source: Europäische Kommission (2013).

Figure 1.5 Daimler's F 125! F-CELL Plug-in HYBRID, Audi A7 h-tron and Symbio FCell's Kangoo ZE-H_2 (range extender).
Sources: Daimler AG (2015e), Audi AG (2014) and Symbio Fcell (2015).

consumption. As the power and energy content of the larger battery is considerable, it is also possible (but not necessary) to reduce the size of the FC system moderately. Figure 1.5 shows some examples of plug-in FCEVs from different car manufacturers. At this point, only a few OEMs are working on this concept.

1.3.4 Range extender FCEV

The components integrated into a range extender FCEV (RE-FCEV) are in principle the same as those used in a plug-in FCEV. However, their dimensioning is very different. The range extender FCEV is derived from a BEV. In the extreme case, a small power FC system (5–30 kW) is simply added to the drive train of an already developed

BEV, while other concepts are more adjusted to the combination of a large battery and a small FC system. An example of an RE-FCEV is shown in Figure 1.5.

A range extender FCEV is not really an adequate FCEV. The FC system provides a longer driving range for the BEV and helps to overcome the drawbacks of BEVs. In particular, when the FC system power is in the lower range, the driving characteristics of the vehicle change significantly as soon as the state of charge of the battery is at the minimum limit. Usually the battery in a BEV provides enough power to the electric motor to have reasonable acceleration and top speed. If the battery of a range extender FCEV is empty and the electricity is supplied by the small FC system only, the acceleration and top speed of the vehicle are much lower. This is a significant disadvantage of this concept. However, range extender FCEVs do provide an easier way to integrate a fuel cell system in a vehicle, as the requirements on the FC system are lower than for a full-power FCEV. Usually companies and research institutes who develop RE-FCEVs do not need to develop the complete electric drive train. It is possible in principle to add a small FC system to a BEV that is already available on the market. However, in the long run, the cost of RE-FCEVs will be higher than full power FCEVs.

1.4 Technical approaches for the main components of FCEVs

1.4.1 Fuel cell stacks

In the early days of FCEV development, a number of different FC technologies were installed in road vehicles. Karl Kordesch used an alkaline fuel cell stack in his famous car in 1970, and others used phosphoric acid fuel cell stacks (Eichelseder and Klell, 2012). Proton exchange membrane fuel cell stacks have been most commonly used in FCEVs, and today this is the only technology considered by the industry for FCEVs. The reasons for this include the very high power density of the technology, which is a prerequisite for packaging the stacks in the limited space available, as well as some severe problems with the other technologies in a road vehicle that uses air to supply the cathode side of the fuel cells. High temperature fuel cell stacks, like solid oxide fuel cells and molten carbonate fuel cells, are far too large and heavy. Further, they need to be heated up to rather high temperatures when starting the vehicle, which is considered to be a show stopper. The materials used for the components of the PEM fuel cells in FCEVs are described in other chapters of this book.

1.4.2 Air supply systems

Fuel cell stacks for high-power applications in passenger cars and buses usually are supplied with pressurized ambient air. The air supply system consists of filters, sensors, flow meters, humidification system and a compression system, which is the main component of the air supply system (Töpler and Lehmann, 2014). For compression, several solutions have been realized in FCEVs so far. The most important ones are screw compressors and electric turbo chargers (ETCs).

Figure 1.6 Electrical turbo chargers (ETCs).
Source: Daimler AG (n.d.).

ETCs are driven either by a small electric motor or additionally by a turbine using the energy of the cathode exhaust. The advantage of the ETC is the lower power consumption and the smaller size and weight compared to a screw compressor (Wallentowitz and Freildenhoven, 2011). An example of an ETC is shown in Figure 1.6. Humidification is necessary in a PEMFC as the membrane materials used need a minimum degree of water for proton conduction. Further, the lifetime of the membrane material is shortened if the membranes are too dry (Kurzweil, 2013).

Most FC concepts use an external humidifier, where the air supplied to the FC stack is humidified before it enters the stack. As the external humidifier adds cost and takes up space, it is desirable to avoid this component in the FC system. Several groups have worked on FC systems without an external humidifier for more than 10 years (Williams et al., 2004). Recently, one car manufacturer presented an FCEV without an external humidifier. In this case, the membranes need to be humidified internally. Water produced in the stack is directly used to humidify the membrane, that is, by active recirculation of the anode exhaust. This concept causes technical challenges for a homogeneous humidity distribution in the fuel cells.

Several approaches for external humidification have been applied. The current state of the art uses a gas-to-gas humidifier. The exhaust gas from the cathode, with a high water content, is used to humidify the air supplied to the stack. Gas-to-gas humidifiers are based on membranes that are impenetrable to liquid water while allowing water vapour molecules to pass through. Plate and frame humidifiers, as well as humidifiers using hollow fibres, have been developed and used in automotive FC systems. Examples of both concepts are shown in Figure 1.7.

1.4.3 Hydrogen supply

PEMFC stacks need pure hydrogen in order to operate. Thus, it is necessary to either store pure hydrogen on board the vehicle or to produce it from another fuel carried

Figure 1.7 Gas-to-gas humidifier cartridge.
Source: Daimler AG (2015f).

on board. On-board production of hydrogen has been used mainly in the last century in some FCEV prototypes, such as Daimler's NECAR 3 and NECAR 5 (Mohrdieck et al., 2014). These vehicles used methanol as a fuel and an on-board steam methane reformer to produce hydrogen from methanol. This adds a high degree of complexity to the drive train. This strategy is no longer followed by any car manufacturer; today, all FCEVs use pure hydrogen stored in an on-board hydrogen storage system.

Due to the very low volumetric energy density of hydrogen, quite a design effort is needed to store the amount of hydrogen necessary for a reasonable driving range. Several hydrogen storage technologies for FCEVs have been developed and tested. Up to now, the only suitable ways to store significant amounts of hydrogen in a vehicle have been either liquid hydrogen storage or compressed hydrogen storage. A new approach, called cryo-compressed storage, is a storage technology that combines both. However, this technology has not yet reached the same technological maturity as the others. Despite long and intensive design and development work at many research institutes and in industry, no other hydrogen storage technology (such as storage in hydrides) has been developed to the maturity level needed in automotive applications (Sorensen et al., 2005).

A complex and expensive storage system is required to store liquid hydrogen at temperatures below 20.28 K. Even with very good insulation systems using a combination of materials with low thermal conductivity and vacuums, evaporation of hydrogen cannot be prevented totally, leading to considerable hydrogen losses. The hydrogen storage system will be empty after some weeks, even with the best available insulation. This is one of the major drawbacks of on-board liquid hydrogen storage. Another disadvantage is the very high energy consumption needed for the liquefaction of hydrogen, which is in the range of 10 kWh/kg of hydrogen, compared to a total energy content of 33.3 kWh of 1 kg of hydrogen (LBST, 2015).

Today all road vehicle manufacturers are using composite storage tanks to store compressed hydrogen with a pressure of 350 bars or 700 bars. 700-bar storage is used in most passenger cars as the available space is limited. For buses, 350 bars of pressure provides a suitable option due to the greater space available in or on the bus. The compressed hydrogen storage systems consist of one or more composite cylinders, valves, sensors and regulators. Type IV cylinders consisting of an inner liner made from plastics and a carbon fibre reinforcement filled with resin are used most commonly (Eichelseder and Klell, 2012).

1.4.4 Electric motors and power electronics

As with all EVs, FCEVs are driven by one or more electric motors. Several different electric motor technologies can be used in principle. Most FCEVs today use permanent magnet synchronous motors due to their high power and torque density (Wallentowitz and Freildenhoven, 2011). As these need to be supplied with AC voltage, an inverter that transforms the DC voltage supplied by the FC system into AC voltage is needed. Additionally, at least one DC/DC converter is needed to convert the voltage supplied by the battery to the voltage of the FC system. Some approaches even use two DC/DC converters, one for the battery and one for the FC system. The technology for automotive electric motors and power electronics is, in the meantime, very mature, with efficiencies in the area of 95% for these components (Wallentowitz and Freildenhoven, 2011).

1.4.5 Batteries

As has been described previously, adding a battery to the FC drive train leads to a number of advantages and opens new opportunities for the operation of FCEVs. In the case of FC hybrids, the batteries are optimized to provide high power (approximately 30 kW), but the energy content of the battery is rather low (in the range of 1 kWh). The same battery type is also used in HEVs with an ICE. Plug-in FCEVs and range extender FCEVs usually use batteries optimized for high energy density rather than high power, as the much larger size of the batteries in these vehicles provides enough power and the additional energy in the battery can be used to advantage.

As in ICE-HEVs, the technology used initially for the hybrid batteries was nickel-hydride, whereas the latest generation of FCEVs uses lithium–ion (Li–ion) batteries, as all modern EVs do. The Li–ion technology provides a very high energy density, which is needed especially for BEVs. The energy content of the batteries used in FCEVs varies from 1 kWh to several tens of kWh; see also Section 1.2 (Braess and Seiffert, 2013).

1.5 Challenges for FCEVs – Consideration of main markets

FCEVs are not yet commercially available. The reasons are mainly the lack of the necessary hydrogen infrastructure and the high costs of the FC drive train (McKinsey, 2010). FCEVs compete with ICE driven vehicles as well as with other electrified vehicles. The maturity of FCEVs has already been shown in a number of demonstration projects with FC passenger cars and FC buses. Millions of kilometers have been driven successfully with FCEVs. Reliability and availability of the vehicles are both high enough for commercialization. However, some technological advancements are still needed: for example, an increase in the lifetime of the FC system. An even more challenging issue is to reduce the cost of the FC drive train. All types of EVs are still significantly more expensive than ICE vehicles, which is a barrier for commercial success for all EVs. Thus, it is not only the FC system and H_2 storage system which are not yet within the necessary cost range, but also other components like electric motors, power electronics and batteries. The cost for an FCEV has to approach, if not meet, the cost of today's ICE vehicles.

Three measures are needed to achieve the cost goals for commercialization. Firstly, additional technological advances will lead to further cost reduction. Secondly, the buildup of a broader and more capable supplier landscape for FC components will provide another step in cost reduction. Thirdly, an increase in mass production is another step towards achieving the cost goal. In 2013, a number of cooperative agreements between car manufacturers were announced. Toyota is cooperating with BMW; Daimler, Ford and Nissan have agreed to work together; General Motors and Honda announced their cooperation; and Ballard is supporting Volkswagen. All these agreements aim toward increasing production volume, sharing development efforts and motivating the suppliers to increase their activities in developing and producing components for FCEVs.

Today's FCEVs are fueled with pure hydrogen, but hydrogen refuelling stations (HRSs) are still very scarce. About 200 HRSs have been installed worldwide; around 85 of these are located in Europe, approximately 80 in the United States (mainly California) and about 50 in the Asia Pacific region, with a focus on Japan (FCH JU, 2014). Without an adequate number of refuelling stations to cover at least the first main markets for FCEVs, at a minimum, FCEVs cannot be introduced successfully on the market.

Thus, hydrogen infrastructure has to be built in a timely manner. The production of hydrogen is less of a problem, and the same is true for the transport of hydrogen to the HRS. Both technologies are already commercial for other applications of hydrogen. It is the HRSs which are needed. As the number of vehicles that need to be filled with hydrogen is still very low, the business case for HRSs is negative at the moment and is in the first phase of market introduction of FCEVs, mainly because of the low station utilization.

As a consequence, infrastructure companies are hesitant to invest in HRSs. On the other hand, FCEVs will not be commercially successful until potential customers are confident that they will find enough places to fuel their vehicles. To overcome this chicken-and-egg problem, new and joint approaches are needed. In 2009, car manufacturers, oil companies and gas companies started the H_2 Mobility Germany initiative with the goal of building up and operating HRSs in Germany. In the meantime six companies – Air Liquide, Daimler, Linde, OMV, Shell and Total – signed an agreement and founded a joint venture with the goal to build up and operate 400 HRSs in Germany until 2023. This is an important milestone in the history of FCEVs. Other countries in the world followed the German example and have also started similar initiatives. The most notable initiatives have been undertaken in the United Kingdom, Norway, Denmark, Sweden, France, Switzerland, Japan and the United States.

Additional measures to promote commercialization of FCEVs are needed. All European major players in the field of fuel cell and hydrogen technology have joined forces in the fuel cell and hydrogen joint undertaking (FCH JU), which was started in 2007. The Council of the European Union and Parliament as well as industry and academia have decided to continue it until at least 2020. The partners of the FCH JU are the European Commission, the industry grouping (NEW-IG) and research grouping (N.ERGHY).

The goal of the FCH JU is clearly commercialization of fuel cell and hydrogen technologies until 2020. The available funding to support the industry and research institutes in publicly funded projects from 2014 until 2020 is 645 million €. About one-third is dedicated to projects for FC and H_2 technologies in road transport, including HRSs (FCH JU, 2014).

Direct funding of research, development and demonstration projects is one important, but not the only, element for achieving commercialization. Further measures, like subsidizing the purchase of FCEVs, subsidies for production plants, customer incentives, tax reductions and temporary privileges (like access to bus lanes, free parking, etc.), are considered to be viable instruments. Some countries, like Denmark and Sweden, are exempting all EVs from registration taxes, which is significant in both of these countries. Manufacturers who sell FCEVs in these markets already benefit from these measures. Similar efforts are being carried out in other regions of the world, notably the United States and Japan, where FCEVs and HRSs are either co-funded or subsidized.

1.6 Summary and future trends

Road transportation is in a phase of significant changes. The future will see a growing share of FCEVs on the roads. The first products will be passenger cars, vans, buses and two-wheelers. The technology for the vehicles as well as for the hydrogen infrastructure is ripe for the market. Significant cost reductions have been achieved (Wallentowitz and Freildenhoven, 2011). Many industrialized countries are considering FCEVs as one of the most promising solutions for sustainable mobility of the future. Leaders are Germany, the Scandinavian countries, United Kingdom, France, Japan and the United States.

The current generation of FCEVs has achieved a technical maturity and performance that fulfils all customer expectations. Nevertheless, there are still significant efforts needed to achieve full commercialization. The FC drive trains need to become significantly lower in cost, lighter, smaller, more reliable and have an even longer lifetime in the next 10 years. This will be achieved through further research and development in industry and research organizations as well as through mass manufacturing of components and FCEVs.

The most important areas for technological work are identified and research and development programs, like the FCH JU with the European Commission, are already in place. Table 1.2 shows the most important targets for FCEVs as defined by the FCH JU.

Table 1.2 **Most important targets for FCEVs as defined by the FCH-JU**

Application	Parameter	Unit	2012	FCH-JU target		
				2017	2020	2023
Fuel cell electric passenger cars	Specific FC system cost	€/kW	>500	150	100	75
	Assumed number of units (per year) as cost calculation basis			*20,000*	*50,000*	*100,000*
	FC vehicle cost (C-segment)	k€	200	70	50	30
	Tank-to-wheel efficiency(vehicle in New European drive cycle)	%	40	42	45	48
	Availability	%	95	98	98	99
	FC system lifetime	h	2500	5000	6000	7000
Fuel cell electric buses	Specific FC system cost	€/kW	<3500	<1,800,750	1,000,500	800,400
	FC Bus System Lifetime	h	10000	15,000 2×8000	20,000 2×10,000	25,000 2×12,500
	FC Bus cost	k€	1300	700	650	500
	Fuel consumption (vehicle, average of SORT1[a] and SORT2[a] cycle)	kg H_2/100 km	9	8.51	8	7.59
	Availability	%	85	90	95	99
	Assumed number of units (per year) as cost calculation basis			<50	200	>500

Hydrogen storage	Hydrogen storage system cost	€/kg H_2	>3000	800	600	500
	Volumetric capacity(H_2 tank system)	kg/L	0.02	0.022	0.023	0.025
	Gravimetric capacity(H2 tank system)	%	<4	4	5	6
Hydrogen supply	Cost of hydrogen delivered to HRS*	€/kg	5.0[b]–>13[c]	5.0–11	5.0–9.0	4.5–7.0
	Hydrogen refuelling stations cost[d]	M€	1.5–3.5	1.0–2.5	0.8–2.1	0.6–1.6

[a] SORT: Standardised on road test cycle, defined test cycle for buses.

[b] Hydrogen from centralized steam methane reforming. Achieving these targets will be influenced by the evolution of the cost of natural gas. This parameter should be taken into account when assessing progress against this key performance indicator.

[c] Renewable hydrogen, either on site or centralized.

[d] 700 bar HRS with 200–1000 kg/day capacity including on-site storage.

* *Source:* Based on the data from the report "A portfolio of power-trains for Europe: a fact-based analysis, McKinsey 2010."

These targets are considered by the partners of the FCH JU to be the prerequisites for full commercialization of FCEVs.

Concerning costs for FCEVs, it is expected that a cost range similar to HEVs will be achieved around 2020. Given that the necessary minimum HRS infrastructure is also available, full commercialization can start in this timeframe. Numbers of FCEVs will then grow significantly, contributing considerably to the reduction of GHG emissions and energy consumption. FCEVs will then be a part of the production portfolio of every vehicle manufacturer. The variety of available FC drive trains will grow steadily and reach a status where all necessary power classes for vehicles can be supplied. The customers will be able to choose the FC drive train as one of the options for every type of vehicle. According to several studies, the growing number of vehicles will also lead to a positive business case for the HRS operators. The investment in HRS technology in the early years will pay off and a dense network of energy refuelling stations offering hydrogen will be a familiar part of road infrastructure.

1.7 Sources of further information and advice

Books

Eichlseder, H., Klell, M., Wasserstoff in der Fahrzeugtechnik: Erzeugung, Speicherung, Anwendung; mit 26 Tabellen. 3., überarb. Springer Vieweg: Aufl. Wiesbaden, 2012.

Töpler, J., Lehmann, J., Wasserstoff und Brennstoffzelle: Technologien und Marktperspektiven. Imprint: Springer Vieweg: Berlin, Heidelberg, 2014.

Wallentowitz, H., Freiladenhoven, A., Strategien zur Elektrifizierung des Strategien zur Elektrifizierung des Antriebsstranges: Technologien, Märkte und Implikationen. 2., überarbeitete Auflage. Vieweg+Teubner Verlag/Springer Fachmedien Wiesbaden GmbH: Wiesbaden, 2011.

Korthauer, R., Handbuch Lithium-Ionen-Batterien. Imprint: Springer Vieweg: Berlin, Heidelberg, 2013.

Stolten, D., Emonts, B., Fuel Cell Science and Engineering Materials, Processes, Systems and Technology, vol. 1, Wiley-VCH Verlag & Co. KGaA: Weinheim, 2012.

Stolten, D., Emonts, B., Fuel Cell Science and Engineering Materials, Processes, Systems and Technology, vol. 2, Wiley-VCH Verlag & Co. KGaA: Weinheim, 2012.

Major trade/professional bodies

NOW Nationale Organisation Wasserstoff- und Brennstoffzellentechnik is responsible for the coordination and control of the National Innovation Programme for Hydrogen and Fuel Cell Technology (NIP) and the Programme Electromobility Model Regions of the Federal Ministry of Transport and digital infrastructure (BMVI) of Germany.

CEP Clean Energy Partnership is the largest demonstration project for hydrogen mobility in Europe and a flagship project of the National Innovation Programme for Hydrogen and Fuel Cell Technology (NIP) in the transport sector.

H_2 Mobility H_2 Mobility is a hydrogen refuelling network to grow to about 400 filling stations by 2023. The six partners in the "H_2 Mobility" initiative – Air Liquide, Daimler, Linde, OMV, Shell and Total – have set upon a specific action plan for the construction of a nationwide hydrogen refuelling network for FCEVs.

FCH JU Fuel cell and hydrogen joint ndertaking (FCH JU) is a unique public private partnership supporting research, technological development and demonstration (RTD) activities in fuel cell and hydrogen energy technologies in Europe. Its aim is to accelerate the market introduction of these technologies, realizing their potential as an instrument in achieving a carbon-lean energy system. The three members of the FCH JU are the European Commission, fuel cell and hydrogen industries represented by the NEW Industry Grouping and the research community represented by Research Grouping N.ERGHY.

Websites

http://www.fuelcells.org/
http://www.netinform.net/H2/
www.fch-ju.eu
http://www.ukh2mobility.co.uk/the-project/
http://www.nrel.gov/hydrogen/
http://www.hydrogen.energy.gov/
http://ieahia.org/
http://h2usa.org/
http://cleanenergypartnership.de
http://www.dwv-info.de/
http://www.now-gmbh.de/de/
http://www.fuelcellpartnership.org/
http://www.iphe.net/
http://www.h2euro.org/

References

Audi AG, 2014. Audi A7 Sportback h-tron quattro. http://www.audi.de/de/brand/de/vorsprung_durch_technik/content/2014/11/audi-a7-sportback-h-tron-quattro.html (2015-03-24).

Beissmann, T. BMW abandons internal combustion hydrogen technology. http://www.caradvice.com.au/51099/bmw-abandons-internal-combustion-hydrogen-technology/.

Braess, H.-H., Seiffert, U., 2013. Handbuch Kraftfahrzeugtechnik. Springer Fachmedien, Wiesbaden.

Daimler AG, 2009. Mercedes Benz B-Klasse F-CELL. Stuttgart.

Daimler AG. Mehr Schub für die Brennstoffzellentechnologie: strategische Kooperation der Daimler AG und Renault-Nissan Allianz trifft Abkommen mit Ford. http://www.daimler.com/dccom/0-5-7171-49-1569731-1-0-0-0-0-0-8-0-0-0-0-0-0-0-0.html (2015a-02-12).

Daimler AG. NECAR 1. http://media.daimler.com/Projects/c2c/channel/images/656398_1177482_400_267_08C537_26.jpg (12.02.2015b).

Daimler AG. Antriebssysteme für elektrische angetriebene Fahrzeuge. http://www.dlr.de/tt/Portaldata/41/Resources/dokumente/ess_2012/Wind_Antriebssysteme_elektrische_Fahrzeuge.pdf (12.02.2015c).
Daimler AG. B-Klasse F-CELL. Emissionsfrei mit Elektroantrieb und Brennstoffzelle. http://www.mercedes-benz-freiburg.de/content/germany/retailer-7/niederlassung_freiburg/my_library/media_library/downloads/rep_Anlage2_zur_PI_Fahrzeuguebergabe_Dieter_Salomon_pdf.object-Single-MEDIA.tmp/Anlage_2_Broschre_zur_B-Klasse_F-CELL.pdf (12.02.2015d).
Daimler AG. F 125! F-CELL. http://media.daimler.com/Projects/c2c/channel/images/884881_1657055_400_267_11C1220_015.jpg (26.02.2015e).
Daimler AG. Alternative drive systems. http://www.daimler.com/dccom/0-5-1324141-1-1324804-1-0-0-1324800-0-1-8-7145-0-0-0-0-0-0-0.html (12.02.2015f).
Daimler AG. Electric turbo charger (ETC). Nabern.
Daimler, A.G., 2008. Mercedes Benz B-Klasse F-CELL cross sectional model. Daimler AG, Stuttgart.
Die Welt. BMW glaubt nicht mehr an den Wasserstoffmotor, http://www.welt.de/motor/article5450945/BMW-glaubt-nicht-mehr-an-den-Wasserstoffmotor.html.
Eichelseder, H., Klell, M., 2012. Wasserstoff in der Fahrzeugtechnik. Vieweg+Teubner Verlag, Wiesbaden.
Europäische Kommission, 2013. TANK-TO-WHEELS Report Version 4.0. Publications Office of the European Union, Luxemburg.
Fuel Cells and Hydrogen Joint Undertaking (FCH JU), 2014. Multi-Annual Work Plan 2014–2020. European Commission, Brussels.
Ginlex, D.S., Cahen, D., 2011. Fundamentals of Materials for Energy and Environmental Sustainability. Cambridge University, Cambridge.
Honda AG. Neues Brennstoffzellenfahrzeug enthüllt. http://www.honda.de/news/carnews_neues_brennstoffzellenfahrzeug_enthuellt_21-11-2014.php (12.02.2015a).
Honda AG. Honda FCX clarity. http://world.honda.com/FCXClarity/design/image/photo06.jpg (26.02.2015b).
Honda AG. Honda FCX clarity 1. http://automobiles.honda.com/fcx-clarity/specifications.aspx (12.02.2015c); http://www.hondazeroemissions.com/.
Honda AG. Honda FCX clarity 2. http://world.honda.com/FCXClarity/specifications/index.html (12.02.2015d).
Hyundai Motor Company. Hyundai ix35 Fuel Cell 1. http://www.hyundai.de/CMSPages/GetFile.aspx?guid=c0668f42-aa52-4482-906a-65c8169882ef&width=0&height=330 (26.02.2015a).
Hyundai Motor Company. Hyundai ix35 Fuel Cell 2. http://www.hyundai.de/downloads/modell_prospekte/Hyu_ix35_Fuel_Cell_20140515.aspx (12.02.2015b).
Joint Research Centre, 2011a. EUCAR; CONCAWE: Well-to-Tank Report Version 3c, Brussels.
Joint Research Centre, 2011b. EUCAR; CONCAWE: Well-to-Wheels Report Version 3c, Brussels.
Joint Research Centre, 2014. EUCAR; CONCAWE: Well-to-Tank Report Version 4.a, Brussels.
Kurzweil, P., 2013. Brennstoffzellentechnik. Springer Vieweg, Wiesbaden.
LBST. Hydrogen Filling Stations Worldwide. http://www.netinform.net/H2/H2Stations/H2Stations.aspx?Continent=AS&StationID=-1 13.02.2015).
McKinsey, 2010. A portfolio of power-trains for Europe: a fact-based analysis – Die Rolle von batteriebetriebenen Elektrofahrzeugen, Plug-In Hybridfahrzeugen und Brennstoffzellenfahrzeugen.

Mohrdieck, C., et al., 2014. Mobile Anwendungen. In: Töpler, J., Lehmann, J. (Eds.), Wasserstoff und Brennstoffzelle. Technologien und Marktperspektiven. Springer Vieweg, Berlin, Heidelberg, pp. 59–112.

Opel AG. GM/Opel HydroGen 4. http://media.opel.de/media/de/de/opel/photos.detail.html/content/Pages/galleries/intl/en/opel/vehicles/alternative-mobility/hydrogen4.html (2015-02-26).

Rosbach, M., 2012. Untersuchung der Bildung und Oxidation von Ruß in Motoren mit Direkteinspritzung. KIT Scientific Publishing, Karlsruhe.

Sorensen, B.M., Srensen, B., Anon, B., 2005. Hydrogen and Fuel Cells. Emerging Technologies and Applications. Academic Pr Inc, Oxford.

Stolten, D., Grube, T., 2010. HydroGen4 the first year of operation in europe. In: 18th World Hydrogen Energy Conference 2010 – WHEC 2010, Essen.

Symbio Fcell. Symbio FCell's Kangoo ZE-H2, received directly from Symbio FCell (2015-03-24).

Thomas Reuters Deutschland GmbH. Daimler, Nissan und Ford bündeln Kräfte für E-Autos. http://www.handelsblatt.com/unternehmen/industrie/brennstoffzellen-daimler-nissan-und-ford-buendeln-kraefte-fuer-e-autos/7697594.html (12.02.2015).

Töpler, J., Lehmann, J., 2014. Wasserstoff und Brennstoffzelle. Springer-Verlag, Berlin.

Toyota Motor Corporation. Toyota Mirai: die Mobilität der Zukunft! http://www.toyota.de/innovation/mirai.json (12.02.2015a).

Toyota Motor Corporation. Toyota Mirai 1. http://newsroom.toyota.co.jp/en/download/4224903 (26.02.2015b).

Toyota Motor Corporation. Toyota Mirai 2. http://newsroom.toyota.co.jp/en/download/4224903 (12.02.2015c).

Wallentowitz, H., Freildenhoven, A., 2011. Strategien zur Elektrifizierung des Antriebsstranges. Vieweg+Teubner Verlag, Wiesbaden.

Wallentowitz, H., Reif, K., 2011. Handbuch Kraftfahrzeugelektronik. Springer+Teubner Verlag, Wiesbaden.

Wallentowitz, H., Johannaber, M., Schüssler, M., 2003. Hybrid–Elektro–Antriebssysteme–eine Übersicht. Deutsche Gesellschaft für elektrische Straßenfahrzeuge e.V, Aachen.

Williams, M.V., Kunz, H.R., Fenton, J.M., 2004. Operation of Nafion®-based PEM fuel cells with no external humidification: influence of operating conditions and gas diffusion layers. J. Power Sources 135 (1–2), 122–134.

Hydrogen-fueled motorcycles, bicycles, and industrial trucks

2

J.R. Anstrom
The Pennsylvania State University, University Park, PA, USA

2.1 Introduction

Both motorcycle and industrial truck markets present an early opportunity for fuel cell manufacturers to build economies of scale by developing and deploying small modular fuel cells, which can later be scaled up for the more challenging automotive market. The global market for motorcycles is forecast to reach $85 billion by 2015 (King, 2013). Likewise, the global market for material handling equipment is projected to exceed $122 billion by 2018, of which industrial trucks make up a large portion (Companiesandmarkets.com, 2014). These two markets are significantly smaller than the global automotive market and therefore present lower barriers to market entry for early adoption of emerging fuel cell technology.

The integration of fuel cells into electric motorcycles provides a rapid refueling capability and extended range, reducing the "range anxiety" that typically limits consumer interest in electric vehicles. Three-wheeled motorcycles may represent a more compatible platform for fuel cell integration because of the additional space for bulky components. Electric and hydrogen fuel cell motorcycles will benefit from a consumer perception of the technologies as green and environmentally friendly. Electric and hydrogen fuel cell motorcycles have zero tailpipe emissions compared to gasoline-powered motorcycles and are also much quieter. Ironically, the loud and distinctive exhaust note of conventional gasoline motorcycles is considered part of their appeal to by recreational riders, as demonstrated by the 1994–2000 litigation between Harley-Davidson and competitors regarding Harley's attempt to trademark the characteristic exhaust note of V-twin engines (Smit and Van Wyk, 2011). Nevertheless, a certain market segment will value quiet and green motorcycle products. This is again demonstrated by Harley-Davidson, which in June 2014 unveiled their first electric motorcycle prototype, dubbed "Project LiveWire," with a new characteristic sound some have likened to a whirring jet engine (Harley-Davidson, 2014). The exhilarating performance of electric and fuel cell motorcycles, especially high acceleration provided by electric motor torque, will continue to motivate new products and attract consumer attention. As a case in point, the Isle of Man TT Zero electric motorcycle class has shown astonishing progress by eclipsing previous combustion-powered records up through the 1980s in just 6 years of racing (Lavrinc, 2013).

A niche market has arisen in the United States for industrial trucks powered by modular fuel cell systems, which mimic the dimensions and electrical connections of the standard lead acid batteries they replace. This market is now expanding into

Compendium of Hydrogen Energy. http://dx.doi.org/10.1016/B978-1-78242-364-5.00002-6

Europe and Asia. As a commercial application, material handling is less price sensitive than the consumer transportation market to relatively high fuel cell system costs of $14,000–30,000 per unit, versus $2600–5500 for lead acid batteries (Jerram, 2014). This is true so long as the overall business case for vehicle fleet purchase, maintenance, fueling, and overhead is lower for fuel cells than for the lead acid battery case. Commercial applications also have greater access to financing than consumers, so the high initial cost of fuel cell systems may not be prohibitive. This includes the cost to install and maintain an on-site hydrogen fueling infrastructure to refuel a captive fleet of industrial trucks.

One of the greatest benefits of hydrogen fuel cell integration into material handling operations is rapid refueling compared to long charge times and/or exchange operations required for batteries. Hydrogen refueling takes an industrial truck operator only about 3 min. The hydrogen fueling dispenser occupies very little space within a crowded warehouse compared to battery charging or exchange facilities. The bulk of hydrogen refueling equipment, including storage and compressor, can be located outside of the warehouse floor space. Another important attribute of fuel cells is their constant voltage output compared to batteries. As batteries discharge, their voltage drops significantly, placing additional strain on electrical components, slowing production, and requiring higher levels of maintenance. Lastly, there is a public relations benefit for commercial and government warehouse facilities that adopt hydrogen fuel cell material handling equipment, as the public perceives hydrogen as a green and environmentally friendly technology. This impression is based on the quiet and clean operation of fuel cells and the lack of hazardous chemicals or emissions.

2.2 Hydrogen motorcycles and bicycles

The attraction of fuel cells for transportation is their potential for improved fuel economy coupled with low carbon footprint and zero tailpipe emissions when compared to conventional petroleum-fueled vehicles. As a useful thought exercise, consider the extreme fuel economy achievable by fuel cell and battery-powered vehicles compared to their conventional counterparts at the limits of light weighting, downsizing, aerodynamics, and powertrain efficiency. Examples of these extremely fuel-efficient vehicles have been demonstrated during several international competitions, including the Progressive X PRIZE, the Society of Automotive Engineering (SAE) Supermileage® student competition, and the Shell Eco-marathon competition. The 2008–2010 Progressive X PRIZE was an open competition that awarded teams a total of $10 million for production-capable vehicles exceeding 42.5 km/L (100 miles per U.S. gallon gasoline equivalent, mpgge) in three classes: Mainstream four seat, Side by Side two seat, and Tandem two seat. These vehicles are intended to be practical and highway capable with accommodations for comfort and safety. The Mainstream class was by definition for four-wheel, four-seat cars of conventional configuration. This class was won by Very Light Car, built by team Edison2, which achieved 43.57 km/L (102.5 mpgge) powered by an ethanol-fueled internal combustion (IC) engine powertrain (Edison2, 2010). The Side by Side class was by rule for cars with two seats abreast. It was won by Wave II, built by team Li–ion motors, an aerodynamic four-wheel electric car powered by lithium–ion batteries that achieved 79.5 km/L (187 mpgge) (X PRIZE, 2010). The Tandem class was for cars with two seats in line. It was won by E-Tracer #79, built by team X-tracer, an electric version of the Swiss-made MonoTracer streamlined, feet-forward motorcycle that achieved 87.15 km/L (205 mpgge) powered by lithium–ion batteries (X PRIZE, 2010). The X-tracer motorcycle achieved the highest overall fuel efficiency partly because of the tandem seating arrangement and inclined driver position, which nearly halved the vehicle frontal area compared to side-by-side arrangements and thereby significantly lowered the aerodynamic drag. In addition, the electric powertrain is very energy efficient compared to IC engines.

Further to the extreme is the SAE Supermileage® competition, an annual North American student competition challenging teams to propel one person the furthest on a liter of high-octane gasoline. The best of these streamliner tricycles achieve fuel economies in excess of 850.3 km/L (2000 mpgge) (SAE International, 2014).

Likewise, the International Shell Eco-marathon competition is an annual challenge to hundreds of secondary and college teams in the Americas, Europe, and Asia to design and build prototype vehicles that achieve extreme fuel economy. The prototype division consists of one-person ultralight and streamlined tricycles classed by their power train choice including batteries, hydrogen fuel cells, gasoline, or diesel. Their fuel economy is measured over a closed course at 15 mph average speed. Remarkable fuel economies have been achieved, which represent a benchmark for comparison to all other vehicle platforms. The author assisted The Pennsylvania State University Shell Eco-marathon team during the 2008–2010 seasons with the loan of a small, air-cooled 1.2 kW Proton Exchange Membrane fuel cell, which powered an 800 W electric hub motor in the single rear wheel. The Penn State team won first place in the 2008 U.S. fuel cell prototype division with 709.1 km/L (1668 mpgge), first place in 2009 with 815.4 km/L (1917.9 mpgge), and second place in 2010 with 767.8 km/L (1806 mpgge). In recent years, the Shell Eco-marathon prototype cars have improved dramatically in aerodynamic efficiency and weight reduction with top fuel cell and IC engine powered vehicles achieving over 3401 km/L (8000 mpgge) in 2014 competitions as shown in Figure 2.1 (Shell Eco-marathon, 2014).

These prototype vehicles built by students are obviously not practical for on-road use and the results from the 2014 global competitions show significant variability. The fuel economy results from Europe are the highest overall and also demonstrate the proper relative fuel efficiencies among powertrain choices with fuel cell powertrains significantly more efficient and electric powertrains up to three times as efficient as conventional gasoline powertrains. These prototype vehicles provide excellent perspective on the extreme limit of achievable fuel economy and indicate where fuel cell motorcycles could fit on that spectrum compared to conventional cars.

In the real world, two- and four-stroke gasoline motorcycles of around 100 cc displacement and their tricycle taxicab and cargo variants have for decades been important forms of transportation in many developing economies around the world. These motorcycles have been popular because they are affordable, dependable, and fuel efficient, achieving between 42.5 and 63.8 km/L (100–150 mpgge). As these economies have developed, however, the increasing number of motorcycles on urban streets resulted in significant pollution and traffic issues. As a result, countries such

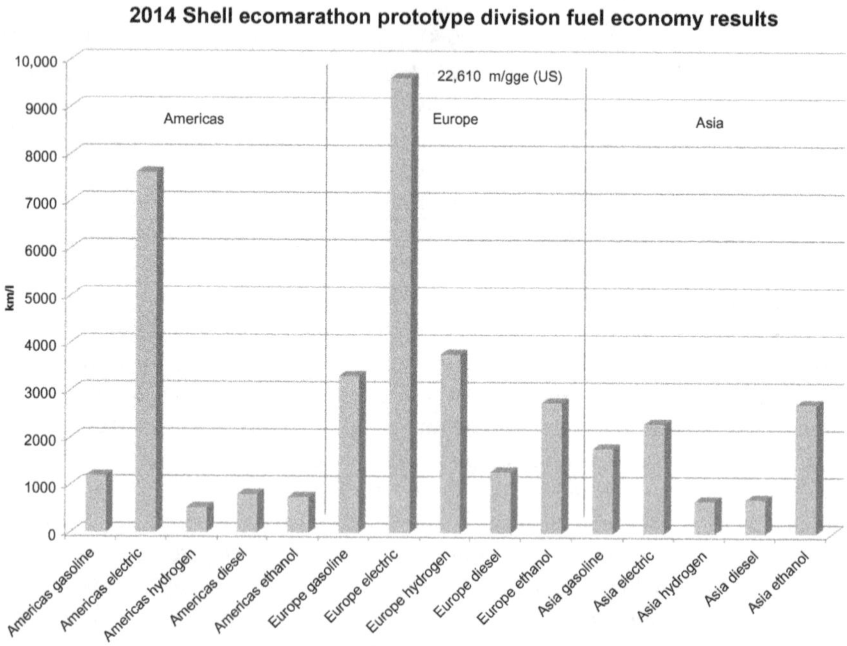

Figure 2.1 Fuel economy results of 2014 shell eco-marathon prototype division.

as India began to significantly tighten emissions restrictions on newly manufactured motorcycles starting in the 1990s (TransportPolicy.net, India). These tighter emissions standards have resulted in an effective production ban on two-stroke IC engine motorcycles and their replacement with less powerful but more fuel efficient and lower emission four-stroke and electric motorcycles. China has also instituted various policies at the national and local levels that have strongly promoted the manufacture and use of electric motorcycles and bicycles over gasoline versions, producing over 21 million electric bicycles in 2008 (Yang, 2010). This global growth of electric motorcycles and bicycles for basic commuting has created low-cost electric platforms that are fuel cell compatible, helping hydrogen fuel cells to penetrate the market more readily.

In 2011, Suzuki became the first motorcycle manufacturer to receive European Whole Vehicle Type Approval to begin manufacture of a hydrogen fuel cell and battery powered hybrid Burgman scooter on a 125-class platform, as shown in Figure 2.2 (MCN, March 2011). The air-cooled hydrogen fuel cell produced by British firm Intelligent Energy can deliver a continuous 2.5 kW while the combination with a relatively small battery pack (compared to other electric scooters) yields a total of 8.2 kW (11 bhp). A 700 bar (10,000 psi) gaseous hydrogen storage tank located at the bottom of the frame stores about 0.5 kg of hydrogen and can be filled in just a few minutes. This configuration works well in heavy urban traffic with low average speeds. At average speeds below 48.3 km/h (30 mph), the total range of the fuel cell Suzuki Burgman is a very impressive 354 km (220 miles). Less range can be expected at higher speeds as the road load will exceed continuous fuel cell power output and rapidly deplete the battery. At 170 kg, the curb weight is only a few kilograms more than the conventional IC

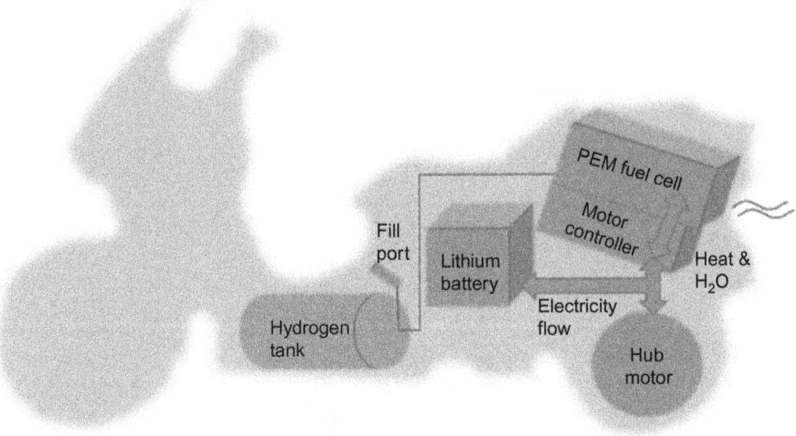

Figure 2.2 Power train schematic of Suzuki hydrogen Burgman (Suzuki Website, 2014).

engine scooter (Intelligent Energy, 2014). Intelligent Energy has also demonstrated the ENV prototype fuel cell motorcycle, which can maintain speeds of 80.5 km/h (50 mph).

There are currently several production battery electric commuter and sport mid-size motorcycles on the market, including the Zero S and Brammo, that have significant all-electric range over 100 km. They are proving that electric motorcycles are now practical. A Zero S demonstrated 117.5 km (73 miles) of highway range at the 2014 21st Century Automotive Challenge hosted at The Pennsylvania State University.

As a second thought experiment, it is worth considering the challenges of designing a fuel cell powered, full-size highway-cruising motorcycle with a range of at least 322 km (200 miles). Since its introduction in 1975, the Honda Gold Wing has been one of the most popular highway cruising motorcycles in North America, Western Europe, Australia, and Japan (Duchene, 2005). Numerous aftermarket trike conversions have also become available, making the Gold Wing a good test case to analyze requirements of fuel cell integration into full-size motorcycles. Table 2.1 shows the measured volume of stock Gold Wing components and the estimated volume for fuel cell hybrid powertrain components, including the hydrogen tank, fuel cell system, batteries, and the motor/controller. A 2014 Gold Wing model GL1800 has a gasoline tank capacity of 25 l (6.6 gal US) and an EPA rated fuel economy of 14.45 km/L (34 mpg US), giving it a range of approximately 363 km (225 miles) (Honda, 2012). All fuel cell components should fit within the available frame or cargo space of the motorcycle. Recent fuel cell vehicle field studies completed by the US DOE determined that they are more than twice as fuel efficient as conventional automobiles (Fuel Cell Today, 2013). Applying these specifications and the fact that 1 kg of hydrogen has the equivalent energy of one U.S. gallon of gasoline, the fuel cell hybrid motorcycle would achieve 28.9 km/L (68 mpg US) and require about 3.3 kg of hydrogen storage to achieve the stock range. A commercially available 3.3 kg tank at 700 mPa pressure would occupy about 80 l, about three times the volume of the gasoline tank (Luxfer, 2014). Also assume a fuel cell power output of 110 kW is needed to match the stock engine because full-size motorcycles are driven at highway speeds, on steep grades,

Table 2.1 Hypothetical fuel cell hybrid road bike volume analysis

Frame	Tank l	Engine l	Trans l	Fuel cell l	Controller l	Motor l	Battery l	Cargo l	Total
2 wheel petrol	25	169	141	–	–	–	–	174	509
2 wheel H_2	80	–	–	268	10	30	50	71	509
3 wheel petrol	25	169	141	–	–	–	–	260	595
3 wheel H_2	80	–	–	268	10	30	50	157	595

with two passengers, and luggage. In addition, they often tow a small cargo trailer. Applying a realistic fuel cell power density of 410 W/L results in a fuel cell volume of 268 L (Rousseau, 2014). Add another 10 L for a motor controller and 30 L for the motor. Finally, a modest battery of 50 L would be needed, assuming that the battery will only assist the fuel cell during startup, braking, and peak power output but provide no significant electric range.

This analysis shows that integrating the components for a fuel cell hybrid powertrain into a full-size highway motorcycle would be rather challenging at today's fuel cell power densities and hydrogen storage energy densities while maintaining stock power, range, and cargo capacity. The two-wheel motorcycle has its cargo capacity reduced by half to accommodate fuel cell components, while the trike version retains the stock cargo capacity of the two-wheel motorcycle. In the United States, factory trikes have recently grown in popularity as the motorcycle riding population ages and older riders look for more comfortable and stable options. Commercial examples such as the Can Am Spyder have more space for fuel cell powertrain components while retaining reasonable cargo space.

While fuel cell implementation within the space and weight constraints of a conventional two-wheel motorcycle may be challenging, there are numerous two- and three-wheel motorcycle platform concepts that may be more compatible. Streamlined and feet-forward motorcycles like the Progressive X PRIZE winner X-Tracer could be easily hybridized with a smaller, less expensive fuel cell. Three-wheeled concepts such as the Toyota i-ROAD and prototype Elio would be similarly good candidates. These fully enclosed trikes provide more room for fuel cell integration and are far more aerodynamic than conventional motorcycles and trikes, leading to higher fuel economy (Car and Driver, 2013).

2.3 Hydrogen industrial trucks

A variety of industrial truck types have been designed to perform different material handling applications. From the perspective of the operator's position, industrial trucks can be either walk behind, stand on, or sit down. Inside a large warehouse operation, the industrial truck fleet tends to be divided into two groups: high-lift for order stacking and picking from shelving and low-lift for order movement and truck loading. Figure 2.3 shows three of the most common small and medium-size industrial truck types, which are typically powered with battery electric drives.

In contrast to motorcycles, industrial trucks have a long history of both battery electric and internal combustion drives, such that battery electric truck types represent about 60% of the market share (Jerram, 2014). In general, smaller industrial trucks have tended to be battery electric while larger trucks have tended to be IC engine powered. Smaller trucks, used inside warehouses and factories where emissions are a greater issue, tend to be electric drive and use large lead acid rechargeable batteries. Larger trucks used outdoors to move and lift heavy loads tend to be IC engine powered using diesel, natural gas, gasoline, propane, and bi-fuel gasoline with propane. Increasingly restrictive air quality standards and improving electric truck performance

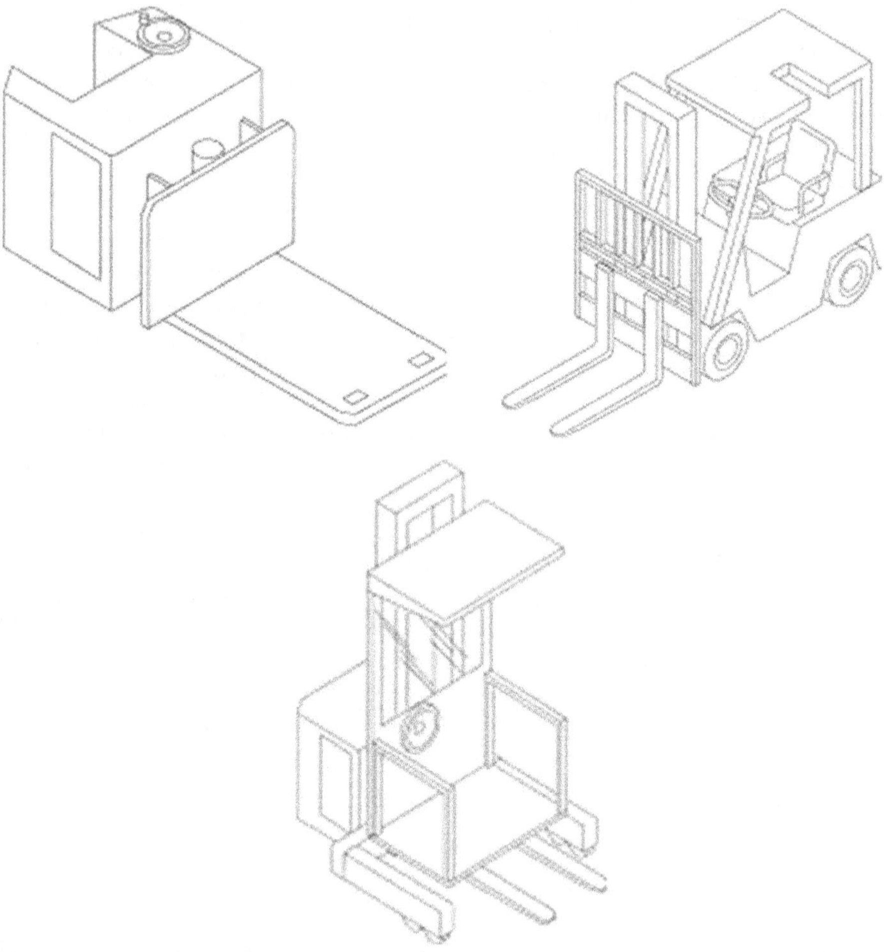

Figure 2.3 Three common types of small to medium-size industrial trucks. Top left—narrow aisle low lift platform, top right—sit down rider, bottom—narrow aisle order picker (OSHA, 2014).

have led to greater percentages of electric drive industrial trucks produced over the last 50 years. Typically, a charged battery will run an entire 8-h shift of continuous use. Battery charging requires several hours and can be done with the battery in or out of the truck. If idle during the next shift, the operator can simply park the truck and plug it into a battery charger. During 24-h operations, however, the industrial truck must remain in continuous service so the heavy batteries must be lifted out and exchanged within a dedicated battery maintenance room staffed by specially trained personnel. In the United States, there has been a recent surge in modular fuel cell systems with dimensions and electrical specifications which mimic the lead acid batteries they are replacing in industrial truck applications. Fuel cell systems can be refueled within a few minutes by the truck operator. This keeps the industrial truck in continuous service and returns valuable warehouse floor space and personnel from battery charging back into productive operations. Industrial truck fuel cell units cost on the order of $20,000 at current volumes. The price of this size fuel cell unit could drop to between \$350 and \$800 at automotive volumes (James et al., 2014). The business case for fuel cell material handling is compelling because it offers improved operational performance, more reliability, less maintenance overhead, and faster refueling compared to battery electric units.

In May 2014, the author visited Wegmans Retail Service Center in Pottsville, Pennsylvania to interview Facility Maintenance Manager David J. Allar Sr. about the company's experience with industrial truck operations using Plug Power GenDrive™ hydrogen fuel cell power units. Wegmans first considered hydrogen fuel cell replacements for its battery-powered trucks in 2009 and began implementation within its grocery and produce operations that same year. The fuel cell systems are a direct drop-in replacement for lead acid batteries and also provide counterweight. Because ample space and weight is allowable in industrial trucks, fuel cell systems can use less expensive and heavier components such as steel fuel tanks to reduce overall system cost. In 2010, Wegmans expanded the program to freezer operations with fuel cells specially modified to operate at below freezing temperatures. By May 2014, the company had converted its entire fleet of 272 trucks to hydrogen fuel cells, completely eliminating all lead acid batteries, their charge facilities, lifts, and maintenance equipment. Elimination of dedicated battery-charging facilities freed up a significant amount of warehouse floor space back into productive operations. Wegmans owns the equipment. Plug Power maintains the fuel cells on site. Air Products and Chemicals Inc. provides the hydrogen fueling equipment as shown in Figure 2.4, most of which

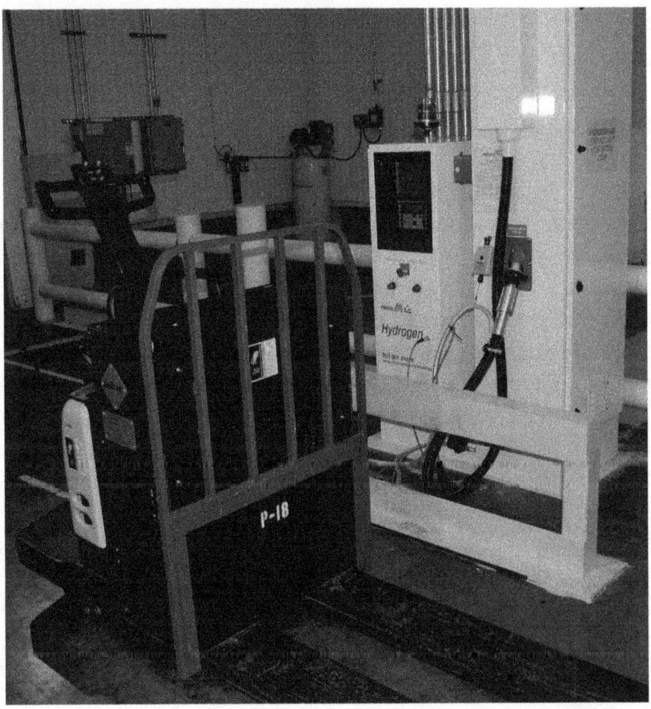

Figure 2.4 Fuel cell industrial truck refueling (Image courtesy of Air Products and Chemicals, Inc.).

is external to the warehouse facilities except for the dispensers. Filling takes only 2–3 min at 350 bar (5000 psi) in quantities of 0.5–1.0 kg. Fuel cell power output ranges from 2.6 to 10 kW, depending on application. Data are collected and waste water is unloaded at the time of fueling.

Wegmans was an early adopter of fuel cell technology for material handling, but not the first in Pennsylvania or the United States. The company experienced some initial challenges, but fuel cell operations are now stable. Wegmans management is satisfied with the overall business case for switching from battery electric to hydrogen fuel cell powered material handling systems. While the capital expenditure for fuel cells is higher than batteries, the combined savings from improved operations, recovered floor space, and reduced maintenance are significant. Lastly, the elimination of battery charging, handling, and chemical safety hazards has both qualitative and quantitative cost savings benefits to operations, including eliminating safety hazards related to battery charging. Several of its neighboring businesses have studied Wegmans's experience and recently followed suit by adopting fuel cells into their warehouse material handling operations.

Plug Power Inc. pioneered this industrial truck application for fuel cells starting in 2008. Plug Power has a supply agreement in place with Ballard Power Systems Inc. for several years to supply fuel cell stacks for its GenDrive™ power units (Ballard.com, 2010). Other Plug Power customers include BMW, Walmart, FedEx, and the U.S. Army, to name a few. Plug Power is now expanding to markets in Europe and Asia, where fuel cell powered industrial trucks have been adopted more slowly.

2.4 Conclusions

Motorcycle, bicycle, and industrial truck markets represent early opportunities for manufacturers to deploy small modular fuel cells and build economies of scale which later can be applied to automotive markets where cost, volume, and other requirements are more challenging. The environmental benefits and rapid refueling advantages of fuel cells are desirable to both markets. The majority of industrial trucks currently produced are already equipped with electric drive and the business case for small fuel cell implementation has proven favorable in certain material handling applications. The additional weight and volume required for fuel cell systems is not an issue within industrial truck designs. Electric drive motorcycles and bicycles have recently demonstrated competitive fuel economy, range, and performance compared to conventional IC engine versions. In China, government policies have promoted significant electric motorcycle and bicycle market penetration. However, the current high price of fuel cells per kilowatt and demanding packaging requirements compared to IC engines are likely to limit the growth of fuel cell powered motorcycles and bicycles to low-power demonstrations for some time until the price drops significantly and the components are further miniaturized.

References

Ballard.com, 2010. Plug power and Ballard extend supply agreement through 2014, July 15. Available at: http://www.ballard.com/about-ballard/newsroom/news-releases/news07151001.aspx (accessed June 2014).

Car and Driver, March 2013, Toyota i-ROAD concept, a strangely cool front-drive, electric three-wheeler—from the Prius Company? Available at: http://www.caranddriver.com/news/toyota-i-road-concept-for-2013-geneva-auto-show-news (accessed June 2014).

Companies and Markets.com, 2014. Material handling equipment: a global strategic business report, May 7. Available at: http://www.companiesandmarkets.com/Market/Logistics/Market-Research/Material-Handling-Equipment-A-Global-Strategic-Business-Report/RPT648281 (accessed August 2014).

Duchene, P., 2005. The New York Times, Honda's gold wing goes the distance, July 4. Available at: http://www.nytimes.com/2005/07/04/automobiles/04BIKE.html?_r=1& (accessed June 2014).

Edison2, 2010. Very light blog, annoy as little as possible, August 20. Available at: http://www.edison2.com/blog/2010/8/20/annoy-as-little-air-as-possible.html (accessed June 2014).

FuelCellToday, 2013. DOE: fuel cell electric vehicles twice as efficient as today's gasoline cars, July 20. Available at: http://www.fuelcelltoday.com/news-archive/2012/july/doe-fuel-cell-electric-vehicles-twice-as-efficient-as-today's-gasoline-cars (accessed July 2014).

Harley-Davidson, 2014. Project LiveWire, http://project.harley-davidson.com/ (accessed August 2014).

Honda, 2012. Gold wing specifications. Available at: http://world.honda.com/GoldWing/spec/index.html (accessed 2014).

Intelligent Energy Zero Emission Fuel Cell Scooter Case Study. Available at: http://www.intelligent-energy.com/automotive/case-studies/suzuki (accessed June 2014).

James, B., Morton, J., Colella, W., 2014. Fuel Cell Transportation Cost Analysis, 2014 Annual US DOE Merit Review, FC018 Available at: http://www.hydrogen.energy.gov/pdfs/review14/fc018_james_2014_o.pdf.

Japanese Automobile Manufacturers Association. Available at: http://www.jama-english.jp/index.html (accessed July 2014).

Jerram, L. 2014, Lead-acid rules as forklift battery, but alternative pursued, SAE Off-Highway Engineering Online, May 1. Available at: http://articles.sae.org/13176.

King, M., 2013. Global motorcycle market, Companiesandmarkets.com, November 28. Available at: https://uk.finance.yahoo.com/news/global-motorcycle-market-000000964.html (accessed July 2014).

Lavrinc, D., 2013. Wired, electric bikes get up speed at isle of man TT, June 5. Available at: http://www.wired.com/2013/06/2013-isle-of-man-tt-zero/ (accessed June 2014).

Luxfer Dynecell hydrogen tank specifications, 2014. Available at: http://www.luxfercylinders.com/products/alternative-fuel-cylinders/504-type-2-hoop-wrap-cylinder-specifications?tags=undefined&units=1 (accessed June 2014).

MCN-Motor Cycle News, 2011. Suzuki's fuel cell scooter gets mass-production approval—but should you want one? March 9. Available at: http://www.motorcyclenews.com/mcn/news/newsresults/general-news/2011/march/mar0911-suzuki-fuel-cell-scooter-gets-mass-production-approval/suzuki/_/R-EPI-129465 (accessed June 2014).

Occupational Safety and Health Administration (OSHA), United States Department of Labor, Sample Daily Checklist for Powered Industrial Trucks, 2014. Available at: https://www.osha.gov/dte/library/pit/daily_pit_checklist.html (accessed July 2014).

Rousseau, A., 2014, Impact of fuel cell system peak efficiency on fuel consumption and cost, 2014 Annual US DOE Merit Review, AN044.

SAE International, 2014. Collegiate Design Series, Supermileage® Competition. Available at: http://students.sae.org/cds/supermileage/results/ (accessed June 2014).

Shell Eco-marathon, 2014. Available at: http://www.shell.com/global/environment-society/eco-marathon.html (accessed August 2014).

Smit and Van Wyk, 2011. Blog, the Harley-Davidson sound trademark, June 29. Available at: http://www.svw.co.za/blog/the-harley-davidson-sound-trademark.html (accessed June 2014).

Suzuki Website, 2014. Environmental Technologies, Adaptation of conceptual diagram of the Burgman Fuel Cell Scooter. Available at: http://www.globalsuzuki.com/corporate/environmental/technologies/ (accessed 2014).

TransportPolicy.net, India, Motorcycle emissions standards. Available at: http://transportpolicy.net/index.php?title=India:_Motorcycles:_Emissions.

X PRIZE Foundation and Progressive Insurance, 2010. Three Teams Awarded Share of $10 Million Purse in Progressive Insurance Automotive X PRIZE for Super Fuel-Efficient Vehicles, September 16, 2010. Available at: http://www.xprize.org/press-release/three-teams-awarded-share-of-10-million-purse-in-progressive-insurance-automotive-x (accessed June 2014).

Yang, C.-J., 2010. Launching strategy for electric vehicles: lessons from China and Taiwan. Technol. Forecast. Soc. Change 77, 831–834.

Hydrogen-fueled marine transportation

3

*F. Vogler**, *G. Sattler*†
*Independent Consultant, Bad Schwartau, Germany, †Independent Consultant, Lübeck, Germany

Abbreviations

AES	all electric ship
AFC	alkaline fuel cell
AIP	air independent propulsion
AUV	autonomous underwater vehicle
BLG	IMO sub-committee "bulk, liquids and gases"
BMBF	Federal Ministry of Education and Research (Germany)
BMUB	Federal Ministry for the Environment, Nature Conservation, Building and Nuclear Safety (Germany)
BMVI	German Federal Ministry of Transport and Digital Infrastructure (Germany)
BMWi	Federal Ministry for Economic Affairs and Energy (Germany)
CNG	compressed natural gas
CO_2	chemical notation for carbon dioxide
DMFC	direct methanol fuel cell
ECA	emission control area
EEDI	energy efficiency design index
EEOI	energy efficiency operational indicator
EQHHPP	Euro-Québec Hydro-Hydrogen Pilot Project
EU	European Union
FC	fuel cell
H_2	chemical notation for hydrogen
HFO	heavy fuel oil
HT	high temperature
HT-FC	high temperature fuel cell
HT-PEMFC	high temperature polymer electrolyte membrane fuel cell
IGF-Code	International Code of safety for gas-fuelled ships
IMO	International Maritime Organization
LNG	liquefied natural gas
LOHC	liquid organic hydrogen carrier
LPG	liquefied petroleum gas
MARPOL	International Convention for the prevention of pollution from ships
MCFC	molten carbonate fuel cell
MDO	marine diesel oil
MEPC	Maritime Environment Protection Committee
MGO	marine gas oil
MSC	Maritime Safety Committee

Compendium of Hydrogen Energy. http://dx.doi.org/10.1016/B978-1-78242-364-5.00003-8

NIP	National Hydrogen and Fuel Cell Technology Innovation Program (Germany)
NM	nautical mile
NO_x	chemical notation for nitrogen oxides
ONR	Office of Naval Research
PAFC	phosphoric acid fuel cell
PEM	polymer electrolyte membrane
PEMFC	polymer electrolyte membrane fuel cell
R&D	research & development
RoRo	roll on roll off
SECA	sulfur emission control area
SEEMP	ship energy efficiency management plan
SSFC	ship service fuel cell
SOFC	solid oxide fuel cell
SOLAS	International Convention for the safety of life at sea
SO_x	chemical notation for sulfur oxides
TBD	to be defined
TCO	total cost of ownership
USCG	United States Coast Guard
XTL	something to liquid (synthetic fuel)
ZBT	Zentrum für BrennstoffzellenTechnik

3.1 Market environment

From experience, the introduction of new technologies often occurs only through economic stimulation or legal forces, particularly when suitable know-how and logistics are already in place for established, conventional systems. The following section introduces the environmental developments in place today in the shipping industry that will influence the design of new vessels and their operation, and which can be seen as one of the main drivers for the use of new technologies in shipping.

3.1.1 Environmental requirements

3.1.1.1 International environmental requirements

In comparison to air- and land-based transport, shipping is still the most efficient and environmentally friendly mode of transport. However, shipping-related CO_2 emissions contributed 3.3% of global CO_2 emissions (International Maritime Organisation, 2009). Furthermore, SO_x, NO_x and particle emissions are of great environmental concern in coastal areas (Corbett et al., 2007–09). Because of this, the International Maritime Organization (IMO) has declared the Baltic Sea, the North Sea, and the Channel from 2006 and 2007 on to be sulfur emission control areas (SECAs), in which strict environmental requirements related to SO_x emissions must be fulfilled. It was further decided to extend these areas in future to emission control areas (ECAs) in which, in addition to the requirements for SO_x emissions, requirements for NO_x emissions, and particulate matter must be fulfilled.

The limits for emissions in shipping related to the ECAs and SECAs are regulated by the Maritime Environment Protection Committee (MEPC) of IMO in the International Convention for the Prevention of Pollution from Ships (MARPOL) Annex VI (International Maritime Organisation, 2006). In the last revision of MARPOL in October 2008 new limit values for emissions were established, to become valid during the following years. The limit value for the sulfur content in fuel was to be lowered in steps. In January 2015 the sulfur content was lowered to 0.1% in SECA areas. By 2020 the intention is to lower the sulfur content globally to 0.5% (Figure 3.1).

In addition to SO_x emissions, NO_x emissions will also be more tightly regulated by 2016. Ships built in 2011 and later already have to fulfill the requirements of Tier II. From 2016 on, the requirements of Tier III as specified by MARPOL Annex VI have to be applied. The NO_x emissions have been defined according to the engine speed (Figure 3.2).

Besides the previously mentioned SECA zones of the Baltic Sea, the North Sea, and the Channel, it was decided by the IMO that, according to a proposal from the USA and Canada, an ECA zone would be established around the North American continent in 2011 (Lloyds Register, 2009–07; US Environmental Protection Agency, 2009–04).

Furthermore, additional ECA zones are currently under discussion. As an example, there are discussions about the Mediterranean Sea, the coastline of Australia, the coastline of Japan, and the territorial waters of Singapore (Meech, 2008).

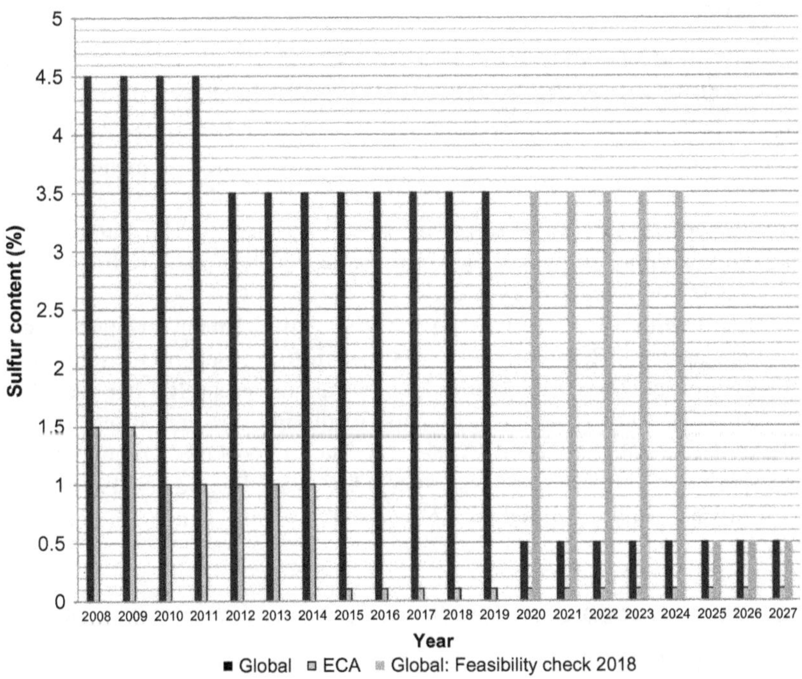

Figure 3.1 Permitted sulfur content according to MARPOL Annex VI.

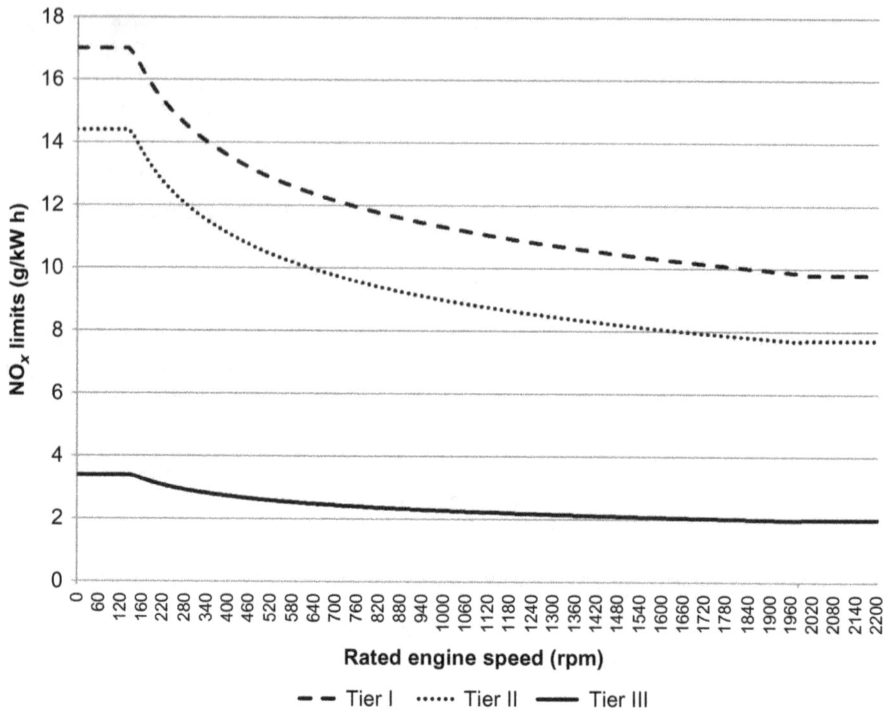

Figure 3.2 IMO limits for NO_x emissions.

3.1.1.2 European environmental requirements

Besides the international environmental requirements of MARPOL Annex VI, the European Union (EU) decided to expand the sulfur emission requirements of the IMO. Accordingly, the maximum sulfur content in vessel fuel for inland navigational vessels and ships berthed in European harbors has been restricted to 0.1% since January 1, 2010. Exceptions apply only for vessels that are berthed less than 2h. For Greece this regulation came into force 2 years later.

According to this EU directive (2005/33/EG) all EU member states are committed from January 1, 2010 on to use only gas oil for vessels on the market with a sulfur content of less than 0.1% (Europäisches Parlament und Rat der europäischen Union, 2005–07). During the latest revision of the EU directive it was decided to align the directive with MARPOL Annex VI. Therefore, the EU requirement from 2020 on will be a sulfur limit of less than 0.5% in all EU waters. Passenger vessels will be furthermore limited to 1.5% sulfur limit in all non-ECA EU waters (200 NM).

3.1.1.3 Further measures for the reduction of ship-based emissions

In addition to all the previously mentioned measures for the reduction of ship-based emissions, the IMO decided to reduce CO_2 emissions in shipping as well. In this regard, the IMO decided on measures that would increase the total ship efficiency of both new and existing vessels. The main instruments in this respect are:

- energy efficiency design index (EEDI)
- energy efficiency operational indicator (EEOI)
- ship energy efficiency management plan (SEEMP)

The EEDI intends a stepwise increase in the requirements on the energy-based efficiency of new vessels. According to predefined parameters the EEDI will be calculated for each specific ship type. The result will then be compared with a predefined limit value. The acceptable limit value of the EEDI will be defined by the IMO according to ship fleet data and will be continuously reduced over time. The EEOI will show the energy efficiency of existing vessels in gram CO_2 per ton and nautical mile. As an additional instrument the SEEMP was introduced as a planning tool for the implementation of energy efficiency increasing measures.

All these changes show that we can also expect more enhancements of environmental regulations for shipping in future. In this short enumeration only global international and European emission regulations were discussed. But besides this, numerous regional and local environmental regulations also exist in different harbors, states, and regions (Den Boer et al., 2009-02). As an example, the NO_x taxes on ship-based emissions in Norwegian territorial waters introduced by the Norwegian government in 2007 should be mentioned. This tax is valid for vessels with a main engine size of above 750kW and applies also to the NO_x emissions of the auxiliary engines, boilers, and turbines. As further examples of special regulations the "Emission Reduction Plan for Ports and Goods Movement in California" and the "Marine Vessel Emission Standard" of Alaska should be mentioned as well.

3.1.2 Regulatory requirements

Figure 3.3 shows a comparison of marine environmental regulations from 2009 to 2020. To comply with these planned international maritime environmental regulations, most of the engine manufacturers are working on engine modifications and technologies for exhaust gas cleaning to reduce engine emissions (Reuß, 2008). Today it is foreseeable that for the planned NO_x limit values from 2016 on an extensive effort in engine development will be required. It is therefore necessary that alternative fuels and technologies come into operation, such as fuel cell (FC) systems or gas engines.

According to international maritime legislation of the IMO, the use of gaseous fuels was prohibited except for limited exceptions for gas tankers. In general it could be stated that fuels with a flashpoint of below 60 °C were not allowed to be used onboard vessels. In the course of the work on environmental regulations for shipping, "INTERIM GUIDELINES ON SAFETY FOR NATURAL GAS-FUELLED ENGINE INSTALLATIONS IN SHIPS" (MSC.285(86)) were developed by the IMO subcommittee bulk, liquids and gases (BLG) (International Maritime Organization, 2009). These interim guidelines came into force in 2010 after the revision of the "Safety of Life at Sea" (SOLAS) convention. But in addition, the BLG subcommittee was assigned to further develop an International Code of Safety for Gas-fueled Ships (IGF

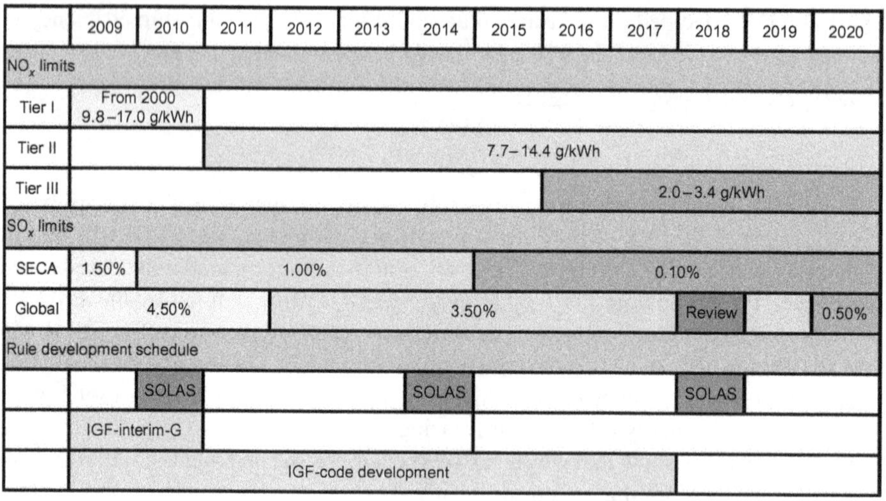

Figure 3.3 Comparison of marine environmental regulations according to MARPOL and the IGF code.

Code), to include not only internal combustion engines and natural gas but all kinds of low flashpoint fuels and other kinds of energy converters, including FCs. According to the current development plan of the code, a specific subchapter for FC systems will be included. The IGF Code is currently under development and it is expected that it will come into force in 2017. With finalization of this code the legal framework for the use of FC systems on board ships will be given an international basis. The current existing MSC.285(86) guideline already allows the use of natural gas as fuel for ships in combustion engines.

3.1.3 Infrastructure requirements

Today shipping is based on a worldwide fuel logistic predicted using traditional liquid fossil fuels such as heavy fuel oil (HFO), marine diesel oil (MDO), or marine gas oil (MGO). The reason for this is the exclusion of marine fuels with a flashpoint below 60 °C. A full logistic chain exists around the world. Basically any of these fuels can be ordered in any harbor around the world.

In the past few years, the use of low flashpoint fuels has come into focus in the shipping industry for environmental reasons. Because of this, natural gas, either in a pressurized state (compressed natural gas, CNG) or liquid state (LNG), came onto the agenda. Slowly an infrastructure development for LNG is on the way, especially in northern Europe. Many bunkering options are under investigation today. But until now no comprehensive fuel supply with natural gas exists. At this stage of development only singular fuel supply solutions have been developed for specific projects.

Bunker possibilities for other gaseous fuels do not exist. Only a few isolated projects are known in which the fuel supply has been developed in parallel to the specific application. As one example, the inland navigational vessel "Alsterwasser" within the ZEMSHIPS project in Hamburg can be mentioned here (Teichert, 2010).

For future FC projects it remains likely that these projects must also develop their own fuel supply either by a singular solution or a regional supply for a first start market. As an example, the introduction of FCs for shipping on a specific lake like Lake Constance or similar should be mentioned. A comprehensive gaseous fuel supply is not expected in the midterm future, although where applicable the use of gas stations for both automobiles and ships could ease the situation for inland waterway vessels.

3.1.4 Market trends

In the past, FC systems in shipping were deeply investigated in several research projects such as the Euro-Québec Hydro-Hydrogen Pilot Project (EQHHPP), FC-Ship, New-H-Ship, or the Fellow Ship projects. But only a few real demonstration projects were begun. The use of FC systems in shipping is an "up and down" situation. One reason for this is the development status of the FC systems themselves and the resulting costs. Therefore, FCs, especially in shipping, are not as visible today as they were during the years 1995–2005. Today initiatives with pure electrical solutions or gaseous fuels for internal combustion engines are currently more in the focus of the shipping industry.

The EU "Clean Power for Transport: A European alternative fuels strategy" aims to reduce the dependence on oil for transport. In 2010 oil accounted for 94% of the fuel consumed in transport, with 84% of this fuel imported into the EU. Therefore, the EU decided to support market development of alternative fuels and invest in the infrastructure for these fuels for all modes of transport. In addition to the reduction of dependence on oil, the strategy was also the decarbonization of fuels. The strategy pointed out that there is no single fuel solution for the future of mobility. Due to the huge energy demand in shipping the strategy identified LNG, LPG, and Biofuels (liquid) as long-term alternatives for the shipping industry. For small vessels with lower energy demand, like inland navigational vessels, hydrogen was also mentioned as one alternative.

Besides the EU fuel strategy the National Hydrogen and Fuel Cell Technology Innovation Program (NIP) in Germany is an example of national market trends. This program was founded by the German Federal Ministry of Transport and Digital Infrastructure (BMVI) together with the Federal Ministry for Economic Affairs and Energy (BMWi), the Federal Ministry of Education and Research (BMBF), and the Federal Ministry for the Environment, Nature Conservation, Building and Nuclear Safety (BMUB). The program is part of the high-tech strategy for Germany and fits into the Fuel Strategy of the Federal Government. The NIP will cover numerous hydrogen- and FC research projects from research and industry partners. The Public Private Partnership started in 2006 and is planned for 10 years. The German government and the industry will invest 1.4 billion Euro till 2016 for research, development, and demonstration projects (German Federal Ministry of Transport and Digital Infrastructure (BMVI), 2014).

Within the NIP one project for power generation on board ships was started. The "e4ships" lighthouse project aims to improve significantly the energy supply on board large vessels. To realize this, the use of polymer electrolyte membrane (PEM) and high-temperature FCs is being planned; these should enable a considerable reduction in emissions and fuel consumption.

In addition to the technical implantation on different vessel types, another important challenge is to derive technical standards for all system types and performance classes. Moreover, better high-performance energy supply systems need to be planned for the future.

3.2 Requirements for marine FCs

Before discussing the specific requirements for FC systems in vessels, a short overview about operation concepts shall be given. It must be considered for all concepts and applications that not only the FCs themselves will be taken into consideration but also the corresponding fuel storage solution for each single application. In shipping, normally no singular solutions exist. Besides the power demand also the specific type of vessel, the route, and the specific operation profile must be considered for the design, as well as the infrastructure possibilities.

3.2.1 Operational concepts for marine FC systems

According to the different technical characteristics of FCs, different areas of application, and operational concepts will result for the specified systems. In general high temperature fuel cell (HT-FC) systems fueled by liquefied natural gas (LNG), synthetic gasses (XTL), or other fossil fuels are suited for auxiliary power on medium and large seagoing vessels according to their power range and stationary operational profile. PEMFC fueled with hydrogen can be used for small vessels for main propulsion and auxiliary power. From this basis, operational concepts for FC systems on board of vessels could be derived.

3.2.1.1 Basic load coverage with parallel operation

Especially for passenger vessels, FC systems could provide part of the auxiliary power for the safe and autarkic power supply of single ship sections (Figure 3.4). In case of a power loss in single sections this application makes it possible to provide enough power for the rest of the vessel to provide the possibility for a safe return to the next possible harbor (safe return to port).

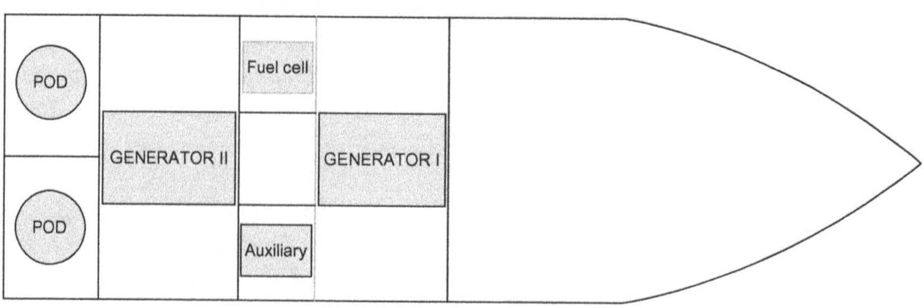

Figure 3.4 Basic load coverage by parallel operation of generator and fuel cell system.

3.2.1.2 Auxiliary power supply with hybrid application

Modern FC systems are able to provide a bigger part of the auxiliary power demand of big vessels when operated as a stationary application (Figure 3.5). Short peak loads and load variations can be covered by other systems, such as battery systems. So, the system can be operated as a hybrid solution.

3.2.1.3 Direct propulsion for small vessel with hybrid application

FC systems can be operated together with a battery system as a hybrid solution. The battery system will buffer and supply electrical power for short peak loads. Such a hybrid system can supply the whole energy demand for auxiliary and main propulsion power of a small vessel (Figure 3.6).

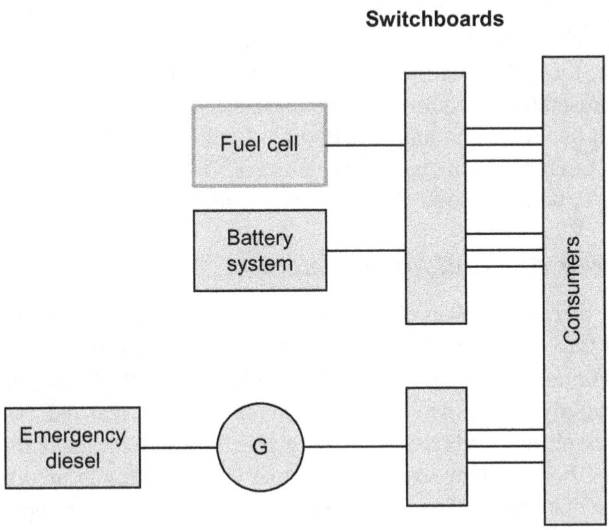

Figure 3.5 Auxiliary powers supply by fuel cell hybrid system.

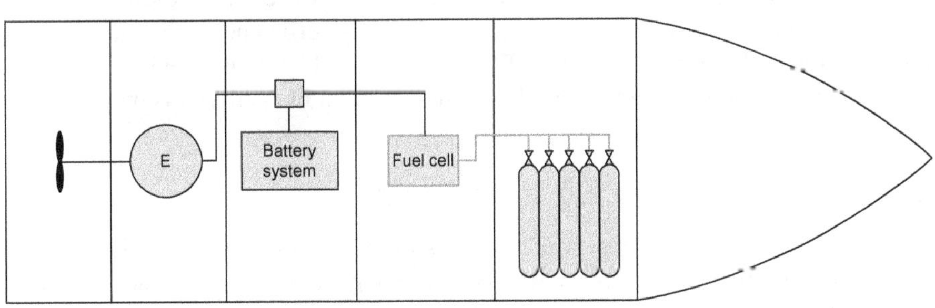

Figure 3.6 Direct propulsion by fuel cell hybrid system (example: ZEMSHIPS project).

3.2.2 General requirements for maritime FC systems

FC systems in maritime applications have to fulfill extended requirements in comparison to stationary land-based applications. In the following subchapters, three main requirements will be given which must be proven by the FC system when used in marine applications.

3.2.2.1 Environmental conditions

The use of systems in marine applications requires compatibility with the environmental conditions of salty, oily, and humid air. Possible effects, like corrosion, have to be considered. This could be achieved, for example, by a proper material selection of suitable coverage and filter systems.

Further requirements result from the ambient conditions in the engine room of a vessel. The FC system has to withstand ambient temperatures of up to 45 °C (electrical components up to 55 °C) at a relative humidity up to 60%. Under these conditions the FC system must be able to provide continuous nominal power.

Furthermore, the FC system must be able to stay fully operable in case of list or accelerations caused by sea motions. In this respect, the FC system must be able to be operable until a heel of 22.5° and longitudinal inclination of 10°. Electrical components must stay operable even up to a heel angel of 45°. Furthermore, the influences of vibrations have to be considered (Germanischer Lloyd, 2009).

3.2.2.2 Power demand and efficiency

During the EU research project "FC-Ship" the required electrical power demand for different vessel types was investigated according to their specific load profile (Viviani et al., 2004). The analysis shows that with module sizes of about 500 kW electrical power, several supply concepts for all sizes of vessel could be realized. For smaller ships like authority vessels or small passenger vessels the FC modules make the whole power supply available. For medium-sized vessels, like ferries, multiple modules can supply the full power or main parts of the energy supply on board. For big seagoing vessels, the FC modules are able to supply the main part of the auxiliary power (Figure 3.7).

To be competitive, the electrical efficiency of the FC system must be higher than that of conventional diesel engines. Therefore, the efficiency of an FC system must clearly be greater than 40% to have an advantage over conventional diesel generator sets. Modern big diesel generator sets today have high efficiencies that are difficult to beat. But also the byproducts of an FC system should be considered in an overall energy balance in order to show the full advantages of an FC system in comparison to traditional diesel engines.

3.2.2.3 System integration into a vessel

For the integration of FC systems into vessels the first criteria will be given by the structural requirements. Besides the limited space in general, also the space for maintenance and repair or a total change of the system/system components have to be considered.

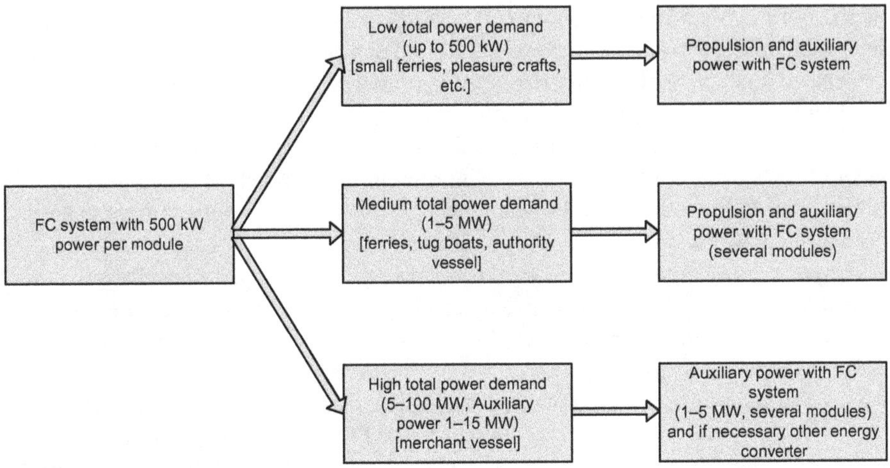

Figure 3.7 Capabilities of maritime fuel cell systems.

Further challenges will be presented by the electrical and thermal integration of the FC system as well as the integration of the protection and control system into the vessel (Germanischer Lloyd, 2002). Especially the electrical integration has to be considered carefully. The onboard system will be operated as an island network with special characteristics (e.g., power density, short-circuit power, frequency stability, dynamics, earthing, and protection philosophy). Furthermore, a potential synchronization possibility with conventional diesel generators and a possible shore connection has to be considered within the design.

3.2.2.4 Supply of essential consumers

A further special topic for vessels is the power supply of essential consumers. For power supply systems for consumers, a respective verification about the reliability of the system is required.

3.2.3 Pleasure craft

FC systems on board pleasure craft will normally be used in two different concepts. Firstly the FC system will be used as a "range extender"; this means that the operational endurance can be considerably prolonged by permanent recharging of the battery. The second concept will be the use of the FC as a hybrid system for the power supply for propulsion and auxiliary purposes.

Generally pleasure craft are used only for limited times during the year; therefore the systems will have only a very limited number of operating hours during one season. As a result of this operation profile, lifetime is not a real criteria of a pleasure craft. The limited operation hours will in general also result in a smaller size of storage tank onboard the vessel. But this must be seen in combination with the existing fuel infrastructure. Basically, no infrastructure for hydrogen supply in harbors exists today.

For the use of direct methanol fuel cells (DMFCs), methanol can be used, which is much more readily available. Several yacht and camping suppliers have special storage cans in their scope of supply. However, the rated power output of DMFC is limited to only a few kilowatts.

More relevant for the user is the availability and reliability of the system. The owner of a pleasure craft uses the vessel normally only for a short time period during his vacation. During that time he wants to rely on his vessel during the season and while staying at sea.

The power range for a pleasure craft is not a very critical item; normally only small FC systems will be required. The size and weight of an FC system including the tank must be seen as a more relevant topic. Pleasure craft owners do not want to reduce the performance of their vessel.

Major benefits of an FC system are the low noise emissions. This will be seen very positively, especially on sailing yachts where the FC system will be used as a range extender or when the vessel will stay at anchor. A further big benefit of the use of the FC system will be its environmental friendliness. Especially in Europe many inland waterways are restricted in the use of combustion engines. Here the FC system can be a real alternative and may also create new markets.

From this perspective a DMFC and a PEMFC seem to be the most suitable types of FCs for a pleasure craft. Since PEMFC will probably be fueled by pure hydrogen, this may cause an infrastructure problem. The DMFC can be refueled by methanol, which can already be ordered from several yacht and camping suppliers.

It should at least be mentioned here that, even if the price is always of some relevance, this may not be such a critical item for pleasure craft, because the pleasure craft market is more of a "hobby market." Of higher relevance for the owner could be the fact that he has a unique feature for his vessel.

3.2.4 Merchant vessels

A merchant vessel will have totally different needs related to the FC system compared to pleasure craft. At least it must be considered that different ship types require different things from the FC system. The requirements presented in the following discussion will relate to a "general" ship.

The main requirement for an FC system of a merchant vessel will be to supply constant power with a high efficiency, especially during part load operation. Therefore, not only the electrical energy should be considered but also the existing byproducts of the FC system, like heat or oxygen-poor exhaust air. For the vessel, the overall efficiency is key issue.

Furthermore, the lifetime of the system is a real topic. In comparison to pleasure craft, merchant vessels will be operated for more than 6000 h per year. Lifetimes in the range of a main engine overhaul must be considered (30,000 h). In relation to the high operation hours, also high availability and reliability is required. Downtimes of the main power suppliers cannot be accepted. In addition, the power supply to essential consumers must be ensured by redundant supply architecture.

Table 3.1 Examples of power demand of different ship types

Ship type	Main engine (kW)	Auxiliary engine (kW)
Passenger ships	25,000	11,000
Mega yachts	10,000	2000
RoRo passenger, cargo vessels	21,000	2500
Other passenger vessels	15,000	5000
Combined cargo/passenger vessels	13,000	5000
RoRo car carriers	10,000	3500
Other RoRo vessels	10,000	4000
General cargo vessels	2800	800
Reefer vessels	11,000	4000
Container vessels 1400 TEU	11,000	3000
Container vessels 5000 TEU	45,000	8000
Container vessels 8000 TEU	70,000	13,000
Bulk carriers	17,000	2500
Oil tankers	8300	3000
Product tankers	8000	3000
Chemical tankers	8000	3000
Tugs	5000	300

It must be considered that merchant vessels have a power demand from about 5–100 MW and therefore also a high energy demand. This is related to the propulsion as well as to the auxiliary power demand. Therefore, HT-FC systems seem to be most suitable. In Table 3.1, an overview of rough power demands of different ship types is given.

In relation to the high power and energy demand and the operation profile of the ship, fuel infrastructure is also a big topic for merchant vessels. The normal vessel will be operated on a worldwide basis. For special applications like ferries, a single point infrastructure may also be sufficient. But the high energy demand will result in the requirement of sufficient fuel storage capacity. This issue is in general a challenge for all mobile applications and therefore a limiting factor. Big storage tanks will require space and weight which then will be lost for the transport of cargo on the vessel. Therefore a fuel with a high energy content and good storage possibilities is needed to allow the smallest possible storage tanks. As discussed in Section 3.3.2.1, most environmentally clean fuels like gasses have quite low specific energy content. And also fossil fuels needing a reformer system will result in weight- and space-consuming technology on the vessel. This discussion shows that the storage tank challenge must be carefully considered in comparison with the available infrastructure on a case by case basis. General recommendations will not lead to a satisfying result.

From the price perspective it should be mentioned that, at the first step, an energy conversion system on board the vessel must be economical to be competitive on the market. This does not necessarily mean that the price of the components must be less expensive but all in all the result for the total cost of ownership (TCO) must fit. If the FC system is not competitive with conventional energy converters, there would be no chance for their use on merchant vessels.

3.2.5 *Navy vessels*

By nature the requirements for navy vessels will differ from all other vessels. The main issue for a navy vessel is to get the right power at the right time. So, availability and reliability will be a tough topic. Lifetime will have a lower relevance. Also, fuel storage capacity may be a topic as a special operation radius is required. Bigger storage capacity will also result in reduced payload of the vessel. For navy vessels this would result in less armament and less intelligence technology.

Furthermore, the protection and risk of the fuel storage must be carefully considered to increase the strength of the vessel.

Costs and efficiency of the systems play only a minor role in this market. But also for this market it is necessary that the benefits dominate the costs. This is of special relevance for underwater vessels.

3.3 Suitable FC systems

While comparing the requirements for marine FC systems and the characteristics of the different FC types, it becomes apparent that DMFC and alkaline fuel cell (AFC) will not be suitable for use in maritime applications. The DMFC can be mentioned as so-called "comfort FC" which will be used up to a power range of 1 kW. The efficiency of this FC type plays a minor role. Lifetime and weight will be more important for portable systems. Because of the low electrical efficiency of below 40% this FC type will not be suitable for maritime FC applications in general. Furthermore it could be expected that the DMFC in comparison to the PEMFC may have a higher noble metal demand, which will clearly increase the investment cost (Heinzel, 2009).

The AFC requires a high quality of process gasses. Besides pure hydrogen on the fuel side, the AFC requires CO_2 free air or oxygen on the air side. The reason for this is that CO_2 will lead to the formation of carbonate within the electrolyte which affects the function of the FC and will lead to a breakdown of the FC in the long term. The required CO_2 separation in case of air operation will increase the technical effort and the space demand and will decrease the electrical efficiency.

The polymer electrolyte membrane fuel cell (PEMFC) is known for its high electrical efficiency of above 50% in the case of hydrogen operation. In the case of operation with natural gas, the efficiency is reduced down to 35–40%, according to the energy demand for the reforming process from natural gas to hydrogen. Currently, systems with a power range up to several hundred kilowatts have been developed. Therefore, a general use for small hydrogen-operated vessels can be derived.

At present the PEMFC technology still has a huge development potential. As an example it should be mentioned that the Israeli technology company CellEra has developed a platinum-free FC suitable for mass production. Platinum is used in PEMFC systems as a catalyst and can be replaced for new developed systems by materials like iron, cobalt, or silver. Accordingly, an enormous cost reduction potential in comparison to conventional PEMFC can be expected (CellEra, 2011).

Another promising technology is the development of high temperature membrane fuel cells (HT-PEMFC), which will be operated as a PEMFC in a temperature range of 120 °C up to 200 °C. According to the operation under these HTs it is possible to clearly decrease the system complexity in comparison to conventional PEMFCs. Due to this reduced system complexity, the system costs will decrease. For example, the water management of the FCs can be simplified (no humidification of the gas required anymore) and the CO tolerance will be much better regarding the high operation temperatures of the FC. This will make the CO precision cleaning unnecessary. This will simplify the heat management of the total system and will therefore lead to a higher total efficiency. Furthermore, the higher temperatures will provide the ability for a heat extraction on a much higher level. At present the power density is not as high as for low temperature PEMFCs (Zentrum für Brennstoffzellentechnik (ZBT), 2009-05). The first successful practical tests were performed during test flights of the first manned FC plane of the "Antares DLR-H2" (Kallo, 2011).

The solid oxide fuel cell (SOFC) and the molten-carbonate fuel cell (MCFC) will be considered as HT-FCs. The advantages of these FC types in comparison to the low- and middle-temperature FCs are the low requirements for the cleanliness of the process gasses and the high operation temperatures. This enables these FC types to be operated directly with natural gas or synthetic fuels which will be converted during the internal reforming process to a hydrogen-rich gas within the FC. A part of the waste heat can be used for downstream processes for the production of electrical and/or thermic energy, for example in microturbine generators or heating and cooling units, which will increase the overall efficiency of the system clearly. Related to the installation of a downstream microturbine process the technical effort will be increased drastically, but according to first investigations it is possible to increase the electrical efficiency from 47% to 55% up to a theoretical 80%.

For maritime applications the use of HT-FCs is of high interest, regarding the fact that these FC types have the highest efficiencies together with the lowest fuel requirements. The minimum requirement for electrical power will be nearly achieved by module sizes of about 400 kW. Present developments are working on module size from 0,5 up to 1 MW electrical power. The MCFC technology was, based on several pilot projects, close to the commercial market, but due to the closing of one of the main manufacturers this development was slowed down. For the SOFC only a few demonstration projects are in operation. For both FC types a high development potential could be expected (Blesl et al., 2007; Gerboni et al., 2008).

The phosphoric acid fuel cell (PAFC) is, based on the numerous field tests, the most mature FC technology. In the short term, the PAFC could be an interim solution for the introduction of FC technology in the maritime market. But due to the low electrical efficiency of up to 42% this technology will not be competitive against conventional diesel engines (Clear Edge Power, 2014).

In the long term, it could be expected that PEMFC, MCFC, and SOFC systems will be used in maritime applications. It could be expected that the PEMFC will be mainly used for hydrogen applications whereas MCFC and SOFC systems will be operated with liquefied gasses or synthetic fuels. For the further development of the PEMFC technology, high potentials could be expected for increase in the efficiency, the use of further fuels and especially the lowering of the cost for the market entry due to the parallel development of platinum free and HT membranes. It could be expected that similar developments in materials research could also lead to a lowering of cost for HT-FC systems.

3.3.1 Required developments for maritime FC systems

Today's existing FC systems for maritime applications are still in a precommercial development status. Therefore further technical development will be required. The following topics must be considered for the further development:

- Power: the module sizes of 400kW today must be increased up to module sizes of over 1MW.
- Weight: The weight of an FC system is 7–19 times higher than that for a diesel generator. The weight per kilowatt must be decreased to a reasonable level. A big part of this development can be achieved by the integration of subsystems into the ship systems. In the long term only, economies of scale will lead to an alignment of the power density with diesel generators. An exception to this are PEMFCs up to 150kW power which already today have about the same power density as diesel generators (Clear Edge Power, 2014; Fuel Cell Energy, 2013).
- Volume: The volume of an FC system is 10–15 times more than for a diesel generator. The volume per kilowatt must be decreased. A big part of this development can be achieved by the integration of subsystems into the ship systems. In the long term only, economies of scale will lead to an alignment of the volume per kilowatt with diesel generators. An exception to this are PEMFCs up to 150kW power which already today need the same space as diesel generators (Clear Edge Power, 2014; Fuel Cell Energy, 2013).
- Lifetime: The lifetime of an FC will be, dependent on the FC type, between 10,000 and 40,000h. The lifetime must be comparable to the main service interval of a conventional diesel generator (about 25,000–30,000h). In the long term, a lifetime between 40,000 and 80,000h should be achieved to clearly reduce the operational and maintenance costs (Wendt, 2006).

Further challenges for the market entry of FC systems will be their reliability and availability. From the operators view also the capacities for the delivery of spare parts and the performance of maintenance and repair work must be similar to the services of diesel engine manufacturers. The FC manufacturers must guarantee the availability of spare parts, short maintenance and repair times and terms of guarantee in line with the industry standard. All this factors are essential requirements for the use of FC systems in shipping.

3.3.2 Hydrogen storage alternatives on board

3.3.2.1 Suitable fuels for maritime FC systems

As discussed in Section 3.1.3, the maritime industry is based on liquid fossil fuels. To use these fuels for FCs a reforming process to a hydrogen rich gas is required. The effort for the reforming is highly related to the sulfur content of the fuel. Principally the use of marine diesel or gas oil for reforming is possible. For technical reasons it is not possible to lower the sulfur content after reforming below 10 ppm, which is too high for a PEMFC. Theoretically it is possible according to the "ionic liquids" process to lower the sulfur content down to 2 ppm but this has not yet been sufficiently investigated. Therefore it is coactively required to use only low sulfur fuels for reforming, which has been desulfurized in a refinery first.

In general the following fuels may be used for FC systems in maritime applications:

- Liquefied natural gas (LNG)
- Liquefied petroleum gas (LPG)
- Methanol
- Hydrogen
- XTL

The use of hydrogen as fuel can be excluded for large vessels regarding the fact that the required storage volume for pure hydrogen will be too big for economic operations of seagoing vessels. However, different information from the internet indicates that the storage of hydrogen in liquid organic hydrogen carriers (LOHCs) has emerged from the pure research area and comes into a phase of application development. LOHC with more than 6 wt% of hydrogen and slightly heavier than diesel fuel can be stored on board like diesel fuel. This might become an alternative for coming applications.

The availability of XTL is currently not given. In the short term, the possible fuel demand of the shipping industry cannot be supplied by the existing production facilities. Therefore this fuel may be an alternative from a long-term perspective.

For the other fuels, LNG, LPG, and methanol, worldwide logistical structures are existing. In the short term, there are possibilities to ensure the supply of these fuels in suitable quantities. The current technical development in shipping is focusing in this respect on LNG as fuel. First LNG infrastructures (small scale terminals) for ship fuel are under development in northern Europe and North America.

Besides the general suitability of the fuel the energy density per volume must be considered. In Table 3.2, the volume increase of different fuels is given related to the energy content of the fuel. This means that if the vessel will shift from HFO to LNG it needs 1.8 times the volume for storage of the fuel without considering the space for the different storage technology.

Table 3.2 Volume increase factor of different fuels

Fuel	Volume factor
HFO	1.0
LNG	1.8
LPG	1.7
H_2 liquid	4.7
H_2 700 barg	8.6

3.3.2.2 Cost perspective

In shipping the choice of propulsion systems does not only depend on the direct costs of the energy converter. Rather, the lifetime costs are to be used as criterion. This means higher priced FCs will not have to compete directly with lower priced combustion engines. For the shipping operator it is not the investment costs that are of main interest but the price per kW/h. Therefore a general cost comparison does not make sense. The economic analysis must be performed on a case by case basis for the specific application. Several different factors play a major role for the cost consideration and must therefore be considered. In the following the most important factors are listed:

- Power demand
- Load factor
- Fuel infrastructure
- Fuel demand
- Fuel price
- Operation distance
- Use of power, heat, and other byproducts of the energy converter
- Environmental regulations
- Harbor fees and discounts for cleaner vessel
- Lifetime
- Maintenance effort
- Other operational cost related to the energy converter

These cost analyses are also true for diesel engines. Therefore, different alternatives such as diesel engines operated with sulfur-free diesel fuel, scrubber systems, or LNG are under discussion for different ship types in different sea areas.

3.4 FC integration in ships

3.4.1 Possible markets

The previous chapters have shown that regional and international environmental requirements for shipping will be further established caused by the social and political pressure of the societies. As an outcome of this situation, the demand for clean energy converters in shipping will increase. The developments related to low flashpoint fuels and their regulation will increase the possibilities for FC systems in shipping. Some FC manufacturers have developed FC systems which already are available. The size of the market will be discussed shortly.

3.4.1.1 Overview of the global ship market

The overall worldwide ship market has a size of about 100,000 vessels. The share of the merchant fleet such as passenger vessels, dry cargo vessels, and tankers has a size of 53% of the world fleet. Other vessels like fishery vessels or others have a share of 47% of the residual world fleet. By ship tonnage, 67% of the world fleet consists of bulk carriers, oil tankers, and container vessels. By quantity, tankers, multicargo vessels, and bulk tankers will dominate the world fleet. In 2012, the international order books had a size of 5550 vessels, approximately 5% of the world fleet.

3.4.1.2 Start markets for FC systems

For the introduction of FC technology in seagoing shipping, several development scenarios could be expected. In this section a development scenario will be considered which was discussed within the e4ships project. According to the project, it would be expected that a first real commercial order will be realized after the finalization of the currently ongoing demonstration projects within the e4ships project. During the first years it might be that only one-module installations will be ordered which only will serve as part of the required power demand of the whole vessel to gain experience by the shipping company. During maturing of the technology multi-module installations will also be expected after the first years of experience with the single-module installations.

The question is, in which market segments will we see the first FC systems in shipping? The past FC projects have shown that FC systems mainly were used in small passenger vessels which were operated in a localized fashion. Because of the challenges with the fuel logistics, either gaseous fuels, or liquid low flashpoint fuels, it could be expected that the first customers will operate their vessels on mainly fixed routes. In this respect RoRo vessels and ferries will be one of the first potential market segments (2.5% market share of the world fleet).

A further possible market segment will be the cruise ship industry. For the background of high environmental requirements in the North and Baltic Sea as well as around the North American continent, this technology is of high interest in this shipping branch. Furthermore, marketing reasons may increase interest in the use of this technology to work on a "green image" for the background of criticism of environmental organizations. Especially in Europe, the cruise vessels are mainly berthed directly in the city centers where the environmental concerns are very high. During their worldwide cruises, cruise vessels often visit environmentally critical areas where strong local environmental legislation is in place. And, not least, the ecological awareness of the passengers increases more and more. Further benefits of the FC system, such as the high overall efficiency and low noise and vibration emissions, are mentioned here only complementarily.

For similar motivation the mega-yacht segment may also be interesting for the market entry. Regarding their worldwide operation these vessels also have to fulfill the highest environmental standards. But besides that, in this ship segment mainly comfort counts for the customer. In this respect the owner has very high requirements regarding noise- and vibration emissions as well as visible and smellable air emissions. The cost of the system will have a subordinated importance. Furthermore this new technology will have a unique selling point, which will be positively rated in this market segment.

Beneath these general markets, special applications like harbor vessels, fishing and offshore vessels, and research and authority vessels should be considered as first potential users of marine FC systems. The special purposes of these vessels, their load profile and their local operation area will have a great potential for the use of FCs. In this respect also navy vessels have a great potential for first applications, especially regarding the fact that in addition to the energy conversion also low signatures like noise and vibration are of high interest.

3.4.1.3 Remaining merchant fleet

For the remaining part of the merchant shipping fleet, it can be expected that the use of FC systems for auxiliary power will be much delayed. The reason for this is that the expected higher cost of the new technology will not be tolerated as much as for the emerging markets. It can be expected that an extensive acceptance by the shipping companies will not be observed before cost equalization with conventional diesel generators and maturity of the technology are reached.

As a special market segment the container feeder vessels should be mentioned here. This segment will also have an interesting potential for FC systems due to their local operation, long harbor lay days, and the high environmental restrictions, especially in the North and Baltic Seas.

3.4.2 Safety

The onboard safety of vessels will be generally regulated by the flag state of the respective vessel. The flag state is responsible for compliance with international (IMO) and national safety standards. Safety will normally be checked by the classification society on behalf of the respective flag state. The relevant regulations were already discussed in Section 3.1.2. In the following, safety background and the basic principles will be discussed.

In the past the use of fuels with a flashpoint below 60 °C was totally forbidden (55 °C for emergency generators) to reduce the risk of fire and explosion onboard. When starting the process for alternative fuels of low flashpoint in 2005, the main requirement for this allowance was connected to the fact that the use of low flashpoint fuels should be as safe as the use of conventional fuels.

One important safety principle in shipping is that no single failure can lead to a critical situation of the vessel (one failure principle). One result of that demand will, for example, be the double barrier principle. That means that every low flashpoint fuel must be covered with at least two independent barriers to avoid the possibility of a leakage of the inner barrier leading to a dangerous situation.

In the following, the functional safety requirements within the IGF Code are listed to give a rough overview of the required safety level in shipping:

- The safety, reliability, and dependability of the systems shall be equivalent to that achieved with new and comparable conventional oil-fueled main and auxiliary machinery.
- The probability and consequences of fuel-related hazards shall be limited to a minimum through arrangement and system design, such as ventilation, detection, and safety actions. In the event of gas leakage or failure of the risk-reducing measures, necessary safety actions shall be initiated.

- The design philosophy shall ensure that risk-reducing measures and safety actions for the gas fuel installation do not compromise the required availability of power generation and propulsion.
- Hazardous areas shall be restricted, as far as practicable, to minimize the potential risks that might affect the safety of the ship, persons on board, and equipment.
- Equipment installed in hazardous areas shall be minimized to that required for operational purposes and shall be suitably and appropriately certified.
- Unintended accumulation of explosive, flammable, or toxic gas concentrations shall be prevented.
- System components shall be protected against external damages.
- Sources of ignition in hazardous areas shall be eliminated to reduce the probability of explosions.
- It shall be arranged for safe and suitable fuel supply, storage, and bunkering arrangements capable of receiving and containing the fuel in the required state without leakage. The system shall be designed to prevent venting under all normal operating conditions including idle periods.
- Piping systems, containment, and over-pressure relief arrangements that are of suitable design, construction, and installation for their intended application shall be provided.
- Machinery, systems, and components shall be designed, constructed, installed, operated, maintained, and protected to ensure safe and reliable operation.
- Fuel containment system and machinery spaces containing source that might release gas into the space shall be arranged and located such that a fire or explosion in either will not render the essential machinery or equipment in other compartments inoperable.
- Suitable control, alarm, monitoring, and shutdown systems shall be provided to ensure safe and reliable operation.
- Fixed gas detection suitable for all spaces and areas concerned shall be arranged.
- Fire detection, protection, and extinction measures appropriate to the hazards concerned shall be provided.
- Commissioning, trials, and maintenance of fuel systems and gas utilization machinery shall satisfy the goal in terms of safety, availability, and reliability.
- The technical documentation shall permit an assessment of the compliance of the system and its components with the applicable rules, guidelines, design standards used, and the principles related to safety, availability, maintainability, and reliability.
- A single failure in a technical system or component shall not lead to an unsafe or unreliable situation.

3.4.3 Fuel supply

In this chapter, a rough overview of the integration of a low flashpoint fuel system on board is given in comparison to the integration of a conventional fuel system. Some of the main criteria are explained to give an understanding about the challenges for the integration of such a system onboard a vessel.

In comparison to other applications it is important in shipping that the integrity of the whole vessel should never be endangered by any installed system on board. Therefore, the potential hazards from the system itself have to be considered as well as the hazards occurring by influences from the outside like fire, collision, flooding, etc. In relation to that, the vessel will not rely on one single system (single failure principle), as normally always a certain level of redundancy is required. Furthermore, the design of the vessel always tries to decouple potential risks to avoid any cascade of effects. For example, the energy conversion system is installed in a different compartment than the fuel storage tanks.

Besides these general design criteria, the low flashpoint fuels will bring some new topics onto the desk which have to be considered for the design of the vessel. The biggest change will be the presence of a possible explosive atmosphere. All installations which come into contact with the low flashpoint fuels must be capable of handling these fuels. This requires measures related to ventilation (premier explosion protection), use of ex-proof components, and avoidance of sparks and critical temperatures (secondary explosion protection) and the reduction of effects of possible incidents like separation of rooms, passive fire protection, active firefighting, etc. (tertiary explosion protection).

It must at least be considered that a vessel by purpose must be watertight and therefore will be mostly gastight. This results in some challenges to get rid of spillages of fuel in case of a leakage. An accumulation of gasses should be avoided under all circumstances.

The detailed layout of the storage tank area and the gas conditioning equipment requires a separate room for storage, fuel conditioning, and the energy converter itself. The rooms must be protected by structural fire protection and a suitable firefighting system must be available. Furthermore, a detection system for gas and fire should also be available. Moreover, the space must meet the requirements for a hazardous environment. To avoid any influences from one room to another all fuel conveying pipes must be able to be automatically separated in case of an incident in one of these rooms. Between the fuel conditioning system and the energy converter typically a so-called double-block-bleed valve configuration is requested. This valve configuration consists of two followed separation valves with a bleed valve between them. In case of shut down or emergency, the separation valves will be closed and the bleed valve will be opened. Even if one of these valves fails, the systems could be safely separated.

For the storage tanks, there are some additional requirements which originally came from the experiences of liquefied gas tankers which have been operating now for more than 60 years. The storage tanks are restricted in their position on the vessel to avoid being damaged during a possible collision. Therefore minimum distances from the bottom and side shell of the vessel are requested. Additionally a safety distance from the bow is required. To avoid an overpressure situation within the storage tank it is also requested to have two independent safety valves on the storage tank which are designed for the fire case and will release the content of the tank to a specific vent mast before the storage tank can burst.

3.4.4 Engine room layout

With respect to design, a FC system has some major benefits in comparison to conventional diesel engines. One main benefit is that a FC system does not require an exhaust funnel which is restricted in its position on board the vessel. The exhaust air of the FC can be handled more easily, such as through ventilation systems, if the respective hazardous areas are considered. This additional degree of freedom enables FC systems to be installed decentralized on a vessel. On conventional vessels the engines are normally installed in such a way that you have only one funnel, or one funnel per engine room. Furthermore, FC systems will supply electrical energy and can therefore be installed anywhere in the vessel. Preferably the FC system will be installed as close as possible to the respective consumer. The decentralized layout also has the potential for highly redundant systems. For example, all main vertical fire zones on a vessel could be supplied independently by a separate FC system.

Unfortunately, the current international regulations in shipping will not allow such a design for the time being. But marine designers and shipyards have realized the benefits of such a system and have started to discuss this topic with their authorities.

The precise layout of an "engine room" for a FC system requires sufficient ventilation, gas, and fire detection, a suitable degree of ex-protection as well as structural fire protection and an adequate firefighting system. Especially, all measures related to explosion protection and gas detection will be additionally required in comparison to conventional energy converters.

3.4.5 Operational aspects

One main safety issue related to the integration of a new technology like the FC system onboard a vessel is related to an adequate training of all relevant crew members. In particular, the presence of hazardous areas and the required behavior in these areas must be taught. The new technologies are more complex than conventional systems and therefore will require intensive training.

Today in shipping there should be a possibility for the crew to manually operate any system if required. Regarding the complexity of FC systems themselves, there are fewer possibilities of manual operation of these systems. Manual overrides will be restricted. It could be expected that these systems will be operated nearly fully automatic. The same will also arise in case of maintenance. The crew will not be able to maintain the FC itself, like they can do for conventional systems. It could be expected that they may change filters or a FC stack, but not work on the stack itself. That will be a difference from conventional internal combustion engines, where the crew is able to change nearly any component itself, e.g., change of piston and liner. But due to the modular design of FC systems, this will not be seen as a critical item. It could be expected that the redundancy of FC systems will be higher than for conventional systems. A loss of one stack will reduce the power supply, but it would not be the same situation as losing the full engine power.

3.5 Marine FC projects

In this section, a short overview of past and current FC projects in the maritime field is given. It should be mentioned here that only those projects are discussed which had a main influence on the development of FCs in shipping. Table 3.3 gives a summary of published FC activities in shipping. Three activity fields can be pointed out as main

Table 3.3 **Summary of FC activities in shipping**

		Merchant vessels		Navy vessels	
	Pleasure craft	Surface	Underwater	Surface	Underwater
Studies/projects	4	17		6	3
Developments	1	2		1	4
Prototypes/ demonstrations	20	7	2		1
Production	3				NN

drivers: prototypes/demonstrators of pleasure craft, studies/projects of surface merchant vessels and production of FC submarines.

NN comprises FCs of ThyssenKrupp Marine Systems submarines and material packages for the types 212A and 214 (estimated 30 systems).

3.5.1 Pleasure craft and small passenger vessels

Naturally, today the majority of marine power FC projects are in the range of pleasure craft. Private and institutional facilities are gaining experience with PEMFCs on board inexpensively. The variety of applications ranges from a few hundred watts to some kilowatts (Table 3.4).

In the following some small-scale demonstration projects are presented to allow an orientation of the current market situation:

From the end of the 1990s, several demonstration boats named HydroxyXX powered by FCs were developed in Switzerland (Affolter, 2000).

FCs on sailing yachts have their advantages especially because of the quiet operation. They are therefore superior to diesel engines. With his mid-study thesis, Steffensen not only describes the first GL classified sailing yacht No. 1 but also the integration of FCs on board sailing yachts in general (Steffensen, 2005).

An emission-free concept for boat operations on environmentally sensitive waters was presented with the Fodiator fuel cell-e-drive (Eichinger, 2011).

One of the first FC pleasure crafts ready for series production was introduced by H2Yacht GmbH, Hamburg (Pelka, 2005). A GL classified boat H2Yacht 675 was presented at the Friedrichshafen Interboat Fair 2013 by the Stadtwerke Ulm.

In the area of small passenger vessels the ZEMSHIPS project should be mentioned. Within this project the FC ship "Alsterwasser" was put into service in 2008 (Teichert, 2010). Since then the 100-passenger zero-emission ship is in regular service on Lake Alster in Hamburg for the Alster-Touristik (Alster-Touristik GmbH, 2014). Due to a current breakdown of the filling station for the vessel, there is now a risk that the FC ship may be converted to a conventional diesel engine driven vessel for cost reasons.

In the mega-yacht market the company H2-Yachts, a division of the Switzerland-based H2-Industries AG, presents the world's first energy autonomous superyacht (Stusch, 2013). The yacht will become available for charter in 2016. A special feature is the storage of hydrogen in a LOHC which enables more than 6000 NM at 10 kn.

Table 3.4 **Fuel cell projects for pleasure craft and small passenger vessels**

No	Project/ship's name	Type	Country	Year	Fuel cell	Power (kW)	Type of project
1	"Hydra"	Leisure Boat	Germany	2000	AFC	5	Prototype
2	"No 1"	Sailing Yacht	Germany	2002–2004	PEMFC	4.8	Prototype
3	Duffy electric boat	Water Taxi	USA	2003	PEMFC	41.5	Prototype
4	"Urashima"	AUV	Japan	2003			Prototype
5	Duffy-Herreshoff DH30	Water Taxi	USA	2003	PEMFC	6	Prototype
6	"Hydroxy 3000"	Leisure Boat	Switzerland	2003	PEMFC	3	Prototype
7	Deep C	AUV	Germany	2004	PEMFC		Prototype
8	"Mamelie"	Sailing Yacht	Germany	2004	DMFC	0.05	Prototype
9	"Have Blue XV/1"	Sailing Yacht	USA	2005	PEMFC	10	Prototype
10	VEGA/Pilot Vaporetto	Boat	Italy	2005–2006	PEMFC	12	Prototype
11	"Xperiance NX hydrogen"	Leisure Boat	Netherlands	2006	PEMFC	1.2	Prototype
12	zebotec	Leisure Boat	Germany	2007	PEMFC	24	Prototype
13	"Solgenia"	Research Boat	Germany	2007	PEMFC	3.6	Prototype
14	SY "Emerald"	Sailing Yacht	UK	2007	PEMFC	1	Prototype
15	"Alsterwasser"	Passenger Vessel	Germany	2008–today	PEMFC	50	Commercial
16	Frauscher 600 Riviera HP	Leisure Boat	Austria	2009	PEMFC	4	Commercial
17	BELBIM	Ferry	Turkey	2009	PEMFC	48	Prototype
18	"Nemo H2"	Canal Boat	Netherlands	2009	PEMFC	60–70	Prototype
19	Protium/"Ross Barlow"	Canal Boat	UK	2010	PEMFC	1	Prototype
20	MF "Vågen"	Passenger Ferry	Norway	2010 ff			Prototype
21	Fodiator	Electr. drive	Germany	2010–today	PEMFC	2.5	Commercial
22	"Hornblower Hybrid"	Ferry	USA	2012	PEMFC	32	Prototype
23	"Hydrogenesis"	Ferry	UK	2012	PEMFC	12	Prototype
24	"Fortuna"	Leisure Boat	Germany	2013	PEMFC	2.4	Commercial

3.5.2 Merchant vessels

Because of the high performance and energy requirements for merchant vessels, FC projects are still mainly limited to studies, like FC-Ship (Altmann et al., 2004), CREATING (de Wilde and Tillemans, 2006) or MC WAP (Schembri, 2010) (Table 3.5).

Beside others, FC-Ship draws attention to the problems of using low flashpoint fuels onboard ships. Consequences of these problems can be realized with the "Hornblower Hybrid" ferry in New York. The ship entered into service in 2012; the FC system was not approved by the US Coast Guard (USGC) until 2014 (Blaisdell, 2014).

Beside the academic studies, also a few demonstration projects have been carried out or are ongoing in the area of merchant shipping. The most important projects in Europe were carried out by Scandinavian companies:

METHAPU is a project for the "Validation of renewable Methanol based Auxiliary Power System for commercial Vessels." A 20kW Wärtsilä WFC20 containerized FC generator set was successfully tested onboard the Wallenius car carrier "Undine" (Pagni, 2008).

FellowSHIP (fuel cells for low emission ships) is a joint industry R&D project experimenting with fully integrated FCs onboard vessels and offshore platforms with the goal of making them commercially viable. Since 2009 a MTU 330kW molten carbonate containerized FC system was operated for more than 18,500h onboard the Eidesvik Offshore vessel "Viking Lady," fueled by LNG (Johnson, 2008).

The German project "e4ships" aims to improve significantly the energy supply onboard large vessels (e4ships c/o hySOLUTIONS, 2009). e4ships is a federal government, industry, and science initiated National Innovation Programme Hydrogen and Fuel Cell Technology (NIP) within the National Organisation Hydrogen and Fuel Cell Technology (NOW) (German Federal Ministry of Transport and Digital Infrastructure (BMVI), 2014). e4ships comprises the three projects Pa-X-ell, SchIBZ, and Toplaterne. Its aim is to develop concepts and gain experience with FC systems up to 500kW onboard seagoing vessels.

3.5.3 Navy vessels

3.5.3.1 Surface vessels

FC applications for naval surface ships go about all electric ships (AESs). Power demands are higher than about 500kW. Thus far, FC systems look more like stationary systems. Instead of PEMFCs, HT-FCs like MCFC or SOFC have to be taken into account. Unlike PEMFC, the development stage of HT-FCs is not yet close to the possibility of realizing concrete applications. Thus far only studies and a very limited number of demonstrators have been realized. In 1997, the US Office of Naval Research (ONR) initiated an advanced development program to demonstrate a ship service fuel cell (SSFC) power generation module. When completed, this program will provide a basis for new FC-based power generation modules for future U.S. Navy surface ships (Privette et al., 1998).

Table 3.5 Fuel cell projects for merchant vessels

No.	Project/ship's name	Type	Country	Year	Fuel cell	Power (kW)	Type of project
1	FCShip case 1	RoRo	Germany	2002–2004	MCFC/SOFC	2000	Study
2	FCShip case 2	Ferry	Germany	2002–2004	PEMFC	400	Study
3	Wallenius/"Orcelle'	Car carrier	Sweden	2004		10,000	Study
4	Felicitas	Luxury yacht	Germany	2005–2008	PEMFC/SOFC	200	Study
5	MC WAP	RoRo	Italy	2005–2011	MCFC	150	Study
6	Methapu/"Undine"	Car + truck carrier	Finland	2006–2009	SOFC	20	Prototype
7	FellowShip/"Viking Lady"	Supply vessel	Norway	2007–2010	MCFC	320	Prototype
8	SMART H2/"Elding"	Whale watching	Island	2007–2010	PEMFC	10	Prototype
9	PaXell	Cruise line	Germany	2009–2015	HT-PEMFC	120	Study
10	SchIBZ	Luxury yacht	Germany	2009–2015	SOFC	100	Study
11	Germanischer Lloyd	Feeder ship	Germany	2012		5000	Study
12	Scandlines	Passenger/car ferry	Germany	2012		8300	Study
13	Fincantieri	Range extender	Italy	2013	PEMFC	260	Delivery contract
14	H2-Yachts	Luxury yacht	Switzerland	2013	PEMFC	TBD	Study
15	CREATING	Inland navigational vessel		2006		TBD	Study
16	RIVERCELL	Inland navigational vessel	Germany	2015-2016	PEMFC	TBD	Study

3.5.3.2 Submarines

FCs for submarines are ready for series production. Similar to space applications the PEMFC must be ready for air independent propulsion (AIP). That means that these FCs must be capable of pure oxygen operation. The circle of manufacturers is for the time being very limited. In addition to the well-known advantages of FCs, like efficient and quiet operation, also no further type of signature (thermal energy, electrical stray field, and others) can be detected. The world leader in conventional submarine design and construction with AIP-systems based on PEMFC is ThyssenKrupp Marine Systems (ThyssenKrupp Marine Systems, 2014). HDW classes 212A and 214 have set new standards especially in the areas of signatures and range. For the time being, hydrogen is stored in metal hydride cylinders. Liquid oxygen is transported in tanks on board. In future submarine applications, alternative hydrogen carriers like methanol or ethanol might be used to prolong underwater endurance. Ongoing reformer developments were reported at the EHEC-Conference 2014 in Seville (Krummrich and Llabres, 2014; Castro, 2014).

FC projects for navy vessels are summarized in Table 3.6.

3.6 Future trends

Future trends for FC systems in shipping are hard to predict. Currently it can be observed that the shipping industry is looking in the direction of gas as fuel, mainly LNG, in internal combustion engines. For smaller vessels, also pure electric vessels have been built (Passenger/Car-Ferry in Norway).

Based on today's knowledge level, it could be expected that hydrogen will not be the preferable fuel solution on larger merchant vessels for FCs in the future, because of the volumetric storage density of pure hydrogen. Nonetheless first concepts have been developed for hydrogen-fueled, FC powered merchant vessels. As an example, the Zero Emission Feeder concept from GL should be mentioned (Sames et al., 2012). Currently ongoing scientific investigations with LOHC show a good perspective for LOHC as a fuel alternative for merchant shipping.

Regarding FCs in shipping, there are minor activities which may be related to the relative high costs for FC systems in comparison to diesel engines. Parallel to that,

Table 3.6 Fuel cell projects for navy vessels

No.	Project/ ship's name	Type	Country	Year	Fuel cell	Power (kW)	Type of project
1	"U1"	Submarine	Germany	1988–1989	AFC	100	Prototype
2	U 212A	Submarine	Germany	1995 ff	PEMFC	300	Commercial
3	"Vindicator"	USCG	USA	1999	MCFC	4×675	Study
4	U 214	Submarine	Germany	2000	PEMFC	240	Commercial
5	SSFC	Navy	USA	2000 ff	MCFC	675	Study

Table 3.7 Comparison of emission reduction potential of different fuels and energy converters

Energy carrier	HFO (2% sulfur)	Methane CH_4	Methane CH_4	Hydrogen H_2
Energy converter	*Diesel engine*	*Gas engine*	*Fuel cell*[a]	*Fuel cell*[a]
CO_2 (g/kWh)	560	444	402	0
NO_x (g/kWh)	14	2	0	0
SO_x (g/kWh)	10	~0	0	0
PM (g/kWh)	0.5	~0	0	0

[a] Efficiency of fuel cell 55% in comparison to 42% of engine (1000 kW system).

the shipping industry is still affected by the ongoing international shipping crisis and shipping companies are under financial pressure. But still some shipping companies in northern Europe are working on hybrid concepts where FC systems are also included. As an example, the zero emission concept from Scandlines (Rohde and Nikolajsen, 2013) may be mentioned here.

Without any further cost-related step and further developments in the area of HT-FCs, it can be expected that activities for FCs in shipping will develop on a lower level.

It must be considered that several other technical solutions in addition to FC systems exist that fulfill the upcoming environmental requirements for air emissions. For a better overview of the environmental situation in shipping, a short comparison of the emission reduction potential of different fuels and energy converters in comparison to the HFOs used today is given in Table 3.7 (Germanischer Lloyd, 2010). It should be stated here that scrubber systems downstream of the engine are also an alternative solution.

Table 3.7 clearly shows that FCs operated by hydrogen are the cleanest possible solution. Furthermore, it must be stated that FC systems operated with natural gas have to compete with gas engines from the environmental perspective.

References

Affolter, J.-F., 2000. Swiss Fuel Cell Passenger and Pleasure Boats. Lucerne, "Fuel, Cell 2000", p. 8.

Alster-Touristik GmbH, 2014. ATG. [Online] Available at: http://www.alstertouristik.de/.

Altmann, M., Weindorf, W., Weinberger, M., 2004. Life Cycle Analysis Results of Fuel Cell Ships. L-B-Systemtechnik GmbH/MTU Friedrichshafen GmbH, Ottobrunn/Friedrichshafen.

Blaisdell, M., 2014. Hornblower hybrid is a model of alternative energy innovation. [Online] Available at: http://www.examiner.com/article/hornblower-hybrid-is-a-model-of-alternative-energy-innovation.

Blesl, M., Ohl, M., Leipnitz, T., 2007. Marktentwicklungen bei stationären Brennstoffzellen. eBWK 59 (12), 1–5.

Castro, Á., 2014. Developments in the Context of the Hydrogen and Fuel Cell-Based AIP System. EHEC: Seville.

CellEra, 2011. Platinum-free membrane fuel cell (PFM-FC). [Online] Available at: http://www.cellera.biz/ (Zugriff am 15.05.14).

Clear Edge Power, 2014. PureCell® Model 400 System. [Online] Available at: http://www.clearedgepower.com/energy/purecellmodel400system/ (Zugriff am 15.05.14).

Corbett, J., Winebrake, J.J., Green, E.H., Kasibhatla, P., Eyring, V., Lauer, A., 2007–09. Mortality from ship emissions. A global assessment. Environ. Sci. Technol. 41 (24), 8512–8518.

de Wilde, H., Tillemans, F., 2006. Fuel cell technology in inland navigation. [Online] Available at: http://www.ecn.nl/docs/library/report/2005/rx05177.pdf.

Den Boer, E., Faber, J., Nelissen, D., 2009. Proposal for an environmental ship index. In: Air Pollutants and CO_2. CE Delft, Delft.

c/o hySOLUTIONS, H., 2009. e4ships—fuel cells in marine applications. [Online] Available at: http://www.e4ships.de/.

Eichinger, R., 2011. http://www.fodiator.de/. [Online] Available at: http://www.fodiator.de/.

Europäisches Parlament und Rat der europäischen Union, 2005–07. Richtlinie 2005/33/EG des Europäischen Parlaments und des Rates. Zur Änderung der Richtlinie 1999/32/EG hinsichtlich des Schwefelgehalts von Schiffskraftstoffen. EU: Straßburg.

Fuel Cell Energy, 2013. 300 kW DFC300. [Online] Available at: http://www.fuelcellenergy.com/products-services/products/300-kw-dfc300/ (Zugriff am 15.05.2014).

Gerboni, R., Pehnt, M., Viebahn, P., Lavagno, E., 2008. Final Report on Technical Data, Costs and Life Cycle Inventories of Fuel Cells. NEEDS: Rom.

German Federal Ministry of Transport and Digital Infrastructure (BMVI), 2014. Nationales Innovationsprogramm Wasserstoff und Brennstoffzellentechnologie (NIP). [Online] Available at: http://www.bmvi.de//SharedDocs/DE/Artikel/UI/nationales-innovationsprogramm-wasserstoff-und-brennstoffzellentechnologie-nip.html (Zugriff am 15.05.2014).

Heinzel, A., 2009. Stand der Technik von Polymer-Elektrolyt-Membran-Brennstoffzellen. Ein Überblick. Chemie Ingenieur Technik, Special Issue: Brennstoffzellen und Wasserstofftechnologie 81 (5), 567–571.

International Maritime Organisation, Sub Committee Bulk Liquids and Gases (BLG), 2009. MSC-86 Resolution (285) Interim Guideline on Safety for Natural Gas Fuelled Engine Installation in Ships. International Maritime Organization: London.

International Maritime Organisation, MEPC, 2006. International Convention for the Prevention of Pollution from Ships. IMO, London.

International Maritime Organisation, MEPC, 2009. Prevention of Air Pollution from Ships. Second IMO GHG Study 2009. Executive Summary. MEPC 59/4/7. IMO, London.

Johnson, P., 2008. Fuel Cells for Commercial Ships. [Online] Available at: http://onlinepubs.trb.org/onlinepubs/archive/marineboard/fall08/johnson.pdf.

Kallo, J., 2011. Antares DLR-H2—Abheben mit Brennstoffzellenantrieb. Deutschen Zentrum für Luft- und Raumfahrt (DLR), Stuttgart.

Krummrich, S., Llabres, J., 2014. Methanol Reformer—The Next Milestone for Fuel Cell Powered Submarines. EHEC, Seville.

Germanischer Lloyd, 2002. Klassifikations- und Bauvorschriften, VI Ergänzende Vorschriften und Richtlinien, Teil 3 Maschinenanlagen, Kapitel 11 Richtlinien für den Einsatz von Brennstoffzellen-Systemen an Bord von Wasserfahrzeugen. GL, Hamburg.

Germanischer Lloyd, 2009. Klassifikations- und Bauvorschriften, I Schiffstechnik, Teil 1—Seeschiffe, Kapitel 2 Maschinenanlagen. GL, Hamburg.

Germanischer Lloyd, 2010. Endbericht zur Marktstudie "Brennstoffzellensysteme in der Seefahrt". Germanischer Lloyd, Hamburg.

Lloyds Register, 2009–07. IMO MEPC 59 Report. Lloyds Register: London.

Meech, R., 2008. Designation of the Mediterranean Sea as a SO_x Emission Control Area (SECA) under MARPOL Annex VI. REMPEC, Malta.
Pagni, J., 2008. Fuel cells—green power for the future. *Wärtsilä—Twentyfour7 Stakeholder Magazine*, 1, 63–64.
Pelka, W., 2005. H2Yacht. [Online] Available at: http://www.h2yacht.com/.
Privette, R., Privette, R., Flynn, T.A., Perna, M.A., Holland, R., Rahmani, S., Woodburn, C., Scoles, S.W., Watson, R.C., 1998. 2.5 MW PEM Fuel Cell System for Navy Ship Service Power. National Energy Technology Laboratory (NETL): SL.
Reuß, H.-J., 2008. Motorenhersteller zwischen Wunsch und Wirklichkeit. Hansa Int. Maritime J. 145 (7), 44–47, Jahrgang.
Rohde, F., Nikolajsen, C., 2013. Zero-emission design concept for a scandlines ferry. In: STG—International Conference on Ship Efficiency 2013, Hamburg.
Sames, P., Plump, R., Rohde, F., 2012. A vision for a zero emission container feeder vessel. [Online] Available at: http://www.clean-baltic-sea-shipping.com/uploads/files/GL-zero_emission_cont_feeder_Pump.pdf.
Schembri, M., 2010. MC WAP—molten-carbonate fuel cells for waterborne application. [Online] Available at: http://www.fch-ju.eu/sites/default/files/documents/ga2009/Marco%20Schembri.ppt.
Steffensen, M., 2005. Brennstoffzellen-Systeme auf Segelyachten. Technische Universität, Hamburg-Harburg.
Stusch, M., 2013. World's first energy autonomous motor yacht concept by H2-Yachts. [Online] Available at: http://www.charterworld.com/news/62m-motor-yacht-concept-h2yachts-worlds-energy-autonomous-superyacht.
Teichert, T., 2010. Market Potential of the Zemships Technology—An Indicative Empirical Examination. BSU, Hamburg.
ThyssenKrupp Marine Systems, 2014. Submarines. [Online] Available at: https://www.thyssenkrupp-marinesystems.com/en/submarines.html.
US Environmental Protection Agency, 2009. Proposal of Emission Control Area Designation for Geographic Control of Emissions from Ships. EPA, Miami.
Viviani, N., Kähler, N., Brevik, D., Skjölsvik, K., 2004. Summary of Ship Design and Areas of Further R&D, FC Ship. FCShip: SL.
Wendt, H., 2006. Stationäre Brennstoffzellen. Stand der Entwicklung, Kostensituation, Marktaussichten. BWK 58 (10), 46–50.
Zentrum für Brennstoffzellentechnik (ZBT), 2009. Bericht zur Marktstudie Brennstoffzellen in der Seeschifffahrt. ZBT, Duisburg.

Hydrogen-fueled aeroplanes

4

A. Godula-Jopek[*,†], *A. Westenberger*[‡]
[*]Airbus Group Innovations, Munich, Germany, [†]Institute of Chemical Engineering, Polish Academy of Sciences, Gliwice, Poland, [‡]Airbus Operations GmbH, Hamburg, Germany

Abbreviations

ACARE	Advisory Council for Aviation Research and Innovation in Europe
APU	auxiliary power unit
ATAG	air transport action group
CS	Certification Standards
EASA	European Aviation Safety Agency
EQHHPP	Euro-Québec Hydro-Hydrogen Pilot Project
H_2ICE	hydrogen-fueled internal combustion engine
HALE	high altitude long endurance
ICE	internal combustion engine
LCA	life cycle assessment
MBB	Messerschmitt Bölkow Blohm, later Deutsche Airbus
MTOW	maximum take-off weight
OWE	operating weight empty
PEMFC	proton exchange membrane fuel cell
RAT	ram air turbine
UAV	unmanned aerial vehicle
WTW	well-to-wake
WAI	wing anti ice

4.1 Introduction to hydrogen vs. traditional technologies: Differences and similarities, advantages, and disadvantages

Traditionally liquid hydrocarbons are used as fuel for aircraft propulsion. Turbo engines are typically fueled by kerosene, while piston engines run on gasoline or diesel. Main criteria for the selection of kerosene in the past have been based on the chemical and physical properties, for example flammability and ignition energy and the volumetric high specific energy content. In addition, economic and practical aspects like low price, handling, infrastructure, and availability have led to the selection of kerosene as an aircraft fuel. The very narrow flammability window of 6–7.5% in air has been accepted as an advantage.

Compendium of Hydrogen Energy. http://dx.doi.org/10.1016/B978-1-78242-364-5.00004-X

Based on these facts, hydrocarbons are still today the most suitable energy carriers for general and commercial aviation.

Compared to kerosene, the physical and chemical properties and behavior of hydrogen are very different. As well as being able to be burned in any kind of combustion engine, such as piston or turbo engines, hydrogen can also be used as energy to power conversion to electricity by chemical reactions in fuel cells. But even the higher specific energy content by weight of hydrogen—it is about four times higher than kerosene—does not completely outweigh the disadvantage of its higher volume in its liquid or compressed state.

Hydrogen stored as a liquid at 20 K has a higher energy density, but the liquefying process requires about a third of the energy content in the fuel. The cryogenic tanks must be kept in a cryogenic status in order to avoid major heat flux into the storage system, which would lead to an unacceptable boil-off of the liquid hydrogen.

Table 4.1 presents some properties of compressed and liquid hydrogen versus kerosene and other fuels (methanol, diesel, and petrol).

The variety of power and propulsion chain components opens up the spectrum for different applications. These have changed in the recent past with changes in the operational intentions of airborne systems. The most important arguments have been economical suitability to the aircraft production, operation, and environmental compatibility.

More efficient energy-to-power transformation with technologies such as fuel cells, along with environmental aspects, could lead to an advantage over fossil fuels. On the other hand, today's underdeveloped infrastructure has prevented hydrogen from being an alternative fuel for commercial aircraft or even small aircraft, including motor gliders. Only for specific applications like testing and demonstration of aircraft components have other fuels, including hydrogen, been taken into account. For example, several tests have been performed on one of the three turbo engines of

Table 4.1 Mass and volume density of hydrogen compared to other fuels

	Gravimetric energy density		Volumetric energy density		Flammability limits	Explosive limits
Fuel	**MJ kg^{-1}**	**kWh kg^{-1}**	**MJ kg^{-1}**	**kWh kg^{-1}**	**vol.%**	**vol.%**
Hydrogen compressed 200 bar	120	33.3	2.1	0.58		
Hydrogen liquid	120	33.3	8.4	2.33	4–75	18.3–59.0
Methanol	19.7	5.36	15.7	4.36	6–36.5	6–36
Petrol	42	11.36	31.5	8.75	1–7.6	1.1–3.3
Diesel	45.3	12.58	35.5	9.86	0.6–7.5	0.6–7.5
Kerosene	43.5	12.08	31.0	8.6	0.6–4.7	0.7–5

a Tupolev 155, which has been modified accordingly for that purpose (Tupolev 155, 2012). Another example of a demonstration aircraft is the motor glider Antares-H2, developed and tested by the Institute of Technical Thermodynamics of the German Aerospace Center (DLR e.V.) and using proton exchange membrane (PEM) technology for energy conversion into electrical power. On Antares-H2 the measurements showed that the average fuel consumption amounted to about 1 kg of hydrogen per flight hour at level flight. The hydrogen tank had a capacity of 5 kg of gaseous hydrogen at a pressure of 35 MPa; thus a flight time of up to 5 h was possible, resulting in flight distances of 500–700 km over ground, depending on the ambient conditions and flight maneuvers. Several flights have been accomplished covering an overall distance of about 1500 km with an overall flight time of more than 10 h (Rathke et al., 2013). It could be observed that a niche market is emerging for aircraft, such as HALE (high altitude long endurance). However, various power trains for the conversion of H_2 into power are still required (Noll et al., 2004).

This may change in the future, as some scenarios are based on the prediction of crude oil depletion and the fight against global warming due to increasing CO_2 concentration in the atmosphere (http:// www.atag.org/our-activities/climate-change.htm, 2014). Between 1989 and 2011 aviation traffic grew 4.6% per year (Pereira et al., 2014). This had consequences in the emissions of greenhouse gases (GHGs) globally and local pollutant emissions as well, including particulate matter (PM), sulfur oxides (SO_x), carbon monoxide (CO), hydrocarbons (HCs), and nitrogen oxides (NO_x). It is forecast (Eurocontrol, 2014) that at the European level, from 2015 onwards, growth of air traffic is expected to be at around 2.7%. For the entire 2014–2020 period, flight growth is forecast to average 2.5% per year. This will lead to the increase of emitted GHGs and other pollutants, which must be reduced in order to meet the goals of vision 2020 and 2050. In 2011, aviation already represented 14% of the CO_2 emissions of the transportation sector in Europe. It has been assessed that the radiative forcing resulting from CO_2, H_2O vapor, NO_x, CO, PM, and SO_x emitted to the atmosphere could be in the range of 2–4 times higher than CO_2 emissions by aircraft (Pereira et al., 2014). According to Janic (2014) commercial air transportation contributes about 2–3% of the total manmade emissions of the GHGs. The majority comes from the transport sector (about 20%), of which the air transportation contribution is about 12%. It is questionable whether domestic air transportation is the most air-polluting transport mode. The quantity of fuel consumption per passenger kilometer has improved in the last two decades by *ca.* 1–2%. The question remains whether hydrogen (LH_2) or natural gas (LNG) can be alternatives for aviation. Several studies have been carried out to evaluate more sustainable alternatives for aircraft in order to assess the difference in energy consumption by different fuels (kerosene Jet A, liquid hydrogen, and LNG) applied in different aircrafts (well-to-wake, WTW, approach) and in order to assess whether alternatives (Jet A and LH_2) could be used to reduce emissions of pollutants in aviation. For example, an evaluation of energy consumption and emissions of pollutions from production of renewable and nonrenewable technologies has been performed using the model for LCA (life cycle assessment) for hydrogen (Pereira et al., 2014). The aircraft (long-range, above 5000 km, and short-range flights, below

5000 km) were analyzed with three different fuels: standard Jet A fuel, LNG, and liquid hydrogen coming from different sources (e.g., SMR, electrolysis). It was discovered that LH_2 coming from electrolysis with electricity from hydro energy shows the best WTW results for all aircraft (19% and 80% less energy consumption, respectively, than the Jet A fueled aircraft). In terms of emitted pollutants, LNG aircraft showed the worst results (more than 4% and 58% less than Jet A and LH_2 from hydro energy aircraft). In the case of LH_2 aircraft, the NO_x pollution was in the range of 16% and 23% (depending on the energy needed for the production of liquid hydrogen). An observation has been made (Pereira et al., 2014) that even LH_2 produced by SMR has lower energy consumption (8%) than the same aircraft with Jet A fuel. Hydrogen obtained by electrolysis with electricity from hydro energy could reduce the environmental cost by *ca.* 51–60% (depending on the aircraft type and flight duration) in comparison to the same aircraft fueled by Jet A.

4.2 Hydrogen fuel on aircraft—Challenges and requirements

Specific application of hydrogen in the aeronautic sector with its operating conditions requires dedicated technologies and design solutions. This could result in very specific drivers for the final selection of technologies for handling, storage, and energy to power conversion. Among the main key challenges that have to be considered are the type of application, handling, specific weight, and volume and the thermal behavior of the conversion technology.

For military and paramilitary use on relatively small air vehicles, hydrogen offers two different advantages. The first one comes from the high energy content by weight; the second comes from the ability to convert the contained energy at low temperature by electrochemical reaction in fuel cells into electrical power. With long endurance flight, a lightweight fuel like hydrogen is an advantage. As the weight of the propulsion system of the aircraft remains nearly the same, regardless of the mission time it is designed for and operated at, it is obvious that the specific weight of the energy has a major influence on the maximum take-off weight (MTOW) of the entire vehicle. Depending on the size of the aircraft and its mission time, different storage systems offer solutions that could have an advantage over kerosene or gasoline. Another aspect is the option to convert the energy content of hydrogen at low temperatures and low noise to electrical power by using fuel cell technology. In this case the aircraft is nearly undetectable by the usual means such as infrared detection. Civil research aircraft, mainly pilotless vehicles (UAVs, unmanned aerial vehicle), use a similar technology. For those vehicles lightweight solutions are of a high but not the only priority. Architectures including energy recovery technologies like the use of photovoltaic and "reversible" fuel cell technology are extremely beneficial (Noll et al., 2004). The main components for such a power train are a hydrogen and oxygen storage system, a reversible fuel cell, a water reservoir, power electronics, electromotor, and solar cell on the wing. This arrangement operates constantly on the electromotor. Electrical power is provided at night by the fuel cell fueled by hydrogen and oxygen from the storage systems (pressure bottles). The process water is stored in the water reservoir. During the day the electrical power comes from the photovoltaic cells. One part of the electrical power is consumed directly by the electromotor, while the rest of the electrical power goes to the reversible fuel cell, which operates now in an electrolyzer mode, splitting up the process water from the reservoir into H_2 and O_2. Compressor units fill up the storage systems. At the next night phase the process continues with night mode operation.

Looking at larger scale aircraft such as commercial ones, the long-term availability of hydrogen and the environmental aspects are the main priorities and concerns. The first studies in this direction have been performed by "Daniel Brewer" at the air framer McDonnell Douglas. Later on the idea was picked up by MBB (Messerschmitt Bölkow Blohm and later Deutsche Airbus) in cooperation with the Russian air framer Tupolev. Continuation of this activity led to the question of sufficient "clean and green" hydrogen production. The Euro-Quebec Hydro-Hydrogen Project (EQHHPP) was set up to answer part of this question. Electricity from hydro power should cover a significant amount of the hydrogen needed for the "Cryoplane"-type aircraft. Results have shown, however, that the total fuel demand could not be covered by this source. On the other hand, the Desertec concept predicts that several percent of the world's energy is delivered by the sun. During 6h deserts receive the solar energy equivalent to the energy mankind on earth consumes in 1 year. The efficiency of sun to electricity and then to electrolysis for producing hydrogen depends on the conversion, compression, and distribution route. The life cycle analyses could therefore vary greatly. It could be shown that the impact of H_2 production is close to zero on the environment.

As the next step the European project "Cryoplane" has been launched (Figure 4.1). The intention was to achieve long-term continuing growth of civil aviation until every human being on earth can fly as often and as far as desired, and when doing so, create no harm to other human beings or the environment. The project was driven by MBB with the objective to develop a conceptual basis for applicability, safety, and full environmental compatibility. This system analysis covered all relevant technical, environmental, societal, and strategic aspects, providing a sound basis for initiating larger scale activities in order to prepare for the development and introduction of liquid hydrogen as an aviation fuel. It investigated medium/long-term scenarios for a smooth transition from kerosene to hydrogen. On the technical side it could be confirmed that hydrogen as a fuel for large commercial aircraft is feasible. The increase of the wetted surface of the aircraft due to the super isolated liquid hydrogen vessels led to an increase in drag, which caused higher energy consumption by 9–14%. The structure of the storage tank increased the OWE (operating weight empty) by 23%. On the other hand, the high energy by weight ratio of hydrogen led to a variation of the MTOW between +4.4% and −14.8%, depending on the mission range of the aircraft. One major question, which could not be fully answered by testing, was the influence of the emission of water vapor in the higher atmosphere. It could only be assumed that water vapor, which is a strong GHG, remains only a few weeks in the atmosphere, whereas CO_2 remains for about 100 years in the air.

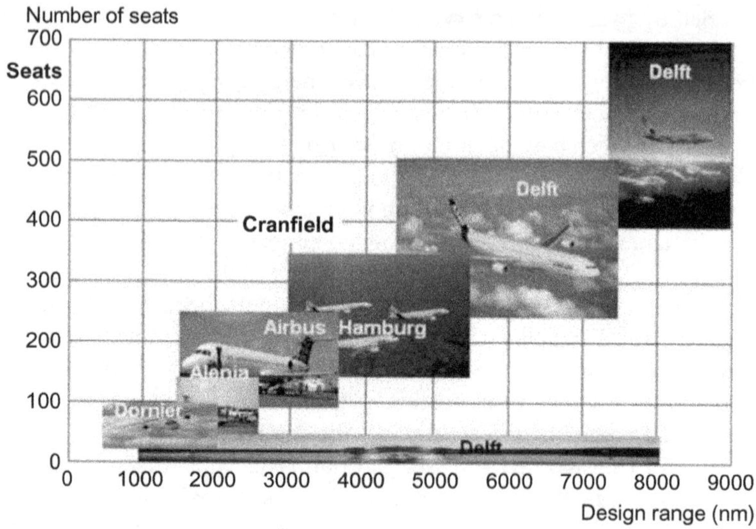

Figure 4.1 Range of aircraft categories that have been the subject of this project.
Source: EU-Project "Cryoplane" CONTRACT No.: G4RD-CT-2000-00192, PROJECT No.: 24.09.2002. RD1-1999-10014, Final technical report—publishable version.

Extra-atmospheric ultra-high speed vehicles like the Sänger, Zehst (Hyperschall-Studie ZEHST, n.d.), and others belong to a special category of aircraft. The initial concepts have been military motivated and driven. The intention for extra-atmospheric flight operation was to increase the speed of the vehicle to maximum in order to perform technical reconnaissance. Civil intentions have been devoted to fast commercial travel. The propulsion concepts have been based on hydrogen as fuel and pure oxygen as the oxidant. This was necessary because of the low partial pressure of oxygen at the operating altitude above 45,000 ft (15,000 km) as a consequence of the low atmospheric pressure. In order to compensate for this effect of low partial pressure, pure oxygen is needed as in the case of space vehicles (Figure 4.2).

4.3 Advantages and disadvantages of hydrogen storage methods in aeronautics

With respect to the hydrogen storage options, several critical features have to be solved. In general the weight, volume, and cost of hydrogen storage systems are too high, as with automotive applications. In addition to this, handling and refueling issues are not yet optimized. For these reasons hydrogen remains in niche applications in all transport sectors, including aviation. Due to that fact, there is no one preferred storage solution to be selected. The best solution depends always on the characteristics of the onboard consumer, its operation and other reasons such as costs, turn-around time, and environmental conditions. On the other hand, these different storage solutions open up

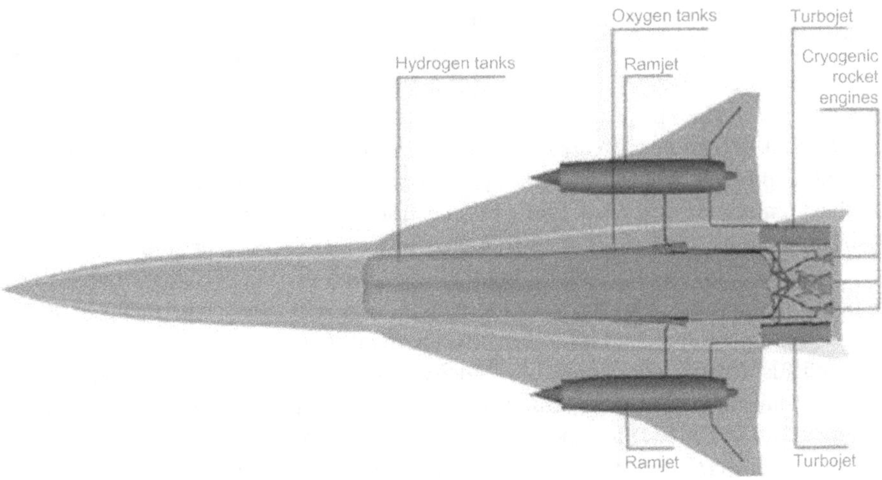

Figure 4.2 Propulsion configuration of the project Zehst.
By permission of Airbus Group Innovation Work.

quite a wide range of suitable designs and architectures. Onboard hydrogen storage systems create severe technical challenges for the aeronautic industry, which are more significant than for automotive applications. Unfortunately there are no suitable components for commercial aircraft available at present.

It is quite clear that the storage system for cars has to comply with different constraints than those for aircraft, like cost, easy and safe handling, and crash-proof design. In general it could be stated that the most critical item in aeronautics is the weight. Because hydrogen is now only used on aircraft or flying vehicles in limited numbers, it can be assumed that only trained staff will do the handling and refueling. This will change when hydrogen-fueled power systems enter the commercial aircraft market. In this case it must be decided whether the hydrogen services will be done by refueling or by just exchanging the whole storage tank, like a battery.

If applications grow to the "Cryoplane" size, the storage system has to follow completely different criteria. With current knowledge, liquid hydrogen is the best choice because of its high energy density. Ground handling and turn-around time has to be considered differently. Even then, the complete fuel system from the tank to the engines will be a major challenge. The "Cryoplane" study shows that all this could be possible, because the chemical and the space industries deliver quite a lot of technical solutions.

Concerning costs and infrastructure, the aeronautical application should be the early adopters of such a new technology. However, since the commercial air transport systems have to comply with high level safety standards and simultaneously with strong competitive forces between operators and investors, air-framers remain more and more conservative and are waiting for other early adopters, which may be the automotive industry.

The current U.S. Department of Energy (DOE) targets for onboard hydrogen storage systems for light duty vehicles require that in 2017 hydrogen gravimetric and volumetric capacities should reach level of 5.5 wt.% and 0.04 kg L^{-1}, respectively, corresponding to usable specific energy from hydrogen of 1.8 kWh kg^{-1} (U.S. Department of Energy, 2009). High-pressure cylinders for compressed gas and other high-pressure elements limit the choice of construction materials and fabrication techniques, within weight, volume, performance, and cost constraints. In addition, hydrogen refueling times are too long (Krishna et al., 2012). Research is also needed on improving hydrogen discharge kinetics and simplifying the reactor required for discharging hydrogen on board the vehicle (like the volume, weight, and operation). For metal hydrides, weight, system volume, and refueling time are the primary issues. In case of high surface area adsorbents, which have low density and low hydrogen binding energies and as such require cryogenic temperatures, volumetric capacity, and operating temperature are important. Nowadays the most proven, tested, and commercially available hydrogen storage is in gaseous or in a liquid form. In order to achieve a satisfying energy density with compressed storage systems, the operating pressure has been raised up to 70 MPa (700 bar) (Anders, 2007). The constraints and challenges for storage systems are, among others, the following: high volumetric and gravimetric hydrogen densities due to weight and space limitations, low operating pressures for safety concerns, fast kinetics for loading and de-loading, reversibility, operating temperatures in the range of −50 to 150 °C and costs of hydrogen storage system (Marnellos et al., 2008). For stationary applications, gravimetric and volumetric densities are not so critical due to the fact that on ground there is enough available space. This is not the case for aeronautical applications.

The main concern with LH_2 storage is an issue of minimizing hydrogen losses from liquid hydrogen storage vessels due to heat leaks, which are a function of the size, shape, and thermal insulation of the vessel used. Compressed hydrogen is still the preferred option for storing and distributing hydrogen, especially for the automotive sector, due to quick refueling and established infrastructure for compression. Compressed hydrogen stored at cryogenic temperatures (CcH_2) is a compromise between liquid hydrogen and compressed gas storage at ambient temperatures. CcH_2 refers to the storage of hydrogen at cryogenic temperatures in a container that can be pressurized to 250–350 bar, as opposed to current cryogenic containers storing liquid hydrogen at near-ambient pressures. Cryo-compressed hydrogen storage can include liquid hydrogen, cold compressed hydrogen, or hydrogen in a two-phase region (saturated liquid and vapor) (Ahluwalia et al., 2010). CcH_2 storage takes advantage of the LH_2-based infrastructure and can be compatible with the 350 bar CGH_2 infrastructure. It seems that solid-state hydrogen storage is potentially better when taking into account energy efficiency, safety, and gravimetric and volumetric storage capacities. With regard to the technical targets and requirements of the US DOE for onboard hydrogen storage systems for light duty vehicles, some materials meet the criteria in terms of gravimetric and volumetric hydrogen storage capacities but it looks like, for the time being, they could not satisfy the targets on a system level (0.055 kg $H_2\,kg^{-1}$ and 0.040 kg $H_2\,L^{-1}$ system in 2017). In addition, there are several other factors that limit or even prevent their practical application. Such material should be safe, lightweight and have an inexpensive storage medium when compared with gaseous and liquid storage. The state of the art of solid-state hydrogen storage focusing on both physical and chemical aspects has been presented by Reardon et al. (2012).

4.4 Available energy conversion technologies

Depending on production, hydrogen as a secondary energy carrier could be very attractive as a clean or zero emission fuel of the future. As the oxidation product of hydrogen, only water or water vapor is formed, therefore there are practically zero harmful emissions. Hydrogen can be utilized in several different modes for power generation with potentially high efficiency. It can be used in an internal combustion engine (ICE)—e.g., a piston engine as in the "phantom eye"—to generate mechanical power for a propeller; gas turbines producing thrust by pass fans and turbine exhaust gas, as proven in the TU 155; or in the fuel cell to produce electrical power for propulsion based on an electric motor in combination with a fan drive or a conventional propeller. The most significant difference from kerosene is the possibility of using hydrogen in combination with fuel cell technology. Fuel cell converts the chemical energy of fuel into electrical energy directly as electrons are released during the formation of water molecules from hydrogen and oxygen. Depending on the operational conditions, the requirements of the propulsion chain may differ and therefore mainly influence the selection of the components of the whole chain. Usually long-range aircraft, like commercial ones, need a very efficient propulsion system which can be operated at high Mach numbers at altitude. This applies as well to HALE aircraft like Helios, with the difference being that the speed of those vehicles is much slower. Other operational conditions may lead to different selections of storage, conversion, and propulsion components, respectively.

Onboard generating systems like fuel-cell powered auxiliary power units (APUs) or RATs (ram air turbines) could provide additional onboard power for aircraft. Considering that, PEMFCs (proton exchange membrane fuel cells) have great potential to emerge as a practical technological solution for power in the transport sector, including aeronautics. For example, in the context of the UAV application, Protonex Technology Corporation had demonstrated (Osenar et al., 2008) in 2008 within the Aerovironment Puma UAV Platform a launchable UAV with low altitude reconnaissance for stationary and moving targets, achieving a flight over 9 h, compared to the typical 2–3 h available on current battery technology. The hybrid power system including fuel cartridge demonstrated above 500 Wh kg^{-1}.

To evaluate the performance of a 200 Wel. PEMFC integrated with an ammonia borane based hydrogen generator (with a tetraethylene glycol dimethylether as a promoter of the reaction), flight tests for 57 min with the UAV platform (Ucon System, RemoEye-006) have been successfully conducted at the Goheung Aviation Center at the Korea Aerospace Research Institute (Seo et al., 2014). The UAV was mainly maneuvered at an altitude of 200 m with a cruising speed of 60 km h^{-1}. The total duration time for H_2 production including the processes for preheating, take-off, cruising and maneuvering, and taxiing was close to 1 h.

4.5 Available infrastructure (production, airport)

Since commercial aircraft today have propulsion or power systems, it is obvious that there is no hydrogen infrastructure available. There are smaller hydrogen filling stations for the supply of ground service vehicles at some airports (Stolten, 2010). Consequently, a suitable size of hydrogen provision needs to be built up. The size and type of supply system depends on the type and size of the consumer—this could be the full propulsion system, a large one or even a small power system. Depending on the aircraft category, that could vary between a large hub with a high traffic density and a small remote airport. Another dimension comes from the aircraft type, which means a long-range type or a short-range type. This applies to both applications, whether primary propulsion or system powered applications (Cryoplane). Due to the wide range of possibilities of hydrogen demand, a separate investigation is mandatory. In the case of very frequent demand, as for power independent galleys which have a generator set consisting of CH_2 storage and a fuel cell as a converter, a suitable CH_2 infrastructure at each airport would be needed. It can be assumed that in this case the hydrogen could be provided by a central production system and distributed by trucks. In the case of synergies, which means that other consumers would like to use ground vehicles, the final demand could increase to a size where an onsite large-scale production would be beneficial. Unfortunately, for airports the impact is minor for the time being. The condition for hydrogen storage is not very different from the storage for hydrocarbons. The storage for liquid hydrocarbon is even more critical, due to the fact that in case of a leakage it could contaminate the soil and the water.

The demand for LH_2 for the propulsion of aircraft would require a major change to the airport. In this case, it could be expected that onsite production would be mandatory. The "Cryoplane" project showed that even this scenario is technically feasible.

4.6 Operational aspects (turn around)

The hydrogen service on ground between arrival at the gate and leaving the gate could vary depending on the size and the operation of the applied power system. An extremely small service could have few to no service requirements—this will be the case if a fuel cell unit with a hydrogen storage system is used as a "sleeping" emergency power system. At the other extreme, a service and infrastructure could be a huge challenge, which would be the case if hydrogen is used for aircraft propulsion.

In any case, technical solutions and requirements for safe handling do exist, since hydrogen is present in industry on different scales. It could be assumed that small storage units are based on solid-state hydrogen storage systems, where the complete power unit is electrically disconnected and exchanged, like the Horizon Aeropack or the hydrogen storage cartage systems from companies Zoz GmbH and Hera. For the next larger storage size, gaseous compressed hydrogen is the best choice (Rau, 2003). For this application AirProducts offers a storage system, which could be qualified for onboard use of a commercial aircraft. Hydrogen energy storage systems could be serviced by either replacement of the entire storage system or by onboard refilling of the pressure bottle. With respect to the short turn-around time needed, an exchange of a complete cartridge-like storage system is more likely. At certain sizes liquefied hydrogen is the best state for storing at the aircraft. Even taking into account that the storage vessel is quite heavy and expensive, it has an advantage in the overall energy density in weight. In the Cryoplane project, different categories of aircraft, where the main engines were fueled by LH_2, have been studied. It was found that the ratio of stored LH_2 to a structural tank weight increased with larger aircraft sizes and longer mission distances. Another storage concept is the cryo-compressed storage (Cryo-Compressed Hydrogen Storage for Vehicular Applications, U.S. Department of Energy Hydrogen Program, 2006). This kind of storage system was developed for the car industry. A benefit for aircraft applications has not yet been found.

4.7 Safety aspects (layout, design, and strategy)

Safety issues in handling of hydrogen are very important at each step, starting from the production through storage, to the end-use application. It is important to remember that hydrogen has been in industrial use for more than 100 years. The world production today is about 54 Mt, equal in terms of energy to 105 Mt kerosene per year (Stolten and Grube, 2010). In 1937 Hans von Ohein operated his first turbo engine running on hydrogen. The car industry began to work on hydrogen in 1967, as one option for an alternative fuel for ICE and electrical drive chain with fuel cells as energy converters. For spacecraft hydrogen is already a traditional energy carrier. Submarine vessels like the U212 are operated by several national armed forces.

Along with all this practical use, regulations have been developed in order to provide safe operation and handling of hydrogen. No significant incident has been recorded since the use of H_2 in aeronautics. The Hindenburg accident in 1936 is nowadays thought to be a synonym for the dangers of hydrogen, but after thorough investigation it was discovered that most probably electrostatics involving the painting and fabric of the airship was the cause, resulting in the burning of the hydrogen (about 200,000 m^3 of hydrogen in 16 cells on board). The accident of the American space shuttle Challenger in 1986 was initiated by the failure of a seal (small o-ring). Although hydrogen was used as a fuel, most likely the same accident would have happened with other fuels. The Challenger catastrophe has been used as a case study in numerous discussions on engineering safety and the workplace. Safety issues for storing and using hydrogen in aeronautical vehicles include the wide concentration range of flammability of hydrogen (4–75% volume) as compared to gasoline (1.0–7.6% volume), and the wide detonation range (18.3–59% volume vs. 1.1–3.3% for gasoline). Because the hydrogen molecule is very small, it diffuses extremely easily. As a consequence, hydrogen reacts with the material of the containers; this is the case especially on the surface and this causes an embrittlement. Other complications come from the fact that hydrogen has no odor and that it can burn invisibly.

It is evident that very careful and comprehensive safety management is mandatory in order to minimize any potential risk of using hydrogen and assure its safe handling and operation. For that purpose the database-driven website supported by the U.S. Department of Energy, The Hydrogen Incident Working Tool (www.h2incidents.org), contains all records on safety aspects needed when using and working with hydrogen and hydrogen-operated systems. The database is based on sharing of lessons learned and other relevant information gained from experiences when using and working with hydrogen from a variety of global sources, including industrial, government, and academic facilities. In addition there is an available online manual, the so-called Hydrogen Safety Best Practices (www.H2BestPractices.org, n.d.) to share the benefits of extensive experience by providing suggestions and recommendations related to the safe handling and use of hydrogen.

4.8 Safety strategy

In order to prevent an uncontrolled reaction of hydrogen and oxygen (or any other oxidant), different safety strategies have been developed by layout and design in order to:

- Keep the possibility of having hydrogen and oxygen in an ignitable ratio absolutely low.
- Prevent any source of ignition.

In cases where an ignitable gas mixture can appear, it must be ensured by design that (i) the amount is small enough to lead to no mayor event, (ii) the resistance time is low and (iii) no ignition is possible.

This has to be achieved respectively by the following key activities: (i) the system architecture and lay out, (ii) design of components, (iii) a suitable integration into the vehicle (sufficient ventilation and prevention of an accumulation of inflammable hydrogen air concentration in an amount that could cause a major incident), (iv) selection of safety systems (detection of H_2 and/or O_2, system health monitoring, fire detection, fire extinguishing) and, last but not least, by (v) handling and operational procedures.

In all cases a similar level of safety has to be achieved. For certification aspects, safety strategy has to be demonstrated by failure analyses, calculations, and tests. The selection of the right strategy depends on the design of the components and integration. The lifetime for safe operation is of great importance for the system components. On the overall system level a failure mode analysis has to be carried out and this should be related to the safe operation on the overall aircraft level.

4.9 Certification aspects

In general for aircraft the official codes and standards are applicable. Design standards are issued in Europe by the European Aviation Safety Agency (EASA) for small and commuter aircraft by CS23 and for large commercial aircraft by CS25. Those safety standards address the usual design and technology features. Novel technology and design has to pass a special rule-making process, at the end of which the aircraft designer has to document by analyses, tests, and typical field operations at least a similar level of safety as for the conventional technical solution. Such analyses have been a subject of several past projects, like the Cryoplane. In summary, it has always been concluded that in principle there is no major difference between hydrocarbons and hydrogen as concerns safe design and operation. According to the different chemical and physical characteristics and behaviors, technical solutions have to be adapted in order to meet at least a similar or even better safety level.

For certification aspects, a safe design and a safe operation must be shown and demonstrated for the whole lifetime of the commercially operated aircraft. Before design work starts, a safety strategy has to be outlined. This means that consequences from abnormal situations and incidents have to be classified and assessed, which is the common rule for the whole aircraft design. For classified incidents, which have to be taken into account, emergency landings, rapid cabin pressure drop, bird strike, lightning strike, volcano ash and others should be included. Other events typical to the technology are also to be taken into account and must be analyzed by appropriate failure assessments, based on failure by design, aging, handling, and operation by trained/untrained staff. The safety philosophy has to be defined at the beginning of any design work and design principles are to be derived from that philosophy. For different levels of incidents, their probability per flight hour and their consequences, rated as a hazard classification by the certification standards (CS) for aircraft, have been defined and are presented in Table 4.2.

This means that malfunctions are tolerated at different levels of consequences. If a detectable malfunction with no consequences appears and a safe flight can continue, it would be classified as minor; however, at the same time it could lead to an unacceptable event for the operator. This leads to a mixture of real safety issues and unacceptable operational situations. In any case, both requirements have to be taken into account. This shows that certification by the airworthiness authorities is mandatory, but full customer acceptance could require a higher standard.

Table 4.2 Level of incidents for aircraft

Hazard classification	Development assurance level	Maximum probability per flight hour
No effect	E	—
Minor	D	—
Major	C	10^{-5}
Hazardous	B	10^{-7}
Catastrophic	A	10^{-9}

4.10 Environmental and economic aspects and public acceptance

It seems that there may be a long way to go to overcome the technical barriers to developing a functioning hydrogen economy. The deployment of totally new infrastructure for transportation including the aeronautic sector is one of the main challenges on the technical, public acceptance, economic, and financial levels. The socioeconomic aspect of the process greatly depends on public acceptance. This issue has been addressed, among others, in the German HyTrust project launched in 2009, which accompanies the German Federal Government's National Innovation Programme. HyTrust analyzed the current state of public acceptance in hydrogen technology in the transportation sector, focusing on three pillars of social acceptance: global, local, and market acceptance. Based on the HyTrust results from interviews within focus groups, citizen conferences, and representative surveys, it can be said that the German population has a very positive attitude toward hydrogen-powered cars (Zimmer et al., 2013). However, it is not easy to predict their acceptance of commercial aircraft with hydrogen as an additional or only fuel on board. Following the discussion of the acceptance of hydrogen-powered road transportation, it could be speculated that a similar attitude could be expected for aeronautics; however, a specific survey will be required in the future. An interesting discussion and conclusions on public acceptability of hydrogen and principal concerns about the introduction of hydrogen energy and policy implications are presented by Ekins and Bellaby (2008).

In terms of environmental pollution, the full chain from the hydrogen production step, removal of impurities that could be a source of pollution, storage, and end use will be addressed. There are a number of significant issues that must be solved before hydrogen can play a substantial role in addressing the nation's and the world's energy security and global warming challenges. The most mature hydrogen production pathways available at present are not necessarily environmentally friendly and green. They include conventional hydrogen production methods by reforming of hydrocarbon fuels or gasification, thus producing pollutants and emitting them into the atmosphere, contributing significantly to the global warming process. Hydrogen produced from renewables could significantly reduce full fuel-cycle emissions by nearly 100%. However, it is estimated that at least two decades are needed before hydrogen can begin to make a significant contribution to reducing global warming pollution, improving air quality, and reducing oil dependence (Herzog and Tatsutani, 2005). Response to environmental stresses such as climate change and the problems of centralized electricity generation based on fossil fuels brought to the forefront the need for alternative energy products, such as nonpolluting renewable energy, particularly solar energy.

Characteristics and impact of GHGs emitted to the atmosphere by commercial air transportation has been recently assessed (Janic, 2014). CO_2 from burned Jet A is typically constant and equals 3.18 g $CO_2\,g^{-1}$ of fuel and increased from 8.8% in 1991 to 12.4% in 2010. Water vapor from burning Jet A is emitted at a constant rate of 1.23 g $H_2O\,g^{-1}$ fuel. This emission could cause the formation of water clouds in the troposphere, where the commercial aircraft perform the cruising phase of their flights.

No_x emissions result from burning the fuel and as well from their presence in the fuel. The SO_x emissions from Jet A is considered almost constant—0.8 g SO_x kg^{-1} of fuel (Celikel and Jelinek, 2001).

In principle, liquid hydrogen can be considered an alternative fuel for commercial air transportation. It is estimated that about three times less LH_2 than Jet A is needed to cover the same flying distance. This would lead to a significant weight reduction and complete reduction of CO_2 emissions. However, H_2O and a certain amount of NO_x would still be present. Because liquid hydrogen is much cleaner than Jet A it is anticipated that the engines would have longer life (about 25% longer) and would lower maintenance costs (Janic, 2014). However, there are several disadvantages which must also be taken into account. Changing the fuel would require new aircraft designs to accommodate liquid hydrogen, which has about four times the volume of Jet A fuel. This would in consequence increase the weight of the fuel tanks and fuel system, which in addition needs a very precise insulation system to prevent potential leakage of hydrogen. For the time being, there are no specific safety procedures for handling of hydrogen and no fuel supply infrastructure at airports. The latter would need complicated logistics including production, storage at cryogenic conditions as well as a proper distribution chain from the airport refueling system to the end-user–airplane. As already mentioned, there is public acceptance for hydrogen-fueled cars, but the question remains whether a similar trend could be observed for cryogenic aircraft. It is estimated that the total costs of producing liquid hydrogen and its delivery to the airlines in Europe could be in the range of 6.9 US\$ kg^{-1} of hydrogen (Janic, 2014). This is about 1.8–2.0 times more than kerosene today at the same energy content. A key prerequisite for introducing hydrogen for airplanes is obviously also that hydrogen delivery must be cost-competitive against kerosene. Facts like the cost development of energy over time and the different production technologies of hydrogen, which depend also on the opportunities in different regions, makes it quite difficult to precisely predict the future situation. Geopolitical developments and political measures may have significant influences on the future situation as well. Figure 4.3 shows the dynamics of the jet fuel price in the United States (www.eia.gov, n.d.).

4.11 Future trends

The European air transport system is facing new challenges regarding its competitiveness, performance, and sustainability. The Advisory Council for Aviation Research and Innovation in Europe (ACARE) has presented in *Europe's Vision for Aviation, Flightpath 2050* highly ambitious goals for European aviation transport and industry by 2050. This includes views on protecting the environment and the energy supply, safety and security, meeting societal and market needs as well as prioritizing research testing capabilities and respective education (Flightpath, 2050). The vision was developed by key stakeholders of European aviation from the aeronautic industry, air traffic management, energy providers, and the research community. This section is written

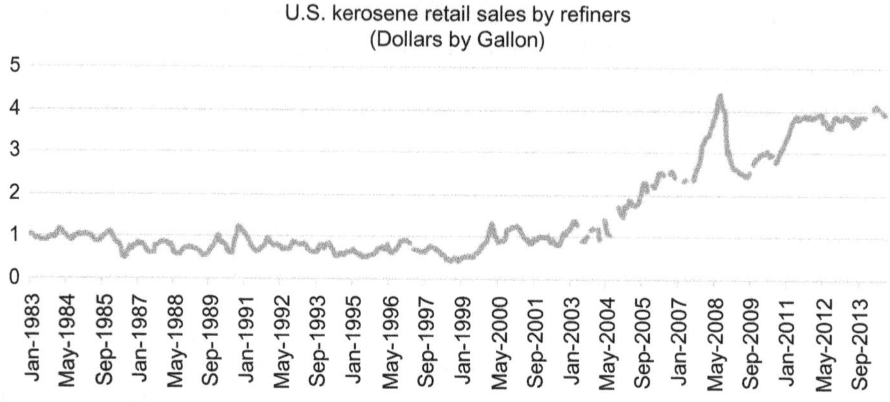

Figure 4.3 Jet fuel retail sales by refiners in the United States.

based on this report and highlights the main challenges and expected directions. It is anticipated that in 2050, the European air transport system will be integrated and interconnected to the global aviation system and interoperability between Europe and other regions will be complete. It is assumed that Europe is leading implementation of international standards covering all aviation issues, including security and safety.

Within Europe, the number of commercial flights is up to 25 million in 2050 compared to 9.4 million in 2011. Production of liquid fuels and energy from sustainable biomass will become an important part of the energy supply. In 2050, the effect of impurities and gases emitted to the atmosphere will be fully understood and the public will be aware of efforts toward environmentally friendly air transportation. It is expected that by 2050, new emerging technologies and procedures will allow the reduction of CO_2 emissions per passenger kilometer to 75% with a 90% reduction in NO_x emissions. The expected noise emission of flying aircraft will be reduced by 65%. In 2050, European aviation will have achieved very high levels of safety with less than one accident per ten million commercial aircraft flights. For specific operations (search and rescue) the number of accidents will be reduced by 80% compared to 2000. Weather and other environmental hazards will be well understood and risks associated with both will be properly mitigated. European aeronautics research will be defined, organized, and funded in a coherent and coordinated way (Flightpath, 2050).

Similar facts and goals are formulated by ATAG 2050 (Air Transport Association Group), which represents all sectors of the air transport industry, about 50 members worldwide. They include airports, the major air-framers, and their suppliers. The global goal of this group is to speak for the aviation industry with one voice and to promote aviation's sustainable growth for the benefit of the social society.

In 2009, the ATAG Board developed a set of environmental targets, which were approved by the IATA Board and the IATA Annual General Meeting. They included:

- improving fuel efficiency by an average of 1.5% per year from 2009 to 2020;
- stabilizing emissions from 2020 with carbon-neutral growth;
- aspiring to the goal of reducing net emissions from aviation by 50% by 2050 compared to 2005 levels.

4.12 Summary

For a large commercial aircraft, several studies have shown that, from a technical point of view, it is possible to use hydrogen as a fuel for propulsion by turbo engines. It was discovered that hydrogen in liquid form can be considered to be the most suitable alternative fuel for commercial air transportation. This has also been proven by some hardware tests and demonstrators. However, changing the fuel would require new aircraft designs to accommodate liquid hydrogen, which has about four times the volume of Jet A fuel. This would in consequence increase the weight of the fuel storage and onboard fuel transfer system, which in addition needs a very efficient insulation system. Otherwise extensive heat transfer into the cryogenic system could lead to boil-off losses of hydrogen.

Technologies to be used for system components can be derived from other sectors, like industry and space. Those studies have not found any show stoppers concerning handling, safety, and operation of hydrogen. Because liquid hydrogen is much cleaner than Jet A, it is anticipated that the engines could have a longer life (about 25% longer) and would require less maintenance cost. But even in this case there are some disadvantages due to the high volume by energy content and the unfavorable storage system, which is a cryogenic pressure vessel integrated in or on top of the fuselage. The low weight by energy cannot fully compensate for the increase of drag and weight of the hydrogen storage system. Apart from the results coming from LCA for different hydrogen production methods, the impact of the water vapor coming from combustion, which is about 2.5 times more than that coming from kerosene combustion, to the atmosphere is not yet fully understood. It can be assumed that water vapor in general is a strong GHG, which causes no harm to the troposphere because it remains only a few days in it. This may change at higher altitudes, such as the usual cruise altitude (10 km, 34,000 ft) and above.

For small manned aircraft compressed hydrogen (pressure cylinders at 200–350 bar) have been selected. The hydrogen is provided to PEMFCs which convert the energy to electrical power. Other air vehicles use different components for their propulsion chain. The selection of those always depends on the kind of operation and the required characteristics such as long endurance, low noise, low thermal signature, or others.

Fuel cell technologies have been a subject of some studies with the intention to provide onboard power to aircraft systems. In this case, a hydrogen-fueled PEMFC was meant to be operated in a multifunctional way, meaning that the fuel cell provides electrical power to the systems and thermal power for heating, e.g., wing anti-icing. The process water could be extracted from the exhaust and after treatment be reused as potable water on board. Even the remaining dry and oxygen-reduced exhaust gas could be used for fire prevention. Such a system could replace the APU, the RATs, and the traditional fire prevention system (Enzinger, 2011). For the time being, there are no specific safety procedures for handling of hydrogen and no fuel supply infrastructure at airports. The latter would need complicated logistics including production, storage at cryogenic conditions as well as proper distribution chain from the airport refueling system to the end-user–airplane. As already mentioned, there is public acceptance for hydrogen-fueled cars, but a question remains whether a similar trend could be observed for cryogenic aircraft. It is estimated that the total costs of producing liquid hydrogen and its delivery to the airlines in Europe could be in the range of 6.9 US$ kg^{-1} of hydrogen. This is rather speculative and depends strongly on the method of hydrogen production. It was discovered that hydrogen obtained by electrolysis with electricity derived from hydro energy could reduce the environmental cost by *ca.* 51–60% (depending on the aircraft type and flight duration) in comparison to the same aircraft fueled by Jet A. An obvious prerequisite for introducing hydrogen as a fuel for airplanes is a cost-competitive hydrogen delivery price (compared to kerosene) and the implementation of a supply infrastructure.

References

Ahluwalia, R.K., Hua, T.Q., Peng, J.-K., Lasher, S., McKenney, K., Sinha, J., 2010. Technical assessment of cryo-compressed hydrogen storage tank systems for automotive applications. Int. J. Hydrogen Energ. 35, 4171–4184 Bde.

Anders, S., 2007. Thermal loading cases of hydrogen high pressure storage cylinders. In: International Conference on Hydrogen Safety, 11–13 September, San Sebastian, Spain.

Celikel, A., Jelinek, F., 2001. Forecasting Civil Aviation Fuel Burn and Emissions in Europe (Interim Report). Eurocontrol Agency, Brussels. EC Note N-8/2001.

Ekins, P., Bellaby, P., 2008. Hydrogen futures: emerging technologies for hydrogen storage and transport. [Buchverf.] G. Walker. Solid-State Hydrogen Storage. Materials and chemistry. Woodhead Publishing Limited, s.l. ISBN 987-1-84569-270-4.

Enzinger, M.F., 2011. Airbus Multifunktionale Brennstoffzelle, High-Tech für die nächste Generation öko-effizienter Flugzeuge. Praxis Seminar Luftfahrt, HAW Hamburg, s.l. 29.09.2011.

Eurocontrol, 2014. EUROCONTROL Seven-Year Forecast February 2014. Flight Movements and Service Units 2014–2020. http://www.eurocontrol.int/sites/default/files/content/documents/official-documents/forecasts/seven-year-flights-service-units-forecast-2014-2020-feb2014.pdf (accessed 23.09.14).

Flightpath 2050, Europe's Vision for Aviation. Publications Office of the European Union, Luxembourg, 2011. ISBN 978-92-79-19724-6.

Herzog, A., Tatsutani, M., November 2005. A Hydrogen Future? An Economic and Environmental Assessment of Hydrogen Production Pathways. Natural Resources Defense Council, s.l.

http://www.atag.org/our-activities/climate-change.htm [Online] (Zitat vom: 17.07.14).

www.flugrevue.de/Sebastian Steinke, Hyperschall-Studie ZEHST [Online].

Janic, M., 2014. Greening commercial air transportation by using liquid hydrogen (LH_2) as a fuel. Int. J. Hydrogen Energ. 39 (29), 16426–16441. http://dx.doi.org/10.1016/j.ijhydene.2014.08.011.

Krishna, R., Titus, E., Maryam, S., Olena, O., Sivakumar, R., Jose, G., Rajkumar, A., Sousa, J.M.G., Ferreira, A.L.C., Campos Gil, J., 2012. Hydrogen Storage for Energy Application. [Buchverf.] Jianjun Liu. Hydrogen Storage. s.l.: Intach, ISBN 978-953-51-0731-6.

Marnellos, G.E., Athanasiou, C., Makridis, S.S., Kikkinides, E.S., 2008. Integration of hydrogen energy technologies in autonomous power systems. [Buchverf.] E.I. Zoulias with N. Lymberopoulos. Hydrogen-Based Autonomous Power Systems. Techno-economic Analysis of the Integration of Hydrogen in Autonomous Power Systems. Springer-Verlag, London. ISBN 978-1-84800-246-3.

Noll, T.E., Brown, J.M., Perez-Davis, M.E., Ishmael, S.D., Tiffany, G.C., Gaier, M., 2004. Investigation of the Helios prototype aircraft, mishap, vol. 1.

Osenar, P., Lauder, N., Launie, P., October 2008. Optimizing PEM Fuel Cell Systems for High Power Density: Enhanced Performance for UAVs and UGVs. Fuel Cell Seminar and Exposition, Phoenix, AZ.

Pereira, S.R., Fontes, T., Coelho, M.C., 2014. Can hydrogen or natural gas be alternatives for aviation? A life cycle assessment. Int. J. Hydrogen Energ. 39, 13266–13275.

Rathke, P., Kallo, J., Stephan, T., Weiss-Ungethüm, J., 2013. Antares DLR-H2—fuel cell testing under aeronautical conditions. In: 4th European PEFC and H2 Forum, 2–5 July 2013. Luzern, Switzerland.

Rau, S., 2003. Dynetek Europe GmbH. Deutscher Wasserstoff-Energietag, Eseen, 12-14 November.

Reardon, H., Hanlon, J.M., Hughes, R.W., Godula-Jopek, A., Mandal, T.K., Gregory, D.H., 2012. Emerging concepts in solid-state hydrogen storage: the role of nanomaterials design. Energ. Environ. Sci. 5, 5951–5979.

Seo, J.-E., Kim, Y., Kim, Y., Kim, K., Hee Lee, J., Hyung Lee, D., Kim, Y., Jae Shin, S., Kim, D.-M., Kim, S.-Y., Kim, T., Won Yoon, C., Woo Nam, S., 2014. Portable ammonia-borane-based H_2 power-pack for unmanned aerial vehicles. J. Power Sources Bde 254, 327–339.

Stolten, D., Grube, T. (Eds.), 2010. Strategic analyses/safety issues/existing and emerging, In: Proceedings of the WHEC. 18th World Hydrogen Energy Conference 2010—WHEC 2010, Essen, May 16–21. ISBN 978-3-89336-655-2.

Tupolev, 2012. Cryogenic aircraft. [Online] http://www.tupolev.ru/English/Show.asp?SectionID=82.

U.S. Department of Energy, September 2009. Targets for onboard hydrogen storage systems for light-duty vehicles. http://www1.eere.energy.gov/hydrogenandfuelcells/storage/pdfs/targets_onboard_hydro_storage_explanation.pdf.

Applications, cryo-compressed hydrogen storage for vehicular. U.S. Department of Energy Hydrogen Program. October 30, 2006.

www.eia.gov [Online].

www.H2BestPractices.org [Online].

www.h2incidents.org [Online].

Zimmer, R., Welke, J., Kaiser, M., 2013. Full steam ahead? Public acceptance of hydrogen technology. In: 4th European PEFC and H2 Forum, 2–5 July, Luzern, Switzerland.

Hydrogen-fueled spacecraft and other space applications of hydrogen

5

N. Frischauf
CERN, Geneva, Switzerland

5.1 Introduction: The potential of hydrogen-powered spacecraft

Hydrogen is the most abundant element in the universe, representing 77% of its constituents directly after the Big Bang. Today, 13.7 billion years later, stars like our Sun have burned *ca.* 5% of it in their fusion processes. Stars are big and powerful, yet the processes that govern their internal fusion reactions are governed primarily by two factors, temperature and density. Temperature is important as it requires a certain amount of energy to overcome the repelling positive charge of the protons and hence to initiate the nuclear reactions between the atomic nuclei, while density is the driving essence behind the reaction rate. Together both factors keep up the nuclear fire at a rate at which the resulting outward directed nuclear pressure perfectly counterbalances the inward centered gravitational force, enabling the star to shine in a stable manner.

An Earth rise as seen by Apollo 8, a human spaceflight mission powered by hydrogen. *Source:* NASA, Apollo 8 Mission.

Compendium of Hydrogen Energy. http://dx.doi.org/10.1016/B978-1-78242-364-5.00005-1

One can therefore also say that hydrogen is nothing else but the ultimate "fuel of the cosmos."

While stars demonstrate the most powerful way—the nuclear way—that hydrogen can be utilized for energy generation (a way that humanity tries to mimic with the thermonuclear fusion reactor ITER in Cadarache, France) there is also a much simpler way. This is the chemical reaction, which comes in two different flavors. The first one, the direct reaction, is well known and manifests itself in the form of fire. The other one, the indirect one, uses a membrane to separate the constituents, thereby controlling the reaction rate and tailoring the underlying oxidation in a way that electric power is generated. Regardless whether the energy is unleashed in a direct or indirect manner, both chemical reactions are very powerful and have therefore established themselves as methods of choice to power our spacecraft, be it in the form of rocket engines or fuel cells.

Although no match for solid rockets, which have existed in one form or another since the thirteenth century (Gruntman, 2004), fuel cells can still be considered as a rather old system, as they were invented in 1839 by Sir William Robert Grove (http://inventors.about.com/od/fstartinventions/a/Fuel_Cells.htm). It was, however, only in the twentieth century that fuel cells saw their first application in aerospace, when they were utilized first within NASA's Gemini and Apollo programs in the 1960s, eventually becoming the main power generation means in the Space Shuttle, until the end of this program in the year 2011.

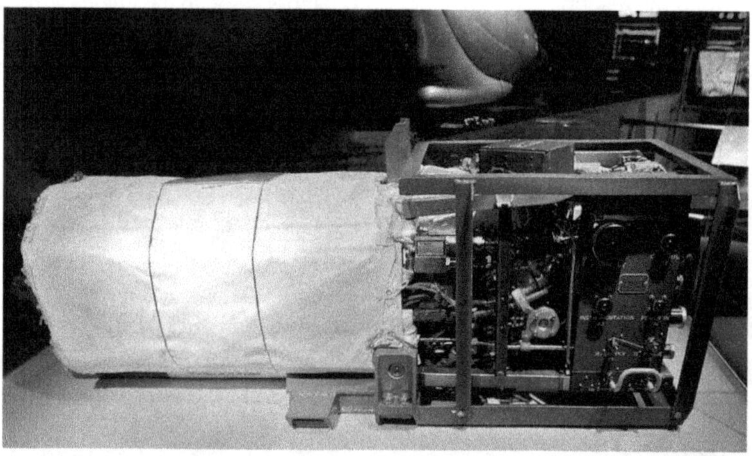

Space shuttle fuel cell.
Source: NASA and Steve Jurvetson.

With the end of the shuttle program, no space systems exist that utilize fuel cells. It is, however, only a matter of time until hydrogen-powered spacecraft will re-emerge, as power generation based on hydrogen is simply too powerful to be neglected. Ambitious space missions requiring high energy and/or high power levels will only be possible if one can rely upon a high-density power system—to date this requirement can only be satisfied by nuclear or fuel cell systems.

In the meantime, while fuel cell systems have been discontinued in space systems (with the final landing of space shuttle Atlantis on July 21, 2011), hydrogen continues to be used as a propellant and thus serves as a prime constituent to fuel humanity's space ambitions. High-performance rocket engines, such as the Vulcain of the European Ariane 5 rocket, utilize the chemical combustion of hydrogen and oxygen to generate the thrust necessary to lift off into space. As these engines demand large amounts of hydrogen with very high flow rates, the storage of cryo-liquified hydrogen has established itself as the state of the art; consequently one speaks of LH_2 (liquid hydrogen) and LOX (liquid oxygen) tanks, which contain the rocket fuel(s).

Artist impression of an Ariane 5 launch. The Vulcain engine is the one between the two boosters.
Source: ESA.

Whether one uses hydrogen in the nuclear fashion, as rocket fuel or as consumable for a fuel cell, storing the hydrogen for its later use remains a—if not THE—major challenge because of its low density. If size does not matter, then one can use gravitational pressure to store the hydrogen. This is exactly what stars do and the reason why even an average star like our sun is so huge, having a diameter which is 109 times larger than that of the Earth.

Naturally, humanity cannot readily copy that "stellar" approach, neither on a spaceship nor here on Earth, and so numerous methodologies for hydrogen storage have been developed over the years, encompassing storage as metal hydride or compressed gas, in the cryo-liquified state, in carbon nanotubes or gas microspheres, as liquid carrier or chemically bonded. Which storage methodology is finally utilized is driven primarily by the concerned application and the associated questions such as storage duration, mass flow rate, available volume and mass budget.

As space projects are always faced with mass and volume constraints, density is imperative. This is true both for rocket engines and for energy systems, even though the underlying requirements and constraints may differ. Rocket engines, which burn for a few minutes with very high mass flow rates, are naturally confronted with very large fuel tanks. In aiming to minimize tank size to the greatest extent, propulsion system designers therefore utilize cryo-liquified tank systems, even though these are very complex systems.

In contrast to rocket engines, energy systems do not feature high mass flow rates; still that does not mean that tank sizes are a nonissue. If power levels are high and/or mission durations are long, hence if one—or both—of these factors exceeds a certain threshold, then tank size also becomes a dominant factor for energy systems. Consequently, it can be that, even for a short duration mission, cryogenic storage solutions need to be built in to allow for a volume efficient energy system. This trade-off was showcased as early as the 1960s and 1970s when NASA's Apollo Programme, which brought man to the Moon, relied on cryo-liquified hydrogen and oxygen to supply a fuel cell to generate power and produce potable water for the spacecraft. An Apollo mission was rather short, lasting "only" for a period of 1 week. If one wants to rely on hydrogen/fuel cell technology for longer time periods than that, such as required when flying to Mars, cryogenic storage methods are only feasible if a near zero boil-off can be assured.

Out of the storage methods depicted in Figure 5.1, the easiest way to ensure this long-term capability is to rely on high pressure gas tanks instead of cryogenic systems (slush/solid—as promising as it is—is still in its infancy). Although gaseous storage is possible, at this time high pressure gas systems are disregarded for space applications due to the technological issues associated with the refilling of the high pressure tanks.

This, however, might change if the technological progress of hydrogen storage within the automotive sector continues. Contrary to space, the automotive industry started to discover the benefits of hydrogen only in the 1990s. Since then, hydrogen has been recognized as a major technology to enable the decarbonization of the transport sector. Consequently, the automotive industry started to implement hydrogen-based power

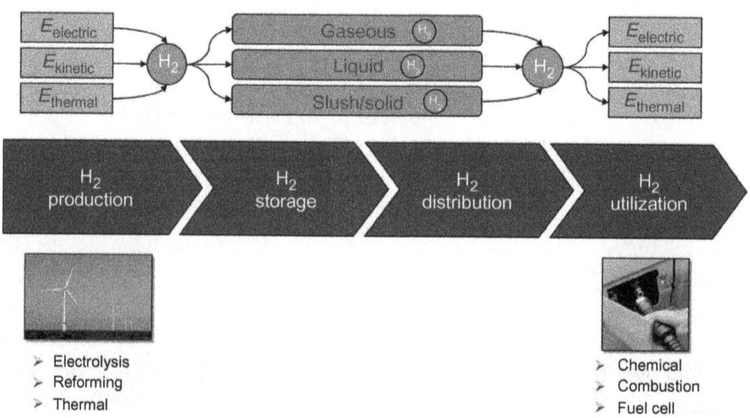

Figure 5.1 The hydrogen value chain.

Figure 5.2 The "hydrogen economy" vision.

systems as the primary drive train system for the next-generation cars. After an initial launch as prototype fuel cell-powered cars in 1997 by Daimler and Toyota (http://inventors.about.com/od/fstartinventions/a/Fuel_Cells.htm), the next-generation commercial fuel cell electric vehicles are now expected to see their market roll-out by 2015 (http://www.fch-ju.eu/sites/default/files/Power_trains_for_Europe_0.pdf; Thomas, 2015). In the years to come thereafter, through a gradual changeover, cars will not burn gasoline or diesel but rather convert hydrogen into electricity to power a high-performance electric motor, provided the infrastructure to produce, store, and distribute the hydrogen exists. Once it does, the world's energy infrastructure will change from its current oil economy to a "hydrogen economy" following the example already set by space (Figure 5.2).

One can expect that the further development of fuel cells will receive a major boost by the R&D efforts in the automotive sector and thus it is only a matter of time until we will see a comeback of fuel cells for/in space—possibly by then relying on other storage systems, which have been spun-in from the automotive domain.

5.2 Advantages and disadvantages of hydrogen-fueled spacecraft

Ten years ago, storage and distribution of hydrogen in the terrestrial sector represented a huge issue, both for the upstream (from the production to the fuel station) as well as for the downstream part (fuel station to the automobile); refer to Figure 5.1. Today, the issue has been partly closed, with the automotive industry agreeing among its members to use the storage of gaseous hydrogen in high pressure tanks as the technical baseline for the automobiles of the future (Car Congress, 2013). With the downstream storage and distribution issue being closed, the research and design focus is now put on the upstream part and the question of how the hydrogen will be best transported from the production facilities (e.g., nuclear power plants, wind parks, etc.) to the dispensers of the refueling stations. It is not a matter of technology per se, but rather a question of cost effectiveness. Only time and continued R&D will tell whether this role is to be fulfilled by hydrogen pipelines, hydrogen trailers, and/or localized hydrogen production facilities.

As far as aerospace is concerned, one can foresee that hydrogen will gradually move into energy and propulsion-related applications. This is due to the fact that hydrogen energy systems scale better than solar cells or batteries when bigger power levels are needed, which means that their mass increases at a lower rate per power increment, leading to a lower mass at a certain power level, and that hydrogen constitutes the best propulsive medium that can be used. No other medium is lighter in mass and can therefore provide readily for high-specific impulse (ISP)[1] levels, which are paramount to save on fuel consumption.

In aiming to picture, the major advantage that comes along when using hydrogen as rocket propellant, let us take the example of a telecom satellite, such as the ASTRA 1 K, with a mass of 5250 kg and an anticipated lifetime of 15 years. As a typical telecom satellite, ASTRA 1 K will be parked in the geostationary belt, 35,786 km above the Earth's equator. It will be brought to a geostationary transfer orbit by a heavy lift launch vehicle (HLLV) like an Ariane 5 or a Proton, and once it has been released there it will fire its main engines to arrive at its designated slot in the Geostationary Orbit (GEO). This maneuver has already taken some fuel, but although the satellite is now in its final parking position, there is more to be spent. The Sun and the Moon, the Earth's gravity field, all sorts of disturbances act on the telecom satellite, trying to move it out of its slot. To maintain in the designated position and out of any collision courses with the adjacent telecom satellites, a roughly 50 m/s velocity increment has to be invested every year. This amounts to 750 m/s in 15 years and, dependent on the chosen propulsion system, the required fuel mass ranges from 1181.40 kg for a classical bi-propellant system with specific impulse (ISP) of 300 s, to 245.04 kg for a plasma engine with an ISP of 1600 s and 99.42 kg for an advanced pulsed plasma thruster with an ISP of 4000 s. The dependency of the required fuel ($m_{\text{propellant}}$) is described by Equation (5.1), whereby $m_{\text{S/C\_Initial}}$ represents the initial spacecraft mass before the maneuver, ΔV is the velocity increment (in our case 15×50 m/s) and g is the standard acceleration of gravity, hence 9.81 m/s^2.

$$m_{\text{propellant}} = m_{\text{S/C\_Initial}} \left(1 - e^{-(\Delta V / g \cdot I_{\text{SP}})}\right) \tag{5.1}$$

Hydrogen would clearly be the best propulsive medium, were it not for the fact that it possesses mediocre storage features due to its low density. At the moment, the fuel savings are more than offset by the mass penalty of the hydrogen tanks, making it impractical to use H_2 as propellant, if the ISP of the engine is too low; except for HLLVs where the thrust-to-weight ratio comes into play as an important factor that needs to

[1] The so-called specific impulse is a figure of merit for the effectiveness of a rocket engine. It is measured in seconds and relates to the exhaust velocity of the engine. In short one can say that the higher the ISP (or the exhaust velocity) the less fuel is needed to achieve a certain velocity increment ΔV (refer to Equation 5.1), which is the ultimate figure of merit as far as orbital dynamics is concerned.

be considered. Once high performance engines become available, however, then things might change and then it will make sense to copy the terrestrial upstream H_2 distribution and storage solutions for space, e.g., by utilizing hydrogen tanker spaceships to re-fuel spaceships and/or satellites in Earth orbit or to supply remote outposts, as envisaged by the Shackleton Energy Company (SEC) (http://www.shackletonenergy.com/, accessed in May 2015) or within the Aurora program of the European Space Agency (ESA) (here as feedstock hydrogen for an in situ resource utilization (ISRU) plant on Mars, utilizing the Sabatier reaction to produce methane as rocket fuel) (ESA S54 Study, 2001–2003).

5.3 Principles: Suitable hydrogen power sources for spacecraft

The retired US Orbiter clearly has written history for utilizing fuel cells in space. Given its power and energy requirements, batteries were not able to do the job; even with today's most advanced lithium–ion technology fuel cells have better performance. It has been simply a question of mass. Batteries at the time of the shuttle design offered no more than 50 Wh/kg if they were depleted by 100% (the so-called "depth of discharge"), not realistic for reliability and lifetime, whereas the now almost 40-year-old fuel cell provided by IFC, a subsidiary of United Technology, offered more than 1000 Wh/kg including all reactant storage including the tanks.

Given the mission profile of the Space Shuttle, which spent 14 days in space at most, the final design choice for the power system was an easy one; the energy system was designed as a primary system, a system where the reactants were not regenerated, hence an electrolyzer was not part of the system (refer to Figure 5.3). An additional advantage of this concept, still holding through today, is that a fuel cell based on hydrogen and oxygen will produce drinkable water—water that can be and was used by the astronauts on board.

Longer space missions, however, clearly require regenerative energy storage, if one wants to avoid bringing along vast amounts of consumables. While batteries may be recharged directly by the solar arrays, in fuel cells, the reactants hydrogen and oxygen need to be recharged by electrolyzing the product water from the fuel

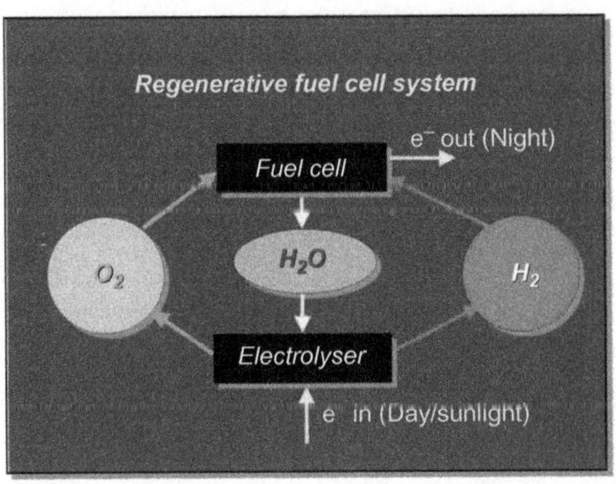

Figure 5.3 A regenerative fuel cell.
Source: ESA.

cell operation whenever solar array power is available. This adds a lot of complexity. Nevertheless, the significantly higher energy density provided by regenerative fuel cells (fuel cell + electrolyzer) compared to batteries would be, technically speaking, a clear choice. A mass savings of 2–3 in power system weight vs. the most advanced space battery system is achievable and also looks very promising in terms of launch cost.

Additionally, a fuel cell can clearly separate power requirements from energy requirements, as the power is sized by the electrode stack (the reactor) and the energy is defined by the size of the reactant storage. In batteries this option is not available, as the reactants are contained in the electrodes; consequently the battery has to be sized for the worst case. The ESA pressurized lunar rover (PLR) study performed a detailed trade between different technologies for energy storage and delivery. Regenerative fuel cell systems (RFCSs) were compared to different battery configurations, supercapacitors as well as a nuclear system. For the case of the Rover, the trade clearly showed that the RFCS provided the lowest system mass by a large margin. This mass savings also justified the choice even with the higher complexity.

Therefore, some future exploration missions, specifically human or robotic missions having high energy demands, will clearly benefit from fuel cell technology. These missions need many new technologies and have technology development programs in place to address cost and risk. Issues that remain to be solved are development risk and cost, complexity, reliability, and especially industrial considerations and conservatism in commercially sensitive areas like telecom satellites. A change within that sector from batteries to fuel cells will require significant modifications of the whole spacecraft with an enormous impact on extra design cost and uncertainties. Only a move to significantly bigger platforms than those being built right now will allow the choice of moving to fuel cells.

With all these aspects taken into account, one can therefore assume that future high power/energy space power systems will be based on regenerative fuel cells (especially PEM fuel cells).[2]

5.4 Advantages and disadvantages of the power sources

Power sources exist in various flavors; most notably within our daily lives are batteries of all types, be they of the primary (nonrechargeable) or secondary (rechargeable) type. Especially, the latter ones may be recharged by plugging in the device, such as a mobile phone or a laptop, to the power grid, while systems with a smaller power requirement may even use a small solar cell to power and recharge (e.g., pocket calculators, remote sensor stations at highways, etc.).

Naturally, even the best system has its limitations and batteries are no exception, although a lot of progress has been made in recent years. This progress has led to an ever-increasing energy density, measured in Wh/kg when assessed for their gravimetric, or Wh/l for their volumetric, performance. Today's battery systems feature energy densities in a range between 40 Wh/kg for lead-acid batteries, 50 Wh/kg for NiH_2,

[2] It shall be noted that all these considerations are purely focusing on nonnuclear power system trade-offs—simply because the automotive role model, whose trade-off applicability is examined within this article, does not consider any nuclear power and/or drive trains.

90 Wh/kg for Nickel–Metal–Hydrate (NiMH), 120 Wh/kg for Li–ion batteries and up to 180 Wh/kg for Li–polymer batteries. As impressive as these numbers are, they are far too low to efficiently propel a normal sized family car from A to B—even a low duty cycle vehicle (LDV) with a mass of 1.5 tons will require a power system, which meets and preferably exceeds an energy density level of 1000 Wh/kg.

As battery performance cannot support such a profile, fuel cells have to step in here and in particular the *polymer electrolyte* or *proton exchange membrane* fuel cell (PEFC or PEMFC) has established itself as the system of choice for mobile applications. This is based on the fact that PEFCs:

- are relatively compact;
- have high power densities;
- have high efficiencies at partial load; and
- feature fairly rapid start-up capabilities.

All these features make the PEFC ideally suited for mobile applications, operating at low temperatures (50–120 °C). In addition, a PEFC can also be utilized for stationary applications, such as uninterrupted power supply (UPS) or auxiliary power unit (APU). As these applications typically feature higher power levels on the order of hundreds of kilowatts up to several megawatts, efficiency considerations become very important and then the PEFC is often substituted with a molten-core fuel cell (MCFC) or a solid oxide fuel cell (SOFC), operating at 650 and 800–1000 °C and featuring an efficiency of 55–60% and 60%, respectively (H_2YDROGEIT, 2015). A showcase for such a "stationary" application is the utilization of a fuel cell as a power source within a submarine, as can be seen on board the German U212A (refer to Figure 5.4).

Although low temperature PEFCs use noble metals, especially platinum, a scarce and hugely expensive resource, and have a water management issue as the electrolyte membrane (e.g., nafion) needs humidification at all times to exhibit high proton conductivity, the other attractive characteristics as mentioned suggest PEFCs as possible power sources in vehicles for both propulsion and as APU. The latter helps to run electric steering and brakes, air conditioning compressors and even entertainment devices.

Similar considerations prevail for the use of batteries and fuel cells in the aerospace sector, with the additional issue that the fuel cell in the spacecraft needs to be of the

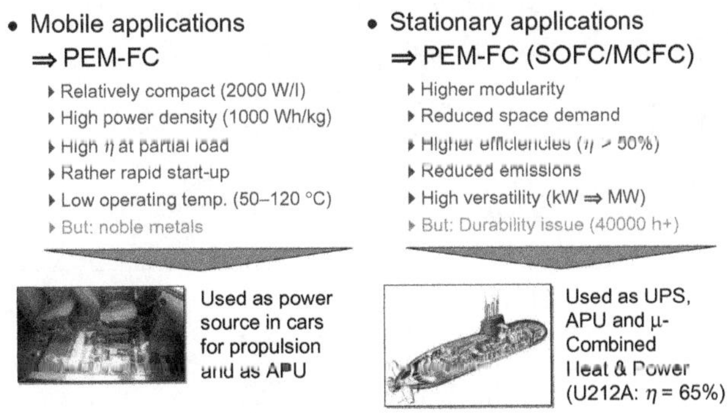

Figure 5.4 Comparison of fuel cells used in mobile and stationary applications. *Source:* JRC-IET (Frischauf et al., 2011).

regenerative type as depicted in Figure 5.3, where the fuel cell's working mode can be reversed to either generate power and H_2O by combining H_2 and O_2 or taking the water and—by working in electrolyzer mode—making H_2 and O_2 out of it to use it as consumables when power is needed again. Due to its future potential, ESA has taken a closer look into this topic by initiating several studies. One particular study related to this topic was the one performed by Thales Alenia Space, dubbed "Fuel Cells for Telecom Satellite System Study," with the objective of determining the overall performance of a selected total RFCS (ESA Fuel Cells for Telecom Satellites System Study, 2008), chosen among the "available" ones, in a closed loop system compared to an advanced Li–ion battery system (G5 technology 180 Wh/cell). The work was focused on the implementation of an RFCS on a large telecom platform (15 kW total satellite power requirement).

The RFCS was subject to the following requirements:

- The FC shall be qualified for >2250 h.
- Efficiency allocation: fuel cell mode > 50%, electrolysis mode > 90%.
- A reliability figure of 0.99 at 15 years is a target. Double isolation shall be implemented.
- The equipment shall be designed to meet space environmental conditions.
- Protections shall be implemented in the RFCS or on board the satellite to avoid any degradation of the mission.
- The RFCS shall respect electromagnetic compatibility for itself and with respect to the rest of the spacecraft. Safety rules shall be respected.
- The supplier shall identify in the user's manual and justify the constraints associated with the powering ON/OFF of the RFCS.
- Maximum voltage range: 0–100 V.

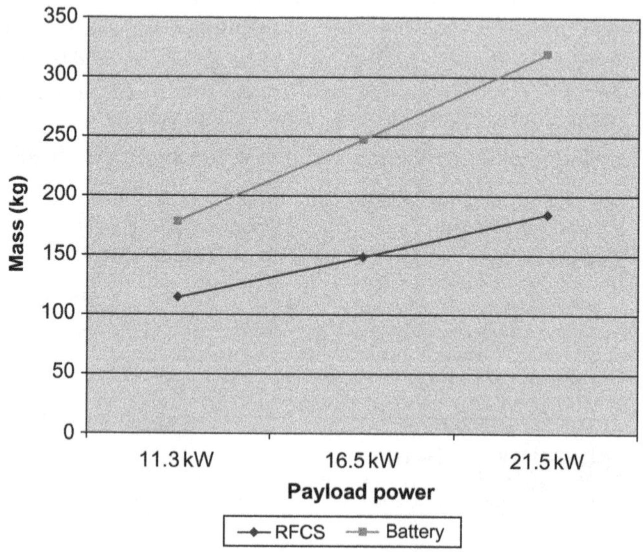

Figure 5.5 Mass trend for the energy storage systems.
Source: ESA Fuel Cells for Telecom Satellite Systems Study.

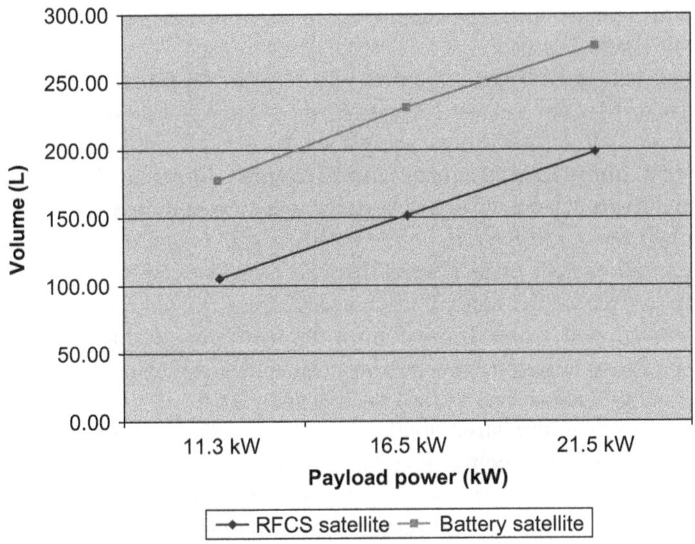

Figure 5.6 Volume trend for the energy storage systems.
Source: ESA Fuel Cells for Telecom Satellite Systems Study.

Trends were derived for power levels from 15 kW up to 30 kW and a system study was carried out for a payload with a power level of 11.3 kW. The following graphics outline the mass (Figure 5.5) and volume trends (Figure 5.6) for the assessed energy storage systems.

In a nutshell, one can say that fuel cells outperform batteries, whenever high power and/or energy levels are required. The transition point within the automotive area is defined by range and size of the car; a typical LDV of 1.5 tons aiming to feature a range of 400 km will be better off relying on fuel cells. For telecom satellites working continuously for 15 years in Earth orbit, generating onboard power by solar arrays with the necessity to maintain onboard power when the satellite passes through the shadow of the Earth, the transition point is around 15 kW of onboard power. Below this point, batteries perform best, while above the 15 kW, a regenerative fuel cell will be the system of choice. A more detailed discussion on the implications of replacing the currently used batteries by fuel cells within a Telecom satellite can be found in Frischauf et al. (2013).

5.5 Challenges for hydrogen-fueled spacecraft

Obviously space is primarily concerned with mass and volume considerations. Bearing in mind that the launch of 1 kg of mass into low earth orbit (LEO) costs between €10,000 and €20,000, both factors play such an important role in the performance of the overall system that a somewhat higher cost of a high-performance fuel cell, carbon hydrogen piping and lightweight cryo-tanks can be readily traded off against the mass/volume savings. Hydrogen production in space is not considered a possibility yet; both launchers and spacecraft that rely on hydrogen take their supplies with them in specific cryogenic tanks. This way, short duration missions can be sustained, while longer missions cannot yet be supported by cryogenic systems due to the boil-off issue.

In the long run, a long-term solution will have to be found, especially when a manned lunar or Martian base is to be established. Such a base will require duty cycle numbers as well as power and energy levels of such a magnitude (50 kW and more; ESA S54 Study, 2002) that battery technologies will not be effective any longer. Consequently, hydrogen-based power systems will step in then and for effectiveness reasons the hydrogen will have to be produced on site. Given the absence of natural gas and high temperature energy wells, hydrogen production by means of electrolysis is likely to become the method of choice, if one does not rely on the utilization of nuclear power. A detailed discussion on the hydrogen production using nuclear energy can be found at http://www-pub.iaea.org/books/IAEABooks/8855/Hydrogen-Production-Using-Nuclear-Energy, accessed in June 2015.

In contrast to space, terrestrial applications focus primarily on safety aspects and on costs. Safety is of prime concern; as the number of systems is high, the likelihood of an accident is therefore even higher and it is untrained personnel that will handle the fuel dispenser, the hydrogen-powered car or the UPS. This is in strong contrast to the space sector, where all systems concerned are under constant supervision and all hydrogen-related activities are performed by well-trained experts, thus significantly reducing the likelihood of major incidents/accidents. The second major driver in the

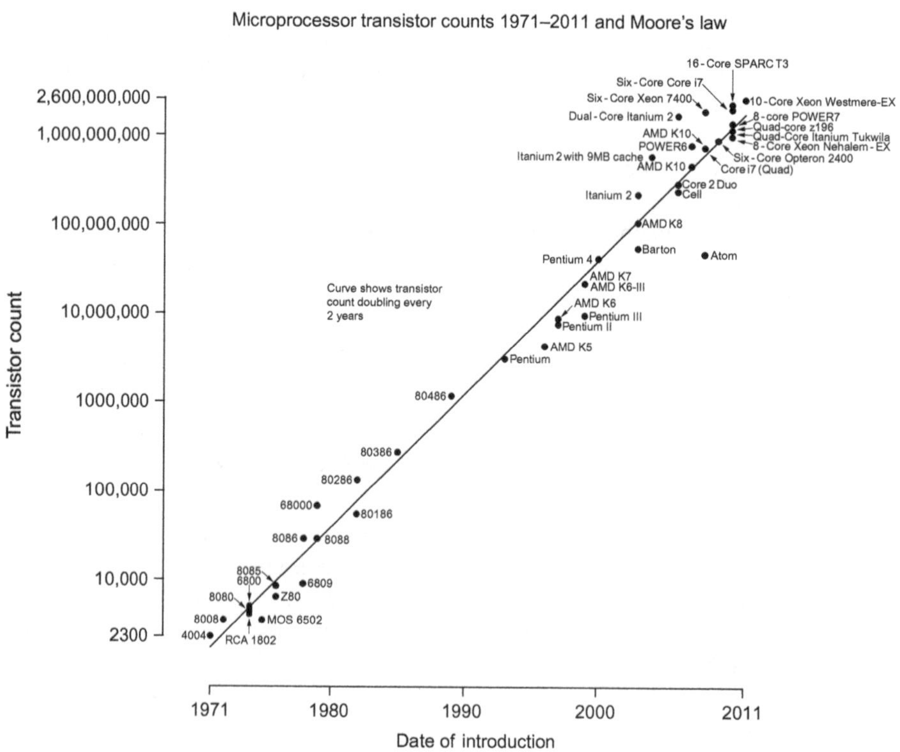

Moore's law, demonstrated by CPU transistor counts.

terrestrial sector is cost—on unit, subsystem, and at system level. In a competing mass market, which is governed by innovation pressure, limiting the time that a system can excel and generate revenues in the market, cost is a major item.

According to Moore's law, electronic systems feature a generation change every 2–3 years (refer to the figure) and due to the high importance of electronics technology for most of today's systems, this leads to a trend of swifter generation changes in all areas. A car model is obsolete after 6 years (10–20 years ago, car models lasted for 9 years or more) and so are its major components—and the fuel cell and its associated hydrogen storage and distribution system are no exceptions. If future generations of spacecraft will increasingly spin-in hydrogen technologies from the automotive sector, this swifter generation change is an issue to be carefully looked at, especially for what concerns general sourcing, the availability of spare parts and the overall obsolescence of key elements and subsystems.

5.6 Other space applications of hydrogen

Fuel cells have been used in NASA's human space missions such as Gemini, Apollo, and the Space Shuttle providing primarily high power levels and, as an added benefit, potable water. These applications were all nonregenerative and demanded the availability of enough fuel and oxidizer to support the entire mission.

Current human exploration architecture and system studies as well as technology developments in the area of RFCSs, a combination of fuel cell and electrolyzer (refer to Figure 5.3), show that this technology is critical for larger human space exploration missions of the future. Typical examples of system studies are lunar habitats/bases and lunar pressurized rovers. But what also has been shown is the overall synergy within a manned architecture between the life support systems, the power management systems (regenerative FC) and the in ISRU systems, which all share the use of O_2 and H_2. Future systems will need to be designed from the start utilizing these synergies, allowing sharing of resources and reduced system mass.

Under a study being performed by CGS for ESA on Energy Provision and Management, an architecture for energy provision for a lunar base has been defined utilizing solar arrays for the power generation, regenerative fuel cells for power conversion, and some batteries for auxiliary power. The design relies on many smaller lunar power plant elements each compatible with a possible lunar lander sized according to the Ariane 5 capabilities. Depending on the power needs and the gradual build-up of a station, additional power plants could be introduced into the system. The solar arrays would provide the power under daylight conditions to electrolyze water into H_2/O_2 which would then be stored for future use. The O_2/H_2 can then either be used at the base during night conditions or transferred to, for example, a lunar rover for use on a specific expedition away from the base.

Under a different ESA contract, Thales Alenia Space looked into the details of a PLR concept, which relied on regenerative fuel cells for its main power generator. The PLR would start its mission with full tanks of O_2/H_2 which would be depleted during the mission. Sizing of the tanks is done to include operational and system contingencies. Batteries would be used for auxiliary power and for redundancy. During all passive

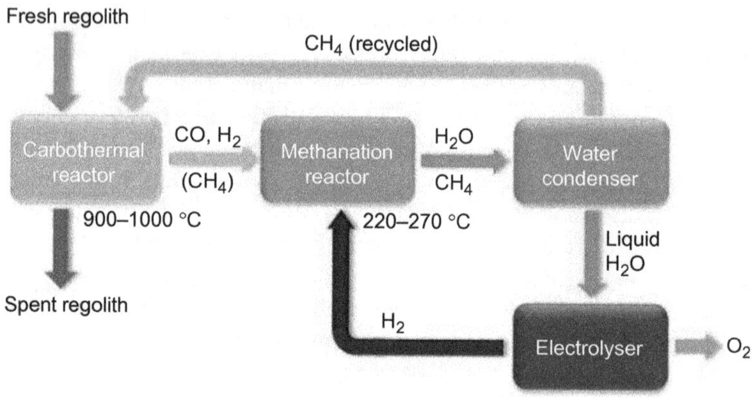

Figure 5.7 Oxygen process ideal process.
Source: CGS/ESA.

operation, i.e., not roving, the PLR would deploy small solar arrays to augment the available power. As with the lunar power elements mentioned previously, this power would be used to electrolyze water into O_2/H_2 for storage. After the mission, the PLR would be able to connect to the lunar base and transfer the water for processing while taking on board O_2/H_2 for the next mission.

Lunar ISRU plants would be possible, which would extract O_2 from the Lunar Regolith and produce water or O_2 directly. Breadboarding activities performed by CGS for ESA have shown that this is feasible using carbothermal reduction of Regolith with methane. The investigated process would use methane and hydrogen in a cyclic manner needing replenishment only to overcome leakage and waste due to inefficient processes (refer to Figure 5.7).

5.7 Market trends

Space is a multibillion euro/dollar business, but is nothing next to the automotive domain, which can be readily seen from the R&D investments outlined later in Figure 5.9.

Based on the different size and the assumption that space will be benefiting from spin-ins rather than driving spin-offs as it used to do, it is important to look into the fuel cell development within the automotive sector.

5.8 Hydrogen storage in spacecraft

With current technologies, hydrogen storage in space is most efficiently achieved on total system level by employing high pressure (100 bar) electrolyzers for reactant regeneration from the water produced by fuel cells during eclipses. This is a significant advantage in terms of complexity and reliability compared with very high pressure (700 bar) tanks as proposed for FC cars, which would require compressors not lasting very long without maintenance. Additionally, the reliability aspects of peripherals like pressure transducers and flow controllers is a significant problem.

Given the current shortfalls of cryo-systems (e.g., boil-off), cryo-storage is not yet the method of choice for long-term space missions. With the advent of high-performance cryo-coolers however, this is likely to change. In contrast, short space missions, lasting no longer than a few weeks may be based on primary (nonregenerative) fuel cells. Liquefying the reactants in space in a regenerative system is not a technical option.

Slush hydrogen (http://en.wikipedia.org/wiki/Slush_hydrogen),[3] as depicted in Figure 5.1, is a possible advanced alternative to cryogenic storage, offering another 10–20% higher storage density. It is, however, at a very early stage and might be used for rocket fuel storage primarily.

Today's state of the art for terrestrial hydrogen storage, such as in the automotive industry, comprises 35 MPa (350 bar) and 70 MPa (700 bar) compressed gas tanks. Carbon fiber fully wrapped, reinforced tanks are already in use in prototype hydrogen-powered vehicles. Two types of inner liners are typically used: metal (aluminum, Cr-alloys) ones in "Type 3" storage pressure vessels and high molecular weight polymer in "Type 4" tanks, as described in ISO 15869 (ISO/CD 15869, 2001). The application of such materials comes from the need of guaranteeing impermeability of the inner liner to the hydrogen molecules while having the tank being as lightweight as possible. Figure 5.8 outlines the key features of the different types of compressed gas tanks.

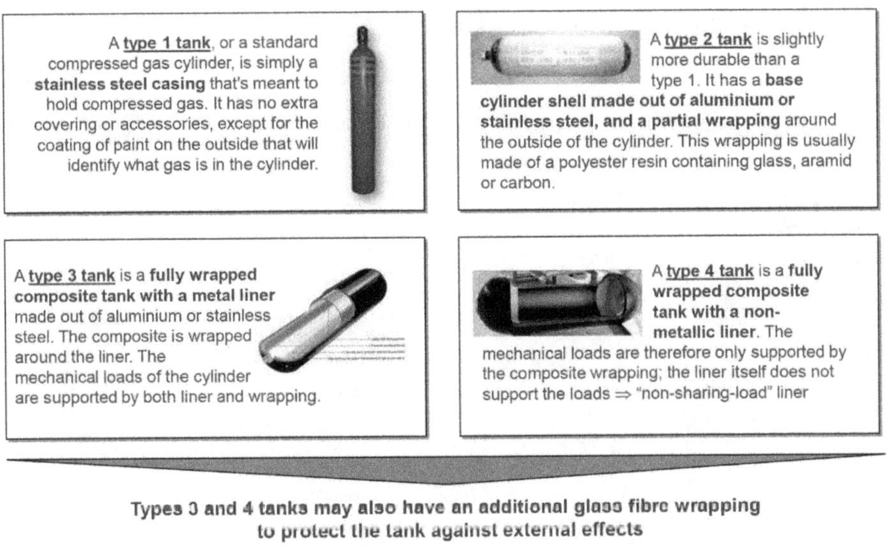

Figure 5.8 Different types of compressed gas tanks.
Source: JRC-IET (Frischauf et al., 2011).

[3] Slush hydrogen is a combination of liquid hydrogen and solid hydrogen at the triple point with a lower temperature and a higher density than liquid hydrogen. It is formed by bringing liquid hydrogen down to nearly the melting point (14.01 K or −259.14 °C) that increases density by 16–20% as compared to liquid hydrogen. It is proposed as a rocket fuel in place of liquid hydrogen in order to improve tankage and thus reduce the dry weight of the vehicle.

Both SAE 2579 and draft ISO 15869 propose the following tests for compressed hydrogen storage tanks:

- Sequential exposure to impact, chemicals and cyclic stresses.
- Sequential exposure to static high pressures (simulates vehicle parking) and fueling stresses.
- Exposure to hydrogen fueling under extreme ambient temperatures.
- Simulated fueling failure (i.e., overpressures) at the end of vehicle service.

It is up to specific research institutes and the industry to perform and further develop these test regimes as scientific knowledge progresses. And it is imperative to do this on a global level so that the "hydrogen economy" vision (refer to Figure 5.2) can be realized.

Given the fact that space agencies around the world can no longer afford a dedicated fuel cell and hydrogen tank storage development for a dedicated space program,[4] a spin-in is the preferred—if not the only—option. And chances are good that such a spin-in will occur. A recent study published in a JRC Technical Note has assessed the R&D expenditures of the automotive industry and the aerospace and defense sectors, confirming that the automotive industry is the largest R&D investor in the EU-27, accounting for one-quarter of total industrial R&D investments. And this observation is likely to remain valid in the years to come, as both car manufacturers and component suppliers show elevated R&D intensities with an increasing trend over the past years.

As is visible in Figure 5.9, the worldwide R&D investment of the automotive sector is larger than that of the aerospace and defense sector by a factor of 5.1 worldwide

	R&D investment (€bn)		**Sales (€bn)**		**Number of employees (million)**	
	World	EU27	World	EU27	World	EU27
Automotive manufacturers	53	20.9	1213	423	2.76	1.26
Automotive suppliers	19.6	9.5	437	156	2.33	0.98
Commercial vehicles and trucks	6.9	2.4	233	66	0.62	0.22
Automotive industry	**79.5**	**32.8**	**1883**	**645**	**5.7**	**2.5**
Aerospace & defence	*15.6*	*7.5*	*379*	*129*	*1.75*	*0.55*
All industries	*431*	*130*	*13897*	*5712*	*45.1*	*21*

Source: Derived from the EU Scoreboard 2009 (DG RTD-IPTS, 2009) (rounded numbers).
Note: The table shows the aggregated figures for 65 EU-based and 140 worldwide companies related to the automotive sector,

Figure 5.9 R&D investments, sales, and total number of employees of companies included in the EU Scoreboard related to the "automotive" sector (2008), "aerospace and defense," and all industries.
Source: JRC Technical Note JRC 5872714 (http://ipts.jrc.ec.europa.eu/publications/pub.cfm?id=3279).

[4] The "golden times" of large-scale space programs like Apollo are gone.

and by 4.4 in the EU-27. This leads to a clear conclusion in the discussion of spin-in vs. spin-off: the sheer difference in R&D expenditure in the aerospace and the automotive sector clearly favors a spin-in of terrestrial fuel cell technology into the space sector. It may therefore be that future spacecraft will more and more adopt the terrestrial standards set forth by the automotive sector. While any space fuel cell must be of the regenerative type, tanks may reach higher and higher pressure levels, just as we see happening in the terrestrial world.

5.9 Advantages and disadvantages of the various potential storage methods

Albeit complex compared to battery-based energy systems, fuel cells excel when high energy and/or high power demands become important. Given their higher specific energy density, which leads to a mass savings of least 2–3 times in power system weight compared to today's most advanced space battery system, fuel cell-based space power systems are an option to be considered for space missions with power levels of 15 kW or more. Today's telecommunication satellites already employ such power levels, certain military reconnaissance satellites using synthetic aperture RADAR (SAR) might do as well, and future exploration missions, going to the Lagrange points, the Moon or Mars will see even higher power levels, especially when humans are on board these spacecraft.

In contrast to the Space Shuttle mission, which lasted a maximum of 14 days and could therefore employ a primary fuel cell, the fuel cell systems utilized in long-term exploration missions and high power/energy Earth observation and telecom satellites will be of the regenerative type, featuring a fuel cell and an electrolyzer (refer to Figure 5.3). In times of excessive power consumption, the fuel cell will use the on-board hydrogen and oxygen supplies to generate power; the water produced (out of the hydrogen and oxygen) will be stored on board. As soon as the power balance is positive again (e.g., when the satellite has left the eclipse and is generating power with its solar arrays again), the electrolyzer starts up and converts the water into H_2 and O_2 again, which is then stored on board for the next discharge cycle. Although this system is more complex than a battery, it offers the great advantage that one can separate power requirements from energy requirements. As such the power is sized by the electrode stack (the reactor), while the energy is defined by the size of the reactant storage. In batteries this option is not available, as the reactants are contained in the electrodes; consequently, a battery has to be sized for the worst case.

From the financial perspective, the higher complexity of fuel cells vs. any battery system will lead to higher initial costs, costs that need to be offset by their better performance. This better performance is achieved by the better scaling characteristic of the respective specific power and energy density. While a fuel system can achieve energy densities in excess of 1000 Wh/kg for primary systems and up to 500 Wh/kg for long-term regenerative systems, Li–ion-based battery systems for satellites can achieve around 150 Wh/kg on battery level in the near future. The quintessence is therefore that it does not make sense to use a fuel cell system for a satellite with limited power requirements, like SGEO, the Small Geostationary satellite, which features a mass of 1500–2500 kg and a maximum payload power of 3000 W (ESA Telecommunications & Integrated Applications, SmallGEO, 2011). The cost penalty would be too high. In this case, it is better to stick to the classical battery system.

These cost-driven considerations are similar to the ones in the automotive industry, which foresees employing fuel cells in bigger cars, requiring more power and/or cars that travel longer distances, therefore requiring more energy. As outlined within Frischauf et al. (2013), this is driven primarily by the different scaling features of the two systems. If one wants to achieve a range of at least 400 km, fuel cells outperform battery-based energy systems in terms of greenhouse emissions, regardless of whether the hydrogen has been produced by solar, wind, ethanol, or natural gas.

5.10 Safety concerns regarding the storage of hydrogen in these vehicles

Safety concerns in spacecraft as far as the storage of hydrogen is concerned are mostly connected to the inflammable/explosive behavior, embrittlement of materials that the hydrogen can be in contact with, and associated storage pressures or temperatures, if hydrogen is stored in either high pressure gas tanks or in a cryo-liquified state. While the aerospace sector has the advantage that mostly experts deal with these types of systems, the stronger spin-in of fuel cells and hydrogen storage systems from the automotive into the space sector will lead to an adaptation of terrestrial safety standards for space. Acknowledging this important linkage, we will take a look into the safety standards and regulations which are applicable to terrestrial systems.[5]

Naturally, a thorough overview of all existing standards is not within the scope of this chapter. It will be enough to mention, in relation to international standards, the work of the Technical Committee on Hydrogen Technologies of the International Standards Organization, which produces, among many others, standards for gaseous and liquid hydrogen storage devices and hydrogen blends (ISO/TS 15869, 2009; ISO, 13985, 2006), These two documents specify the requirements for refillable fuel tanks intended for onboard storage of respectively high pressure compressed gaseous and liquid hydrogen on land vehicles. Similarly, the International Electrochemical Committee has focused, with its Technical Committee 105, on standards regarding fuel cell technologies for all commercial land-based applications. Among its many publications it is worth a special mention of the IEC safety and performance testing standards for stationary fuel cells (IEC 62282–3-1, 2007; IEC 62282-3-2, 2006), and the corresponding safety standard for portable fuel cell systems (IEC 62282-5-1, 2007). Important and pioneering work has been performed by SAE International, for example with the Technical Information Report on vehicular hydrogen systems, which

[5] The following text is an excerpt of my dissertation titled "Communication/Exploration/Navigation Technologies—Applications, Trade-Offs and Possible Transfers Between Space and Ground at the Example of MOA^2, a Novel Pulsed Plasma Accelerator."

first among the standardization bodies adopted a full performance-based standard to guarantee safety operation during the whole life of hydrogen pressurized components. The standard consists mainly of an Expected-Service Performance Verification test and a Durability Performance Verification test.

In parallel with these standardization activities, and often building upon them, a similar effort has been invested at national and international levels for the development of legally binding regulations. The European Commission has prepared a regulation for type approval of hydrogen-powered motor vehicles (REGULATION (EC), 2009), which has been approved by the European Parliament and the Council in 2009, and in 2010 (Commission Regulation, 2010) an additional, more technical document containing the implementing measures such as the individual tests required for the type approval. In this document, all the tests and performance requirements needed to ensure safe and reliable function of all the components of a hydrogen-propelled vehicle are described in detail. Typical tests for high pressure components such as tanks are: burst test, bonfire test (resistance to fire), chemical exposure test, ambient temperature, and extreme temperature pressure cycle tests, accelerated stress rupture test, impact damage test, leakage test, hydrogen gas cycling test. According to the hydrogen gas cycling test, for example, the high pressure tank must be subjected without deterioration to 1000 hydrogen filling and emptying cycles, which simulates refilling at the refueling station, followed by the fuel consumption during travel.

More recently, the United Nations Economic Commission for Europe (UN-ECE) has also concluded the drafting work for a Global Technical Regulation for hydrogen-fueled vehicle (UN-ECE), containing compliance tests for fuel system integrity, test procedures for compressed hydrogen storage and for electrical safety.

5.11 Future trends

Future applications for hydrogen in space are likely to be centered on the spin-in of further advanced fuel cell and hydrogen storage technologies for new purposes. The fuel cells being used in NASA's manned space missions like Gemini, Apollo, and the Space Shuttle already provided for high power levels and potable water. When combining these with the latest technological developments in the terrestrial automotive sector, it would be possible to design a pressurized manned rover for the exploration of the Moon and Mars. Similar to an automobile on the Earth, the combination of high pressure hydrogen gas tanks, buffer batteries, hydrogen leakage detectors, and fuel cells could provide for a manned pressurized rover, allowing 2–4 astronauts to explore planetary bodies like the Moon during a cruise of 5–6 weeks, thereby covering a range of 1000–2000 km. As part of an ESA contract with Thales Alenia Space, the Liquifer Systems Group (LSG) has designed a Rover for Advanced Mission Applications (RAMA Concept) (Figure 5.10).

RAMA has been designed for surface missions with a crew of two or three lasting up to approximately 40 days; its source of energy is a liquid hydrogen/liquid oxygen fuel cell, allowing it to be driven and operated during the day as well as the night. Guidance, navigation, and obstacle avoidance systems are foreseen as standard

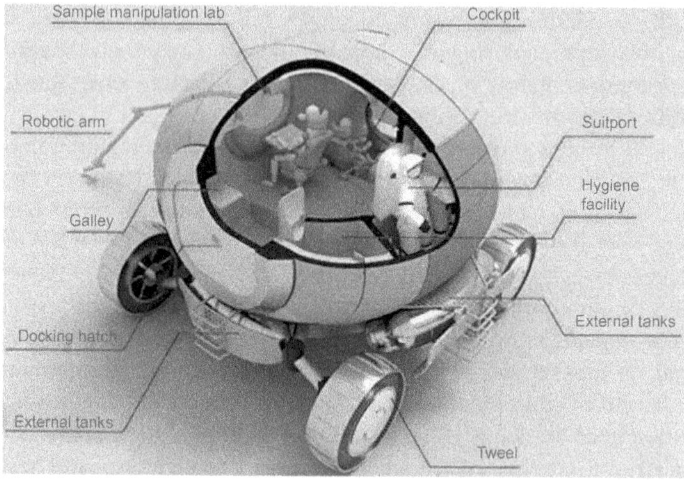

Figure 5.10 Illustration of LSG's RAMA concept.

equipment to allow it to travel safely over rough terrain at all times of the day. The rover allows extravehicular activity and a remote manipulator is provided to recover surface samples, to deploy surface instruments and equipment and, in general, to assist the astronauts' field activities wherever and whenever needed. The vehicle has also been designed to have a very high degree of maneuverability. In addition, RAMA can be operated and replenished from a fixed site base or can cooperate with other rovers of the same type to provide a mobile base. The rover in all cases will be refueled using the products supplied by an in situ resources facility.

Transportation and surface exploration requirements define the size and mass of the rover. RAMA has a launch mass of approximately 7000 kg, a dry mass of about 6200 kg and surface mission masses of between 7800 and 8300 kg. The rover can be launched by a future heavy lift launcher similar to the American ARES V concept. The factor most affecting the mass of the rover, other than the quantities of fuel cell reactants and crew consumables, is the amount of radiation shielding integrated in the design of the rover's pressurized shell. The factor most influencing the rover's external and internal configuration is the launcher's payload envelope and the need for the rover's center-of-mass to be aligned with or close to the launcher's longitudinal axis.

Obviously there is a significant synergy with the fuel cells in the automotive sector, with the major difference being the type of hydrogen storage system: a cryogenic tank in the rover vs. a high pressure gaseous hydrogen tank in the cars. The major reason for this is the current complexity of high pressure compressor systems, which would be required to refill the hydrogen into the rover. With continued progress in the automotive sector, however, this type of compressor will eventually become space qualified as well and then a Moon/Mars rover can transfer-in the hydrogen energy systems and power train to the greatest extent, with all possible cost benefits.

References

Green Car Congress, 2013. Five Hyundai ix35 Fuel Cell vehicles joining London hydrogen project, 21 July. http://www.greencarcongress.com/2013/07/ix35-20130721.html (accessed February 2015).

Commission Regulation No 406/2010 implementing Regulation (EC) No 79/2009 of the European Parliament and of the Council on type-approval of hydrogen-powered motor vehicles.

REGULATION (EC) No 79/2009 OF THE EUROPEAN PARLIAMENT AND OF THE COUNCIL of 14 January 2009 on type-approval of hydrogen-powered motor vehicles, and amending Directive 2007/46/EC.

ESA Fuel Cells for Telecom Satellites System Study, 2008, Contract number: 21349/09/NL/LVH.

ESA Telecommunications & Integrated Applications, SmallGEO, 2011. http://www.esa.int/Our_Activities/Telecommunications_Integrated_Applications/SmallGEO, accessed in June 2015.

ESA S54 Study, 2001–2003. ESA Contract # 14565/00/NL/WK.

ESA S54 Study "Future Power Systems for Space Exploration", Executive Summary, 2002, http://esamultimedia.esa.int/docs/gsp/completed/comp_sc_00_S54.pdf (accessed April 2012).

Frischauf, N., et al., 2011. The hydrogen value chain: applying the automotive role model of the hydrogen economy in the aerospace sector to increase performance and reduce costs. In: JRC-IET, International Astronautical Congress 2011, 04 October 2011.

Frischauf, N., et al., 2013. The hydrogen value chain: applying the automotive role model of the hydrogen economy in the aerospace sector to increase performance and reduce costs, Acta Astronautica 88, 8–24, http://dx.doi.org/10.1016/j.actaastro.2013.01.002.

Gruntman, M., 2004. Chapters 1–2, Blazing the Trail: The Early History of Spacecraft and Rocketry, AIAA. ISBN 1-56347-705-X.

H_2YDROGEIT, February 2015. http://www.hydrogeit.de/bz-typen.htm.

http://en.wikipedia.org/wiki/Slush_hydrogen (accessed February 2015).

http://inventors.about.com/od/fstartinventions/a/Fuel_Cells.htm (accessed September 2011).

http://ipts.jrc.ec.europa.eu/publications/pub.cfm?id=3279 (accessed October 2014).

http://www.fch-ju.eu/sites/default/files/Power_trains_for_Europe_0.pdf.

IEC 62282–3-1. Edition 1.0 (2007-04-24) Fuel cell technologies—Part 3-1: stationary fuel cell power systems—safety.

IEC 62282–3-2. Edition 1.0 (2006-03-21) Fuel cell technologies—Part 3-2: stationary fuel cell power systems—performance test methods.

IEC 62282–5-1. Edition 1.0 (2007-02-23) Fuel cell technologies—Part 5-1: portable fuel cell power systems—safety.

ISO 13985:2006. Liquid hydrogen—land vehicle fuel tanks.

ISO/CD 15869, 2001. Gaseous hydrogen and hydrogen blends—land vehicle fuel tanks. ISO.

ISO/TS 15869:2009. Gaseous hydrogen and hydrogen blends—land vehicle fuel tanks.

RAMA Concept, Liquifer Homepage, http://www.liquifer.at/projects/1xg/RAMA/e_rama.htm.

Thomas, C.E., February 2015. Hydrogen-powered fuel cell electric vehicles compared to the alternatives. http://www.azocleantech.com/article.aspx?ArticleID=214.

UN-ECE, World Forum for Harmonization of Vehicle Regulations (WP.29), Working Party on Passive Safety (GRSP), Informal Group on Hydrogen and Fuel Cell Vehicles. http://live.unece.org/transwelcome/areas-of-work/vehicle-regulations/working-parties-and-documents/working-party-on-passive-safety-grsp/informal-group-on-hfcv-sgs.html.

Part Two

Other applications of hydrogen

Hydrogen fuel cells for portable applications

6

P.P. Kundu, K. Dutta
University of Calcutta, Kolkata, India

6.1 Introduction

Present day portable electronic devices (PEDs) represent a unique class of multi-functionary systems. Hand-held mobile devices exemplify cumulative communicating, computing and entertaining functionalities. Multipurpose PEDs have instantly drawn extreme mass attention and market demand, simply because they have potentially made human life a lot more secure, convenient, and hassle-free. Therefore, it is very much expected that with progress in time, these devices will be available with increased built-in functionalities and improved sophistication. However, such progress will essentially demand equally capable charging and power back-up systems for these ultramodern devices to run efficiently. For example, a third-generation (B3G) multimedia phone will require nearly 3790 mAh per day, which means a need for three 5 h charges for an one-day use (Chang et al., 2009). Soldiers and military personnel are outfitted with high-tech electronics, such as radios, night vision devices, portable computers, and personal cooling systems, in order to significantly increase awareness of the enemy environment. These applications demand compact and long-life power sources. In addition, demand also exists for lower weight and longer operating power sources for mobile notebook personal computers (Chang et al., 2009).

From their very advent, PEDs have used battery power to fulfill their energy requirements. Table 6.1 lists power demands of different PEDs (Kundu and Jang, 2009), while Figure 6.1 schematically demonstrates the power demands of next-generation mobile phones and notebook personal computers (Chang et al., 2009). According to power demand, we can classify the PEDs into small personal electronics (e.g., mp3 player, cameras, etc.) requiring around 3 W of power, and large personal electronics (e.g., laptops, printers, radios, etc.) requiring up to 30 W of power. However, the limited progress that has been made in current battery technology has generally failed to cope with the rapid progress achieved in the field of PED fabrication (Suominen et al., 2011). As a result, researchers and technologists have considered other available alternative energy sources as potential replacements for batteries (Stone, 2007; Cook-Chennault et al., 2008; Shimizu et al., 2004; Achmad et al., 2011; Kamarudin et al., 2009; Rashidi et al., 2009; Urbani et al., 2007; Maynard and Meyers, 2002; Yalcinoz and Alam, 2008). In this respect, fuel cells have strongly claimed their applicability, primarily owing to certain critical advantages they enjoy over battery systems, such as (a) instant recharging via a replacement or a refilled fuel cartridge, (b) independence from electricity, (c) longer cell lifetime, (d) lower operating temperature, (e) higher

Compendium of Hydrogen Energy. http://dx.doi.org/10.1016/B978-1-78242-364-5.00006-3

Table 6.1 **Power demands of different PEDs (Kundu and Jang, 2009)**

Items	Power required (W)
Cellular phone	1
Personal digital assistant (PDA)	1
Notebook personal computer	20–30
Flashlights and toys	1–10
Tablet personal computer	10
Playstation portable (PSP)	2
Digital multimedia broadcast-receiving (DMB) phone	3
iPhone	2
Robot	10–15
Digital camera	1

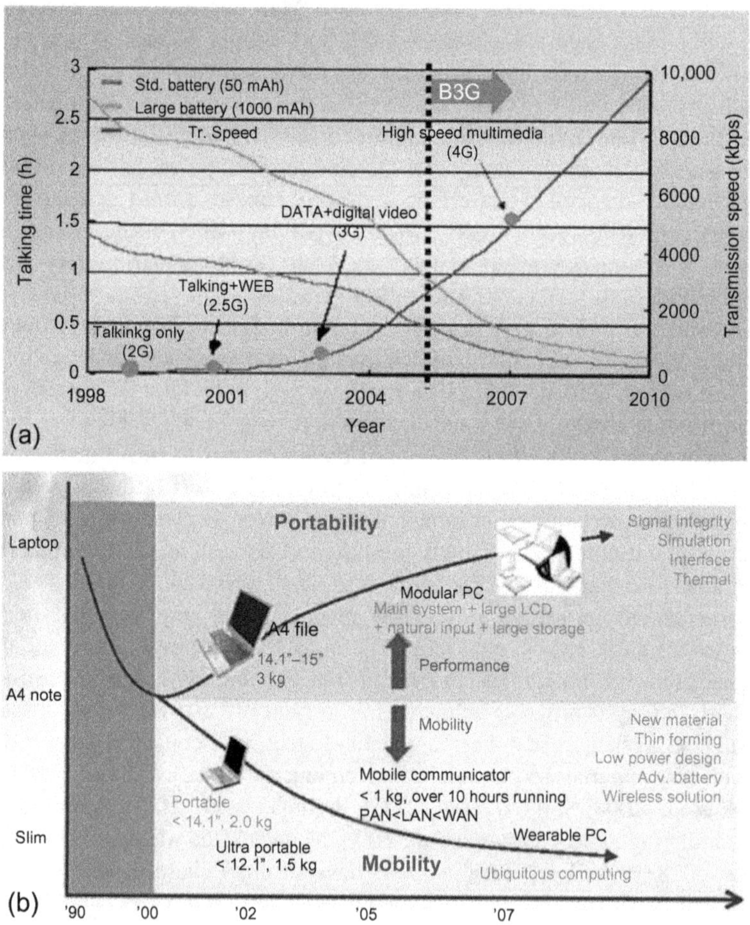

Figure 6.1 A schematic illustration of the power demands of next-generation (a) mobile phone and (b) notebook personal computers (Chang et al., 2009).

free energy content of the fuel, (f) higher energy conversion (chemical to electrical) efficiency, (g) significant weight reduction potential, (h) convenience and reliability, and (i) no point-of-use emissions (Inman et al., 2011; Scott, 2004). In addition, the efficiency of fuel cells, as compared to the batteries, does not depend on the size, since doubling of operating time requires only doubling the amount of fuel and not doubling of the capacity of the unit itself. Therefore, they can be highly efficient even for small systems. Again, unlike batteries, fuel cells have no "memory effect" when they are refueled. Moreover, electrochemical and materials degradation factors that usually reduce the longevity of batteries is not so critical in the case of fuel cells.

Among the various types of fuel cells that have been investigated so far, those using hydrogen as fuel (i.e., hydrogen fuel cells) are far superior in terms of energy density produced (Table 6.2) (Viswanathan and Scibioh, 2006). In order to enhance the prospect of commercialization of hydrogen fuel cells as a PED energy source, their energy efficiency, energy and power densities, as well as cost, have to be competitive with the presently existing rechargeable batteries. It should be noted in this respect that these requirements are different from other types of fuel cells for residential or vehicle application, since the latter can utilize auxiliary compartments as required. However, for portable applications these compartments have to be miniaturized effectively or eliminated totally, which means that the stack materials themselves should be able to perform the functions of these auxiliary compartments.

The design of a hydrogen fuel cell is inherently more complex than that of a battery, owing to the presence of fuel hydrogen and oxidant delivery systems, water removal systems, gas isolation systems, and flexible and low-loss external electrical access (Figure 6.2). These factors are seemingly hindering its miniaturization prospects and subsequent application in miniature PEDs. However, recent progress made in microelectromechanical systems (MEMS) technology and its application in fabricating different fuel cell component parts have given rise to the possibility of miniaturizing the fuel cell devices to make them compatible with miniature PEDs (Kundu and Jang, 2009; Maynard and Meyers, 2002; O'Hayre et al., 2003; Dutta et al., 2014a).

Table 6.2 A comparison of the chemical and electrochemical data of various fuels (Viswanathan and Scibioh, 2006)

Types of fuel	ΔG^0 (kcal/mol)	E^0 (theoretical) (V)	E^0 (maximum) (V)	Energy density (kWh/kg)
Hydrogen	−56.69	1.23	1.15	32.67
Methanol	−166.80	1.21	0.98	6.13
Ammonia	−80.80	1.17	0.62	5.52
Formaldehyde	−124.70	1.35	1.15	4.82
Formic acid	−68.20	1.48	1.14	1.72
Methane	−195.50	1.06	0.58	–
Propane	−503.20	1.08	0.65	–

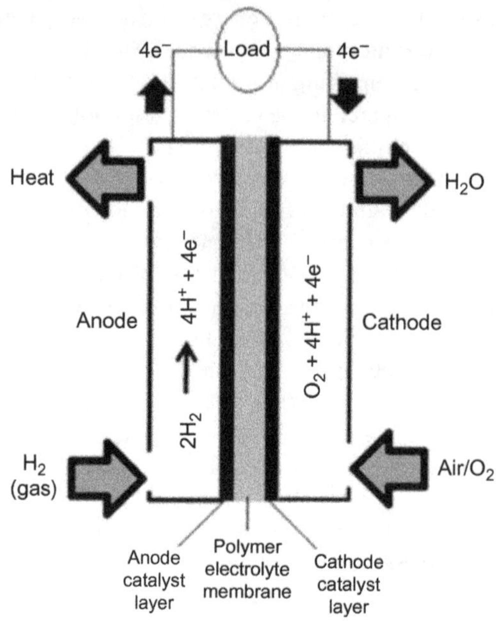

Figure 6.2 A schematic illustration of a hydrogen fuel cell system.

6.2 Drawbacks of hydrogen fuel cells regarding application in portable devices

6.2.1 Storage of hydrogen fuel

The development of new generation PEDs with increased functionality is accompanied by simultaneous reduction in device size and ever-increasing demand for high power density. The power delivered by hydrogen fuel cells (from μW to mW range) can be potentially used in implantable applications and low power MEMS. Advancement of hydrogen fuel cell power systems in portable applications is mainly hindered by problems associated with hydrogen storage. When designing storage facilities for hydrogen, the issues which must be considered are weight, volume, efficiency, safety, and most importantly "the cost." Hydrogen fuel can be stored in pressurized containers or within chemicals or solids capable of reversible absorption/desorption of hydrogen with low energy exchange. State-of-the-art metal hydride cartridges, with storage densities of up to 1–2% weight, may yield energy densities in the range 50–90 Wh kg^{-1} and 140 Wh L^{-1}, which is comparable to modern Li–ion batteries (Fernández-Moreno et al., 2013). Protonex, Inc. and Millenium Cell, Inc. demonstrated fueling by hydrogen generated from sodium borohydride solutions, and produced specific energy values as high as 375 Wh kg^{-1} for systems that operate for 12 h on a single charge.

6.2.2 Cost-related factors

Cost is the most fundamental and critical barrier in the commercialization of fuel cells, and therefore, development of low-cost materials is the need of the hour. Membrane electrode assembly (MEA) components, including the polymer electrolyte membrane (PEM), the catalyst and the gas diffusion layers (GDLs), account for about 40–50% of the cost of a fuel cell. For instance, at present approximately one-third of the stack cost is still contributed by Pt alone (James et al., 2011), although this fraction has come down from over half of the fuel cell cost in 2008 (Zhang, 2008). In addition, Nafion membranes (a product of DuPont) cost in the range of US$ 800 m^{-2}. Hence, MEAs are considered to be the single largest contributor to the total cost of a fuel cell. The use of hydrogen in fuel cells is more efficient when compared to traditional technology, leading to higher quality of energy with fewer waste products. However, the higher efficiency of fuel cells does not always account for their high initial costs. Therefore, it is very clear that market penetration of polymer electrolyte membrane fuel cells (PEMFCs) will only be possible when their capital cost decreases and becomes more competitive with other available power sources (Cottrell et al., 2011; Shaw et al., 2013). Breakthroughs in constituent materials, which can potentially bring the overall costs down toward commercial cost, will assist in achieving the desired performance targets (Pollet et al., 2012). Numerous efforts for developing alternative cost effective and efficient membrane and catalyst materials have been reported over the years, including utilization of alternative PEM materials in place of traditionally used high-cost Nafion membranes and alternative catalyst materials in place of traditionally used high-cost and limited abundance platinum metal (Dutta et al., 2014b,c,d, 2015a,b; Das et al., 2014; Kumar et al., 2014; Zhang and Shen, 2012; Peighambardoust et al., 2010). In addition, some major innovations, such as low platinum catalyst loading and fabrication of thin film electrodes, have been reported. As a result of these continuing efforts, costs have been falling exceptionally rapidly, and are expected to continue doing so for the next 5–10 years.

6.2.3 Performance-hindering factors

For portable applications, both the energy and power density of the device should be high. The power density that current Li–ion batteries supply is ~200 Wh kg^{-1}. Therefore, hydrogen fuel cells should be able to generate higher power density compared to this in order to be competitive in the PED sector. This, in turn, requires a high reaction rate at the electrodes and proton transfer rate across the PEM. A hydrogen fuel cell most preferably employs Nafion membranes as PEMs. The primary function of a PEM is to transport protons generated in the anode chamber of the fuel cell to the cathode chamber, while keeping the reactants and products of one chamber separated from those of the other chamber. The state-of-the-art Nafion, a perfluorosulfonic acid membrane, can achieve a high level of proton conductivity (i.e., ~0.1 S cm^{-1}); however it requires 100% hydration in order to do so. This particular criterion is preventing hydrogen fuel cells from operating at high temperature (>60–80 °C) at normal atmospheric pressure. Moreover, it also requires fully hydrated fuels, i.e., hydrogen and oxygen/air, in order to function properly. Since, practically, hydrogen fuels are often generated from hydrocarbon (such as gasoline and natural gas) reforming using water–gas shift reactions, they often remain contaminated with up to 1% carbon monoxide (CO). This CO causes catalyst poisoning at the anode, particularly at the low operating temperature of the cell. As a result, a high operating temperature (~100 °C) is required in order to prevent the CO poisoning effect and to accelerate the reaction kinetics of the anode. Moreover, at high operating temperatures, the necessity for water management is eliminated. Therefore, development of alternative PEMs that require low hydration conditions for achievement of high proton conductivity and CO tolerant catalysts are required in order to increase the efficiency of hydrogen fuel cells toward practical portable applications.

6.3 Present status

Utilization of hydrogen fuel cells in PEDs has been gaining momentum at a steady rate, which is likely to increase with further progress made in terms of overall cost, and stability and longevity of power supplies from fuel cells. Owing to the ever-increasing functionality and subsequent miniaturization of PEDs, development of small and compact power delivery systems (providing high power-to-size ratio) is necessary. Various research activities on hydrogen fed micro-fuel cells have been presented in Table 6.3 (Kundu et al., 2007). These examples have been so chosen as to give an indication toward development made in the field of hydrogen fuel cells for portable applications on the laboratory scale.

O'Hayre et al. (2003) developed a portable hydrogen/oxygen fuel cell, using printed-circuit board (PCB) technology. Device voltages ranging from 1 V single cells to 16 V planar arrays were obtained, with power output ranging from <1 to >200 W. The lightweight laminate PCB technology was claimed to be the best prototype for achieving >700 mW cm^{-2} area power density and >400 mW cm^{-3} volumetric power density. In addition, the PCB technology was found to be prospective for developing portable fuel cells below 1 kW capacity.

With the objective of aiding in the development of portable fuel cell systems, the International Association for Hydrogen Energy (IAHE) chapter at Oakland University (OU) has designed, constructed, and tested a hydrogen-fueled portable PEMFC stack. The fuel cell created by the IAHE-OU chapter possessed a size dimension of 5.95 cm × 5.95 cm × 1.95 cm, a weight of 0.275 kg and can produce continuous power up to 12.5 W, with a peak power output of 17 W. Any source of pure dry hydrogen (at 2–3 psig) can be used to fuel the reported fuel cell. The total fuel cell system is composed of the portable fuel cell stack, an electrical fan with controller, and 5 and 12 V DC–DC boost converters. An image of the portable fuel cell system in operation and the polarization curve obtained has been presented in Figures 6.3 and 6.4, respectively (Inman et al., 2011).

Penn State's IAHE group designed, built and tested a 72 cm^3 fuel cell stack, and received a rather unsteady maximum power output of 42 W (Figure 6.5). The cell was operated for 1 h at 28.5 W with complete stability. The final design was composed

Table 6.3 **Experimental observations in the literature (Kundu et al., 2007)**

Fuel cells	Authors	Substrate material and feed flow rate	Dimension of cell and MEA	Comments
Direct hydrogen fuel cell	Lee et al. (2002)	Glass and silicon; H_2 and O_2 flow rate control by monitoring inlet pressure (35 kPa for two cell assembly and 100 kPa for four cell assembly)	Two cell assembly in wet etch glass: MEA area of each cell: 5 cm^2 and split type channel with 150 μm deep. Four-cell assembly in dry-etched silicon: The flow chambers were etched 200 μm deep and the square distribution pillars were 100 μm × 100 μm in size, arranged in a uniform rectangular array 100 mm apart	Maximum power density: 20 mW cm^{-2} (glass). Maximum power density: 42 mW cm^{-2} (silicon wafer)
	Yu et al. (2003a)	Silicon wafer: H_2 flow rate: 50 mL min^{-1}; O_2 flow rate: 50 mL min^{-1}	MEA area: 5 cm^2	Maximum power density: 107.3–194.3 mW cm^{-2}
	Yu et al. (2003b)	Silicon wafer: H_2 flow rate: 40 mL min^{-1}; O_2 flow rate: 60 mL min^{-1}	MEA area of each cell: 3 cm^2. Number of cells: 2. Pt loading in cathode and anode: 1 mg cm^{-2}	Maximum power density in twin cell: 190.4 mW cm^{-2}
	Shah et al. (2003, 2004)	Silicon and PDMS; hydrogen flow rate: 1 sccm	MEA area: 1.4 cm × 1.2 cm; channel width of 5 μm and number of channels in each plate: 768	Maximum power density: 0.35 mW cm^{-2}
	Hahn et al. (2004)	Sandwiched metal–polymer foils; hydrogen flow: 0.5 sccm	Size of micro-fuel cell: <10 cm^3 Size of Planar micro-fuel cells 0.2 cm^3; MEA of each cell: 0.18 cm^2 (total number of cells: 3); MEA of one cell: 0.54 cm^2	Maximum power density: 120 mW cm^{-2}; it can be potentially applied in wireless sensor networks, chip cards or autonomous microsystems; air-breathing
	Yamazaki (2004)	Silicon wafer: hydrogen and oxygen	MEA area: 1 cm^2	Maximum power density: 37 mW cm^{-2}; the porous silicon layer was there to support the MEA

Continued

Table 6.3 Continued

Fuel cells	Authors	Substrate material and feed flow rate	Dimension of cell and MEA	Comments
	Hsieh et al. (2004)	Polymethyl methacrylate (PMMA); H_2 flow rate: 10 $cm^3 min^{-1}$ at 4 atm; air flow rate	Membrane: Nafion 117; MEA area: 5 cm^2; flow field plate: parallel serpentine channel (width of 400 μm and depth of 200 μm); Pt loading in cathode and anode: 0.15 $mg cm^{-2}$; size of micro-fuel cell: ~16 cm^3	Maximum power density: 31 $mW cm^{-2}$
	Hsieh et al. (2005)	Cu and SU-8 photoresist H_2 flow rate: 6, 7, 8, 9 and 10 sccm; air velocity: 25 $cm s^{-1}$	–	Air breathing and forced air type; maximum power density: 22 $mW cm^{-2}$ (air breathing) and 29 $mW cm^{-2}$ (forced air)
	Chan et al. (2005)	Polymethyl methacrylate (PMMA); hydrogen pressure: 10 psi; air flow rate: 50 $mL min^{-1}$; oxygen flow rate: 20 $mL min^{-1}$	MEA area: 3 cm^2; flow field plate; spiral channel (width and depth of 220 μm); Pt loading in cathode and anode: 1 $mg cm^{-2}$; size of micro-fuel cell: 3.5 cm^3	Maximum power density: 82 $mW cm^{-2}$ (air in the cathode) maximum power density: 315 $mW cm^{-2}$ (oxygen in the cathode)
Reformed hydrogen fuel cell Fuel: methanol	Yamazaki (2004)	Silicon wafer. Size of reformer: 25 mm × 17 mm × 1.3 mm (micro-channels with a width of 600 μm and a depth of 400 μm)	95% and more of the conversion ratio at the reaction temperature of 280 °C	–
	Holladay et al. (2004)	Two cells in series (area of each cell: 1 cm^2); the volume of fuel processor (a catalytic combustor, two vaporizers, a heat exchanger, and a catalytic methanol reformer): less than 0.25 cm^3 and a mass of less than 1 g	99% conversion of methanol at around ~400 °C	Maximum power density: ~10 $mW cm^{-2}$

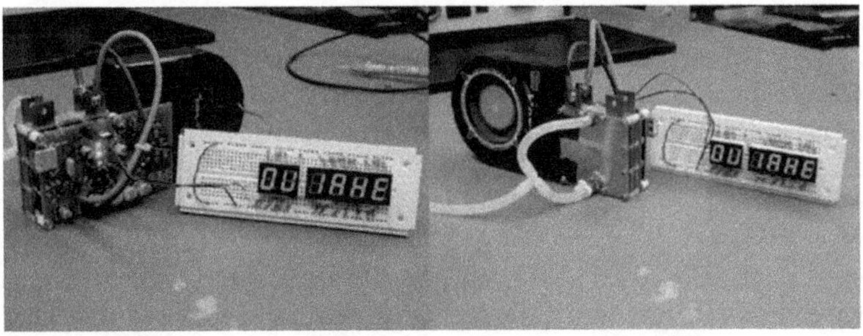

Figure 6.3 Images of the portable fuel cell system in operation (Inman et al., 2011).

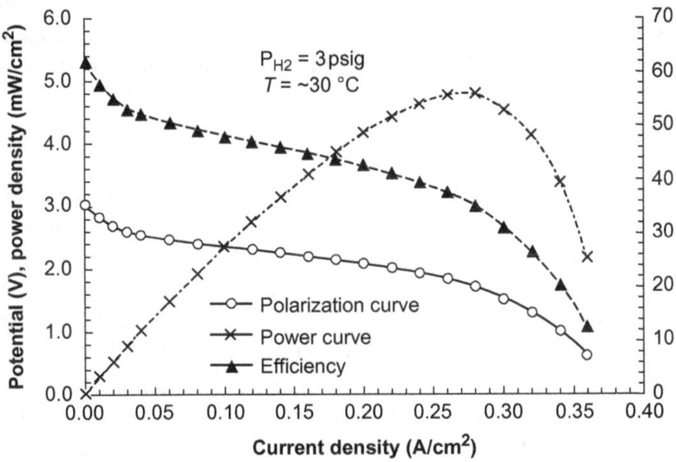

Figure 6.4 Polarization curve of the IAHE-OU fuel cell stack for steady-state operating conditions (Inman et al., 2011).

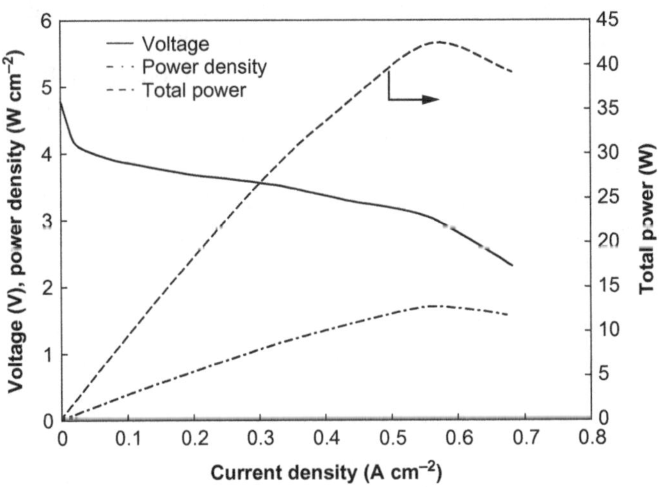

Figure 6.5 Stack performance curves showing voltage, power density and total power versus current density (Manahan et al., 2011).

of a cell stack containing 6 bipolar/end plates, i.e., 5 cells, each of which measures 6 cm × 6 cm × 1.975 cm and with an active area of approximately 25 cm^2. The total open circuit voltage obtained was approximately 4.75 V, amounting to an average value of 0.95 V per cell. With the previously mentioned active area, the maximum power obtained during operation was approximately 1.7 W cm^{-2}. The maximum sustained power over 1 h averaged to 1.15 W cm^{-2}; while the total cost of the project totaled $2245 (Manahan et al., 2011).

Urbani et al. developed a PEMFC for portable applications, which can operate at room temperature (Urbani et al., 2007). The device was composed of a homemade air breathing fuel cell stack, a metal hydride tank for hydrogen supply, a DC–DC converter for power output control and a fan for cooling the stack. The stack was composed of 10 cells, with an active surface of 25 cm^2, and produces a rated power of 15 W at 6 V and 2 A. The stack could successfully run with end-off fed hydrogen, without appreciable performance degradation during the operating time. The final assembled system was able to generate a power of 12 W at 9.5 V, and could continuously power a portable DVD player for 3 h. This power unit was able to operate for about 100 h without maintenance.

In addition to laboratory-based developments discussed above, considerable industrial developments of hydrogen fuel cells and their application in PEDs have also been witnessed in the last decade. Table 6.4 lists some of these developments (Kundu et al., 2007). These examples have been chosen so as to give an indication of developments made in the field of hydrogen fuel cells for portable applications on the industrial scale.

A typical micro-fuel cell has an output of 300 mW, which is only enough to keep a cell phone battery charged. On the other hand, a laptop consuming 30 W of power requires 100 miniature fuel cells in order to sustain continuous operation. Toshiba reported prototype fuel cells for laptops and other portable applications, generating 20–100 W of power. They further reported that at 100 W/L, the units are compact and the specific energy is comparable with a NiCd battery. Panasonic claimed to have doubled the power output from 10 to 20 W, while maintaining similar size of the fuel cell. They further specified a calendar life of 5000 h on the condition that the fuel cell is used intermittently for 8 h/day. Angstrom Power developed a portable fuel cell running on stored hydrogen while taking oxygen from the air. This system is devoid of pump and fan, and is therefore totally silent. The 21 cc cartridge provides the equivalent energy of about 10 AA disposable alkaline batteries, and the runtime between refueling is 20 h. However, for application in a laptop (requiring ~30 W of power), a small fuel cell cannot provide enough output to sustain the demand. Therefore, the system needs a battery as backup, leading to the fuel cell serving only in the capacity of a charger (Carter et al., 2012).

In 2008, Angstrom (which has subsequently been acquired by BIC) integrated a hydrogen fuel cell system into a popular slim-line Motorola mobile phone. Although realizing full integration of fuel cells into commercial mobile phones is still a distant matter, progress has been made with respect to size and cost reduction. In October 2011, patent applications filed by Apple were published, which revealed that the company was working on development of lightweight monopolar fuel cells for the purpose of integration into its ever-slimmer devices. Further applications in December 2011 showed how the company could integrate the units directly into its MacBook laptops

Table 6.4 **Status of different companies and laboratories in the development of micro-fuel cells (Kundu et al., 2007)**

Type of micro-fuel cell	Company's name	Dimension and volume	Maximum power density and power	Potential applications
Micro-reformed hydrogen fuel cell	Ultracell (ultracellpower); fuel: methanol	150 mm × 230 mm × 43 mm (weight = 1 kg; with cartridge of 200 cm^3 and 550 cm^3)	Maximum power 25 W; cartridge duration: 9 h and 24 h at 20 W for 200 and 550 cm^3 volume of cartridge, respectively	Military purposes, portable electronics
	Casio (Yahata et al., 2006); fuel: methanol	Reformer + PROX system size: ~20 cm^3	2.5 W	Charger
	Seiko Instruments Inc. (Iwasaki, 2007); fuel: sodium borohydride	1 W system: 75 mm × 40 mm × 60 mm; 3 W system: 80 mm × 45 mm × 70 mm; 10 W system: 200 mm × 65 mm × 53 mm	Time of operation: 2–5 h	Maximum energy density of the system: 60 $Wh\,L^{-1}$
Hydrogen fed micro-fuel cell	Angstrom Power, Inc. (angstrompower)	25 mm × 66 mm × 100 mm	Maximum power: 2 W (through hydrogen storage); time of operation: 6 h	Charger for cell phones, PDAs, and digital cameras; needs infrastructure for refueling hydrogen into the system; energy density: 72.7 $Wh\,L^{-1}$
	NTT DoCoMo and Aquafairy Co. (Commsdesign)	24 mm × 24 mm × 70 mm; weight: 45 g	Maximum power: 2 W (800 mAh at 3.6 V) fuel cartridge (water plus hydrogen producing catalyst): 10 cm^3; time of operation: 5 h	Charger for lithium–ion battery in 3G handsets. Energy density of the system: 248 $Wh\,L^{-1}$
	Hitachi Maxell Ltd (neasia)	97 mm × 87 mm × 31 mm	Maximum power: 10 W; aluminum: 20 g; water: 40 cm^3; time of operation: 4–5 h	For laptop; energy density of the system: <191 $Wh\,L^{-1}$
	Canon, Inc. (Shibata, 2007)	Fuel cell: 35 mm × 35 mm × 12 mm; metal hydride: 60 mm in length with 10 mm diameter and fuel cell: 55 mm × 77 mm × 3.5 mm; metal hydride: 55 mm × 66 mm × 5 mm	Stack power density: 0.2–0.3 Wcm^3; system power density: 0.1–0.15 Wcm^3	Maximum energy density of the system: ~160 $Wh\,L^{-1}$

Figure 6.6 Horizon MiniPak fuel cell charger (Hart et al., 2014).
© E4tech, 2014, and Horizon Fuel Cell Technologies, reprinted with permission.

and into an external charging system for iPads and iPhones. This latter application is rapidly emerging as a commercial reality for fuel cells (Carter et al., 2012).

At the 2010 Consumer Electronics Show held in Las Vegas, Horizon Fuel Cell Technologies (Horizon) revealed its MiniPak portable PEMFC electronics charger to the world (Figure 6.6) (Hart et al., 2014). With a 2 W USB output, the unit runs from solid-state hydrogen cylinders (refillable from a desktop electrolyzer). The units have been trialed extensively throughout 2011 to positive reviews (Carter et al., 2012). The company continued the global rollout of this charging device, and has been offering customization options for retail partners and bulk buyers (Figure 6.7) (Carter and Wing, 2013).

In February 2011, Sweden's myFC and Japan's Aquafairy demonstrated water-activated PEMFC portable chargers at the Mobile World Congress in Barcelona and the FC Expo in Japan, respectively (Carter et al., 2012). Aquafairy began domestic sales of its AF-M3000 system in April 2011, and since then the demand has been healthy. May 2012 saw the first commercial sales of myFC's PowerTrekk device, which launched across 13 major global markets in a staggered manner to allow the company to meet strong demand for the system. PowerTrekk charger is a fuel cell-battery hybrid rated at either 5 or 6.5 W (depending on the model). It uses a chemical hydride store for supply of the fuel hydrogen (Hart et al., 2014). In August 2012, Horizon began commercial sales of its MiniPak through REI, which is North America's largest retail chain for recreational outdoor gear, as well as many other American, European, and Asian retail outlets (Carter et al., 2012).

Jadoo Systems has commercialized the N-GenTM, a 100 W system that uses hydrogen supplied from a uniquely designed and user-friendly metal hydride canister (N-StorTM). This system can be rapidly refilled using Jadoo's portable fill station, and

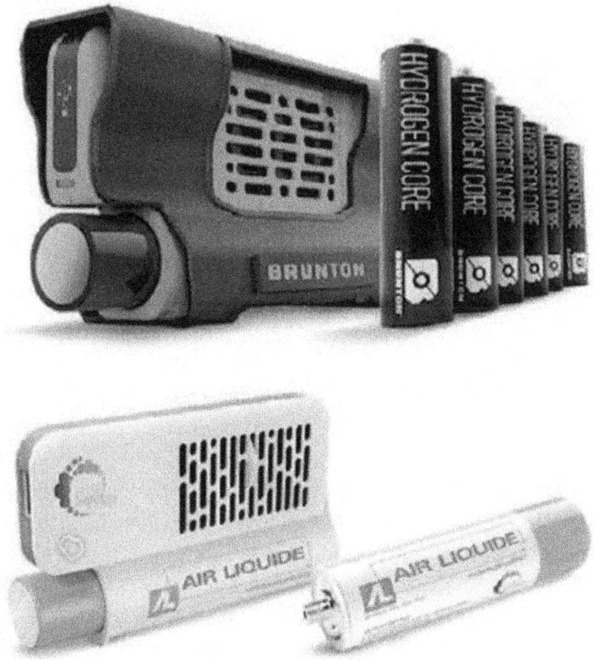

Figure 6.7 Branded MiniPak fuel cell chargers (Carter and Wing, 2013).
© E4tech, 2014, and Horizon Fuel Cell Technologies, reprinted with permission.

has found acceptance as a battery replacement for video cameras and field equipment. This system offers about 3.6 h of operation on a single cartridge, which calculates to 78 Wh kg^{-1} at the system level. The fact that the user can swap out the fuel cartridges to recharge the power source distinguishes this system from other direct hydrogen fuel cells. The cartridges provided with the N-Gen are "hot-swappable" and have a specific energy of 180 Wh kg^{-1}. These portable systems face competition with lithium ion batteries that can provide specific energy values as high as 150 Wh kg^{-1} in large battery packages (Ramírez-Salgado and Domínguez-Aguilar, 2009).

Intelligent Energy's Upp 5 W charger was unveiled at the end of 2013 (Figure 6.8). It utilizes their proprietary air-cooled fuel cell and a metal hydride store of hydrogen. The company targeted at producing and shipping 50,000 units before the end of 2014, by partnering with Brookstone and Sure for distribution (Hart et al., 2014).

6.4 Market penetration

The penetration potentiality of hydrogen fuel cells in the market is extremely important from a commercialization and economic point of view. Since the use of fuel cells is associated with the use of the PED that it is powering, therefore the demand of the latter directly influences the demand of the former (Agnolucci, 2007). However, since other

Figure 6.8 Intelligent Energy's Upp fuel cell charger (Hart et al., 2014).
© E4tech, 2014, and Intelligent Energy Ltd, reprinted with permission.

alternative powering devices exist for PEDs, fuel cells therefore have to be competitive and should be able to be user-friendly both in terms of cost and handling. For example, customers who often do not remember to charge their mobile handsets on a regular basis are more prone to accept hydrogen fuel cell powering technology for their mobile phones over the existing battery technology. On a regional basis, adoption of hydrogen fuel cell powering technology for mobile phones has been more enthusiastic in Asian markets (such as Japan and Korea) over that of the markets in Europe and the USA. Analysis has shown that, for mobile phones and laptops, the higher energy output provided by hydrogen fuel cells is not of prime importance to cause its immediate acceptance by the customers. Cost, user-friendliness, and size-related factors are at present serving as the most important factors in this respect. Again, for recreational users, the clean and silent mode of operation offered by fuel cells in portable generators is more acceptable. Similarly, for utilization in the security sector, the smaller size, light weight, and reliability of fuel cells can increase the speed of soldiers and security personnel. Higher reliability demands simpler logistics for maintenance and spare parts, while the negligible heat and noise produced by fuel cells are generally very difficult to detect by enemies (Agnolucci, 2007). Table 6.5 lists some barriers to fuel cell market penetration (Cottrell et al., 2011). An assessment of energy demand and cost of fuel cells and cartridges has been presented in Table 6.6 (Ramírez-Salgado and Domínguez-Aguilar, 2009). Overall, as long as the cost of fuel cells does not become competitive with the other available power sources, it will not serve as a very attractive alternative. However, once cost is cut down considerably, then in conjunction with the other potential benefits it offers, fuel cells will be the dominant power source at least in the PED market.

Mobile phones serve as the largest potential market for fuel cells, representing 86% of applications in Europe, 85% in Asia, and 76% in North America in 2002. It has been predicted that this prevalent market share in favor of mobile phones will remain similar over the period of the next 5 years, although it will decline in all regions to 83% in Asia, 79% in Europe, and 65% in North America. The primary reason behind this is a relatively slow overall worldwide compound annual growth rate (CAGR) of just 8.0%.

Annual fuel cell shipments in the portable sector grew by a modest 1.5% between 2010 and 2011. However, 2012 witnessed commercial launches of several portable electronics chargers, including the myFC Power Trekk and the Horizon MiniPark,

Table 6.5 Barriers to fuel cell market penetration (Cottrell et al., 2011)

Technical considerations	Even with large amounts of support for fuel cells there is still a need for more technical research and development in most of these applications. Much data need to be collected and shared in order to validate technology. Durability and reliability of components, e.g., improvement in PEM stacks, balance of power (such as sensors, pumps, reformers, and energy converters), must be addressed in experimental and applied environments. Among others, fuel reformation, hydrogen safety, readiness of infrastructure, as applicable, must be considered. Additionally, increased support for R&D in hydrogen storage technologies would address a leading cost factor in middle to large-scale fuel cell applications (10 kW–1 MW), as well as address renewable infrastructure
Cost competitiveness	Clearly, one of the main obstacles to market penetration is the high initial cost associated with stationary fuel cell applications. Market penetration of the PEM fuel cell will only be possible when the capital cost for fuel cells decreases and becomes more competitive with other power sources
Public acceptance	Many consumers who are unfamiliar with hydrogen technology believe that hydrogen is unsafe for use by the average citizen. The public must be made aware that hydrogen is not only safe, but has also proven to be cost effective in many applications in the mass market. If a large majority of the public no longer felt that hydrogen technology is unsafe, then the fostering of innovative government policy would be less cumbersome

Table 6.6 Assessment of energy demand and cost of fuel cell and cartridge (Ramírez-Salgado and Domínguez-Aguilar, 2009)

Application	Fuel cell power (W)	Battery power (W)	Total Wh	Runtime (h)	System size (cc)	Retail fuel cartridge cost (USD)	Fuel cell system selling prize (USD)	
Cell phone	1	2	12	12	10–5	<1	10	7
PDA	1	None	20	20	20	<1.50	10	7
Digital camera	2	3	10–15	5–7	20	3	25	20
Laptop	15	30	120	8	300	5–10	75	50

which resulted in a sevenfold increase of portable fuel cell shipments from 6900 units in 2011 to 50,500 units in 2012. Since 2007, Horizon Fuel Cell Technologies, Heliocentris, and h-tec have been trying to generate public awareness and understanding of fuel cell technology through the commercialization and mass shipment of milliwatt-scale PEMFC toys and education kits. Allowing tomorrow's consumers to gain an understanding of fuel cell technology is important for long-term industry progression, and the commercialization and popularity of these systems has been a precursor to the commercialization of more practical and industrial fuel cells over the last few years. These milliwatt-scale fuel cells currently ship more than 200,000 units per annum with steady growth of approximately 15–25% per year; nevertheless, they contribute little to total megawatts shipped (Carter et al., 2012).

Fuel cells thrive in applications where incumbent technologies are easy to displace and this early market has huge potential revenue—the mobile phone travel charger segment is valued at more than €11 billion, according to myFC. According to the Fuel Cell Industry Review 2014, portable unit shipments and small battery charger-type systems have been gaining traction. This strong growth from 2013 to 2014 has been largely due to increase in the consumer products segment, such as mobile phone chargers. Annual unit shipments figures of fuel cells for the period 2009–2014 have been presented in Table 6.7 (classified into shipments by application, by region, and by fuel cell type). Similarly, annual megawatts shipment figures of fuel cells for the period 2009–2014 have been presented in Table 6.8 (classified into megawatts by application, by region, and by fuel cell type) (Hart et al., 2014).

Table 6.7 Annual unit shipments (by application) of fuel cells for the period 2009–2014 (Hart et al., 2014)

1000 units	2009	2010	2011	2012	2013[a]	Forecast 2014
Shipments by application						
Portable	5.7	6.8	6.9	18.9	13.0	21.8
Stationary	6.7	8.3	16.1	24.1	51.8	45.6
Transport	2.0	2.6	1.6	2.7	2.0	2.9
Total	14.4	17.7	24.6	45.7	66.8	70.2
Shipments by region						
Europe	4.4	4.8	3.9	9.7	6.0	6.1
North America	3.2	3.3	3.3	6.8	8.7	17.1
Asia	6.7	9.5	17.0	28.0	51.1	45.2
Rest of the World	0.1	0.1	0.4	1.2	1.0	1.8
Total	14.4	17.7	24.6	45.7	66.8	70.2
Shipments by fuel cell type						
Polymer electrolyte membrane fuel cell	8.5	10.9	20.4	40.4	58.7	65.3
Direct methanol fuel cell	5.8	6.7	3.6	3.0	2.6	3.1
Phosphoric acid fuel cell	0.0	0.0	0.0	0.0	0.0	0.0
Solid oxide fuel cell	0.1	0.1	0.6	2.3	5.5	1.8
Molten carbonate fuel cell	0.0	0.0	0.0	0.0	0.0	0.0
Alkaline fuel cell	0.0	0.0	0.0	0.0	0.0	0.0
Total	14.4	17.7	24.6	45.7	66.8	70.2

[a] Uncorrected fuel cell today forecast from 2013.

Table 6.8 Annual megawatts shipments (by application) of fuel cells for the period 2009–2014 (Hart et al., 2014)

MW	2009	2010	2011	2012	2013[a]	Forecast 2014
Megawatts by application						
Portable	1.5	0.4	0.4	0.5	0.3	0.5
Stationary	35.4	35.0	81.4	124.9	186.9	147.3
Transport	49.6	55.8	27.6	41.3	28.1	28.2
Total	86.5	91.2	109.4	166.7	215.3	176.0
Megawatts by region						
Europe	2.9	5.8	9.4	17.3	17.3	10.4
North America	37.6	42.5	59.6	61.5	74.7	52.3
Asia	45.3	42.5	39.6	86.1	122.9	112.4
Rest of the World	0.7	0.4	0.8	1.8	0.4	1.0
Total	86.5	91.2	109.4	166.7	215.3	176.0
Megawatts by fuel cell type						
Polymer electrolyte membrane fuel cell	60.0	67.7	49.2	68.3	68.0	69.7
Direct methanol fuel cell	1.1	1.1	0.4	0.3	0.2	0.2
Phosphoric acid fuel cell	6.3	7.9	4.6	9.2	7.9	3.8
Solid oxide fuel cell	1.1	6.7	10.6	26.9	47.0	32.2
Molten carbonate fuel cell	18.0	7.7	44.5	62.0	91.9	70.0
Alkaline fuel cell	0.0	0.1	0.1	0.0	0.3	0.0
Total	86.5	91.2	109.4	166.7	215.3	176.0

[a] Uncorrected fuel cell today forecast from 2013.

6.5 Future perspectives and conclusion

For mobile devices, the target is to increase the device runtime by 5–10 times, without increasing the size or mass of the power source. In order to meet the energy and power density targets for future mobile devices, a specific energy of 500 Wh kg^{-1} or greater, an energy density of 1000 Wh L^{-1} or greater and a specific power not less than 50 W kg^{-1} should be achieved. Typical power levels in mobile devices vary from 1 to 30 W with runtimes varying from 10 to 100 h. Further, any next-generation power source for mobile devices must include battery features, namely, quiet operation, ruggedness, ease of use, operability in any orientation, operability over a wide temperature range (−40°C to +60 °C), and operability in confined environments. These requirements present significant technical challenges. However, in the last decade or so, there have been focused efforts in the commercial and government sectors to develop fuel cell-based power sources operating on high-energy fuels, such as hydrogen, in order to meet the requirements of next-generation portable power sources (Narayan and Valdez, 2008).

The objective to reduce the overall cost of hydrogen fuel cells is currently serving as a novel research topic, and is serving as a challenge for the prospective researchers to lower the cost of MEA, including cost of PEMs and catalysts (without influencing the efficiency of the cell). Since hydrogen fuel cell technology has already become the chosen path for portable applications, therefore developments of future technologies and key components, hardware, controls and manufacturing processes and volumes are being investigated. In addition, leveraged scale-up of manufacture across all the applications will benefit (in terms of cost) the key components of hydrogen fuel cells, like PEMs, catalysts and gas diffusion media. Direct hydrogen fuel cells, with passive operation, are also capable of near-complete fuel utilization. This advantage will avoid the need for micro-mechanics and sensors, thus eliminating added cost, volume and complexity, which could otherwise negatively impact durability and reliability. However, it is too early to assume that the use of hydrogen as a fuel in direct hydrogen fuel cell devices for portable electronic applications has been fully solved. Micro-fuel cells, based on direct hydrogen fuel cells, are the best alternatives to compete with Li-ion battery technologies. Many believe that this fuel cell technology is rapidly progressing toward the point where it will be able to provide the required power and runtime to a new generation of electronic and electrical devices. In addition, micro-fuel cell-based technologies also potentially offer substantial advantages in comparison to existing technology in terms of operation at a competitive price. Therefore, resolving of the issues of fuel storage and supply, in conjunction with generation of consumer demand, will ultimately lead to the success and acceptance of hydrogen fuel cell technology as a prospective alternative to batteries (Stone, 2007). However, manufacturers of micro-fuel cells admit that a direct battery replacement with high power, small size, and competitive price is still several years away. Rather than offering an outright battery replacement, today's micro-fuel cell serves as a charger to provide continuous operation for the onboard battery. It is indeed regretful that the fuel cell has not enjoyed the same breakthrough as microelectronics. It is therefore our hope that the fuel cell will eventually succeed as a clean energy source to extend the range of portable power and reduce pollution.

Acknowledgments

K. D. would like to thank the Council of Scientific and Industrial Research (CSIR), India, for a Senior Research Fellowship (Award no. 09/028 (0901)/2012-EMR-I). P. P. K. is thankful to the Ministry of New and Renewable Energy (MNRE), Govt. of India, for a grant-in-aid (Grant no. 102/56/2009-NT).

References

Achmad, F., Kamarudin, S.K., Daud, W.R.W., Majlan, E.H., 2011. Passive direct methanol fuel cells for portable electronic devices. Appl. Energy 88, 1681–1689.

Agnolucci, P., 2007. Economics and market prospects of portable fuel cells. Int. J. Hydrogen Energ. 32, 4319–4328.

http://www.angstrompower.com.
Carter, D., Wing, J., 2013. The fuel cell industry review. Fuel Cell Today, 1–46.
Carter, D., Ryan, M., Wing, J., 2012. The fuel cell industry review. Fuel Cell Today, 1–52.
Chan, S.H., Nguyen, N.-T., Xia, Z., Wu, Z., 2005. Development of a polymeric micro fuel cell containing laser-micromachined flow channels. J. Micromech. Microeng. 15, 231–236.
Chang, H., Kim, H., Choi, Y.S., Lee, W., 2009. Critical issues in the commercialization of DMFC and role of membranes. In: Zaidi, S.M.J., Matsuura, T. (Eds.), Polymer Membranes for Fuel Cells. Springer, New York, USA, pp. 307–340.
http://www.commsdesign.com/news/showArticle.jhtml?articleID=190400451.
Cook-Chennault, K.A., Thambi, N., Sastry, A.M., 2008. Powering MEMS portable devices—a review of non-regenerative and regenerative power supply systems with special emphasis on piezoelectric energy harvesting systems. Smart Mater. Struct. 17, 043001 (1)-043001 (33). http://dx.doi.org/10.1088/0964-1726/17/4/043001.
Cottrell, C.A., Grasman, S.E., Thomas, M., Martin, K.B., Sheffield, J.W., 2011. Strategies for stationary and portable fuel cell markets. Int. J. Hydrogen Energ. 36, 7969–7975.
Das, S., Kumar, P., Dutta, K., Kundu, P.P., 2014. Partial sulfonation of PVdF-*co*-HFP: a preliminary study and characterization for application in direct methanol fuel cell. Appl. Energy 113, 169–177.
Dutta, K., Kundu, P.P., Kundu, A., 2014a. Fuel cells—exploratory fuel cells | micro-fuel cells. In: Reedijk, J. (Ed.), Reference Module in Chemistry, Molecular Sciences and Chemical Engineering. Elsevier, Amsterdam, Netherlands. http://dx.doi.org/10.1016/B978-0-12-409547-2.10975-8.
Dutta, K., Kumar, P., Das, S., Kundu, P.P., 2014b. Utilization of conducting polymers in fabricating polymer electrolyte membranes for application in direct methanol fuel cells. Polym. Rev. 54, 1–32.
Dutta, K., Das, S., Kumar, P., Kundu, P.P., 2014c. Polymer electrolyte membrane with high selectivity ratio for direct methanol fuel cells: a preliminary study based on blends of partially sulfonated polymers polyaniline and PVdF-*co*-HFP. Appl. Energy 118, 183–191.
Dutta, K., Das, S., Kundu, P.P., 2014d. Low methanol permeable and highly selective membranes composed of pure and/or partially sulfonated PVdF-*co*-HFP and polyaniline. J. Membr. Sci. 468, 42–51.
Dutta, K., Das, S., Kundu, P.P., 2015a. Partially sulfonated polyaniline induced high ion-exchange capacity and selectivity of Nafion membrane for application in direct methanol fuel cells. J. Membr. Sci. 473, 94–101.
Dutta, K., Das, S., Rana, D., Kundu, P.P., 2015b. Enhancements of catalyst distribution and functioning upon utilization of conducting polymers as supporting matrices in DMFCs: a review. Polym. Rev. 55, 1–56.
Fernández-Moreno, J., Guelbenzu, G., Martín, A.J., Folgado, M.A., Ferreira-Aparicio, P., Chaparro, A.M., 2013. A portable system powered with hydrogen and one single air-breathing PEM fuel cell. Appl. Energy 109, 60–66.
Hahn, R., Wagner, S., Schmitz, A., Reichl, H., 2004. Development of a planar micro fuel cell with thin film and micro patterning technologies. J. Power Sources 131, 73–78.
Hart, D., Lehner, F., Rose, R., Lewis, J., 2014. The fuel cell industry review. E4tech, 1–45.
Holladay, J.D., Wainright, J.S., Jones, E.O., Gano, S.R., 2004. Power generation using a mesoscale fuel cell integrated with a microscale fuel processor. J. Power Sources 130, 111–118.
Hsieh, S.-S., Kuo, J.-K., Hwang, C.-F., Tsai, H.-H., 2004. A novel design and microfabrication for a micro PEMFC. Microsyst. Technol. 10, 121–126.

Hsieh, S.-S., Huang, C.-F., Kuo, J.-K., Tsai, H.-H., Yang, S.-H., 2005. SU-8 flow field plates for a micro PEMFC. J. Solid State Electrochem. 9, 121–131.

Inman, K., Ahmad, Z., Shi, Z., Wang, X., 2011. Design of a proton exchange membrane portable fuel cell system for the 1st international association for hydrogen energy design competition. Int. J. Hydrogen Energ. 36, 13868–13874.

Iwasaki, F., 2007. Application of chemical hydride as fuel of a small fuel cell. In: Proceedings of the Third International Hydrogen & Fuel Cell Expo 2007 (FC-11) in the session of "The latest trend of Mobile Fuel Cells (2)—Flood of New hydrogen fuel cells" 2007, 41–79.

James, B.D., Kalinoski, J., Baum, K., 2011. Manufacturing cost analysis of fuel cell systems. In: Proc DOE Hydrogen and Fuel Cells Annual Merit Review, Washington, DC.

Kamarudin, S.K., Achmad, F., Daud, W.R.W., 2009. Overview on the application of direct methanol fuel cell (DMFC) for portable electronic devices. Int. J. Hydrogen Energ. 34, 6902–6916.

Kumar, P., Dutta, K., Das, S., Kundu, P.P., 2014. Membrane prepared by incorporation of crosslinked sulfonated polystyrene in the blend of PVdF-*co*-HFP/Nafion: a preliminary evaluation for application in DMFC. Appl. Energy 123, 66–74.

Kundu, A., Jang, J.H., 2009. Applications—portable | portable devices: fuel cells. In: Garche, J. (Ed.), Encyclopedia of Electrochemical Power Sources. Elsevier, Amsterdam, Netherlands. pp. 39–45.

Kundu, A., Jang, J.H., Gil, J.H., Jung, C.R., Lee, H.R., Kim, S.-H., Ku, B., Oh, Y.S., 2007. Micro-fuel cells—current development and applications. J. Power Sources 170, 67–78.

Lee, S.J., Chang-Chien, A., Cha, S.W., O'Hayre, R., Park, Y.I., Saito, Y., Prinz, F.B., 2002. Design and fabrication of a micro fuel cell array with "flip-flop" interconnection. J. Power Sources 112, 410–418.

Manahan, M.P., Hatzell, M.C., Srouji, A., Chidiac, N.K., Goldberger, B.L., Peck, N.R., Doutt, S.A., Atmur, S.J., Clark, J.T., Salehi, K., Browe, M.A., Mench, M.M., 2011. International hydrogen association for hydrogen energy design competition applied topic A: portable fuel cell. Int. J. Hydrogen Energ. 36, 13875–13879.

Maynard, H.L., Meyers, J.P., 2002. Miniature fuel cells for portable power: design considerations and challenges. J. Vac. Sci. Technol. B 20, 1287–1297.

Narayan, S.R., Valdez, T.I., 2008. High-energy portable fuel cell power sources. Electrochem. Soc. Interface 40–45.

http://neasia.nikkeibp.com/neasia/004518.

O'Hayre, R., Braithwaite, D., Hermann, W., Lee, S.-J., Fabian, T., Cha, S.-W., Saito, Y., Prinz, F.B., 2003. Development of portable fuel cell arrays with printed-circuit technology. J. Power Sources 124, 459–472.

Peighambardoust, S.J., Rowshanzamir, S., Amjadi, M., 2010. Review on the proton exchange membranes for fuel cell applications. Int. J. Hydrogen Energ. 35, 9349–9384.

Pollet, B.G., Staffell, I., Shang, J.L., 2012. Current status of hybrid, battery and fuel cell electric vehicles: from electrochemistry to market prospects. Electrochim. Acta 84, 235–249.

Ramírez-Salgado, J., Domínguez-Aguilar, M.A., 2009. Market survey of fuel cells in Mexico: niche for low power portable systems. J. Power Sources 186, 455–463.

Rashidi, R., Dincer, I., Naterer, G.F., Berg, P., 2009. Performance evaluation of direct methanol fuel cells for portable applications. J. Power Sources 187, 509–516.

Scott, D.S., 2004. Until something better comes along! Int. J. Hydrogen Energ. 29, 1439–1442.

Shah, K., Shin, W.C., Besser, R.S., 2003. Novel microfabrication approaches for directly patterning PEM fuel cell membranes. J. Power Sources 123, 172–181.

Shah, K., Shin, W.C., Besser, R.S., 2004. A PDMS micro proton exchange membrane fuel cell by conventional and non-conventional microfabrication techniques. Sens. Actuators B 97, 157–167.

Shaw, L., Pratt, J., Klebanoff, L., Johnson, T., Arienti, M., Moreno, M., 2013. Analysis of H_2 storage needs for early market "man-portable" fuel cell applications. Int. J. Hydrogen Res. 38, 2810–2823.

Shibata, M., 2007. Proceedings of the Third International Hydrogen & Fuel Cell Expo 2007 (FC-11) in the session of "The latest trend of Mobile Fuel Cells (2)—Flood of New hydrogen fuel cells" 2007, 1–39.

Shimizu, T., Momma, T., Mohamedi, M., Osaka, T., Sarangapani, S., 2004. Design and fabrication of pumpless small direct methanol fuel cells for portable applications. J. Power Sources 137, 277–283.

Stone, C., 2007. Fuel cell technologies powering portable electronic devices. Fuel Cells Bull. 12–15.

Suominen, A., Tuominen, A., Kantola, J., 2011. Analyzing prospects of portable fuel cells with an expert opinion study. Futures 43, 513–524.

http://www.ultracellpower.com.

Urbani, F., Squadrito, G., Barbera, O., Giacoppo, G., Passalacqua, E., Zerbinati, O., 2007. Polymer electrolyte fuel cell mini power unit for portable application. J. Power Sources 169, 334–337.

Viswanathan, B., Scibioh, M.A., 2006. Handbook of Fuel cells: Principle and Applications. University Press, Hyderabad, India, ISBN 81173715572.

Yahata, T., Takeyama, K., Nakayama, S., Kawamura, Y., Igarashi, A., 2006. Air-breathing single cell small proton exchange membrane fuel cell system. In: Fuel Cell Seminar Abstracts, Hawaii, 13 – 17 November, 2006.

Yalcinoz, T., Alam, M.S., 2008. Improved dynamic performance of hybrid PEM fuel cells and ultracapacitors for portable applications. Int. J. Hydrogen Energ. 33, 1932–1940.

Yamazaki, Y., 2004. Application of MEMS technology to micro fuel cells. Electrochim. Acta 50, 663–666.

Yu, J., Cheng, P., Ma, Z., Yi, B., 2003a. Fabrication of miniature silicon wafer fuel cells with improved performance. J. Power Sources 124, 40–46.

Yu, J., Cheng, P., Ma, Z., Yi, B., 2003b. Fabrication of a miniature twin-fuel-cell on silicon wafer. Electrochim. Acta 48, 1537–1541.

Zhang, J., 2008. PEM fuel cell electrocatalysis. http://www.scitopics.com/PEM_fuel_cell_electrocatalysis.html (cited November 2011).

Zhang, H., Shen, P.K., 2012. Recent development of polymer electrolyte membranes for fuel cells. Chem. Rev. 112, 2780–2832.

Large-scale underground storage of hydrogen for the grid integration of renewable energy and other applications

7

U. Bünger, J. Michalski*, F. Crotogino†, O. Kruck†*
*Ludwig-Bölkow-Systemtechnik GmbH (LBST), Ottobrunn, Germany, †KBB Underground Technologies GmbH (KBB), Hannover, Germany

7.1 Hydrogen and the need for energy storage in Europe

The European energy markets are currently undergoing rapid changes to fulfill the energy policy targets as defined by the European Commission, better known as the *Energy 2020* strategy (ES2020, 2010) and the *Energy Roadmap 2050* (ER2050, 2011). The Energy 2020 strategy suggests aspiring to a 20-20-20 target: 20–30% greenhouse gas (GHG) reduction below 1990 levels, share of renewable energies increased to 20% and implement 20% energy savings, all by 2020. Looking farther forward, the Energy Roadmap 2050 sets more ambitious goals to reduce GHG emissions by 80–95% by 2050 (with 1990 as the base). As one consequence, the share of renewable energy should rise substantially according to the energy market scenarios for Europe, achieving at least a 55% share of gross final energy consumption or 64% of electricity provision in the "high energy efficiency scenario" by 2050. The latter could even rise to 97% according to the "high renewable energy scenario."

As the European energy system targets are put into practice, the greater reliance on intermittent renewable energy poses challenges, specifically from temporary and regional mismatches between electricity supply and demand in concert with growing strains on the electricity grid. Depending on regional and policy drivers, various flexibility options can be applied to cope with this challenge, namely grid management, supply and demand side management as well as utilization of large-scale electricity storage such as pumped hydro, compressed air energy storage and hydrogen power-to-gas (PtG) technology.

Several studies claim that grid management by electricity grid extension will render large-scale electricity storage unnecessary in the short to medium term. However, a number of studies have identified relevant nontransmittable electricity between neighboring countries in Europe (SUSPLAN, 2011; Pape et al., 2014; Dena, 2010; or TYNDP, 2012). They estimate the need for grid extension in the order of magnitude of tens of thousands of kilometers, indicating that grid management also poses challenges due to high investment needs and/or missing public acceptance.

Compendium of Hydrogen Energy. http://dx.doi.org/10.1016/B978-1-78242-364-5.00007-5

Supply-side management is considered to be an improvement in flexibility over the conventional generation mix or dispatchable power plants (e.g., gas turbines, biofuel, biomass). These technologies are required to provide a number of different system services such as balancing power, spinning and non-spinning reserves, black start capability, etc. All have in common the need to fulfill several prerequisites, such as the ability for frequent start-up and shutdown, quick response capability, higher ramp rates, and efficient operation during part load.

Demand-side management is understood as an active response of the end-user consumption to the actual grid and intermittent supply situation. This can be achieved mainly by shifting electricity demand from peak load hours to base load hours. Currently, the potential of demand-side management is rather limited to those industrial processes with large cost-reduction potentials, but in the future these mechanisms are also expected to occur for small commerce, services and in the residential sector, including electromobility. However, the overall potential of demand-side management is expected to remain rather limited in comparison to other flexibility options.

For the option of large-scale electricity storage, "cheap" electricity surplus will be the driver in the short to medium term and the need to provide electricity in periods of renewable energy scarcity in the medium to long term. As the possible evolution of future energy, specifically electricity markets, cannot reasonably be overlooked at this time, some European countries have determined their future large-scale storage needs and have arrived at lists of possible scenarios, ranging from only a few to three-digit terawatt-hours, whereas in other countries the topic has not even surfaced as a major area requiring action.

As elaborated by the German Association for Electrical, Electronic and Information Technologies (VDE) study "Energy storage in supply systems with large shares of REN electricity" (VDE, 2008), the only option for large-scale electricity storage at sufficient potentials in Europe is the electrolysis of water producing hydrogen to store underground, where geologically feasible and acceptable to the public. This method can store energy and electricity at TWh-scale for extended periods of time, such as for weeks or months. This energy conversion and storage concept is best known by the term PtG. PtG can be in the form of

- power-to-hydrogen (PtH_2) with hydrogen as energy carrier (Bünger et al., 2014) or
- power-to-methane ($PtCH_4$), generated by a consecutive methanation step from hydrogen with CO_2 (Sterner and Schmidt, 2009).

For the latter option, existing natural gas infrastructures can be used to distribute synthetic methane as a secondary energy carrier.

It is specifically the universality of the PtG options that has moved them into the center of recent discussions on how best to integrate ever-increasing quantities of renewable energy into the energy markets. $PtCH_4$ has a specific advantage in that the transport and distribution infrastructure and relevant end-use technologies such as natural gas turbines or internal combustion engines are widely available worldwide. On the other hand, it is to the favor of the PtH_2 concept that it is characterized by higher process efficiencies in the conversion step (missing the methanation step) and significantly higher efficiencies during end-use (use of fuel cells). Also, the sustainable CO_2 sources needed for methanation in $PtCH_4$ must be taken into account. For the PtH_2 option only, large-scale hydrogen underground storage will be required, since with $PtCH_4$ the existing natural gas infrastructure can continue to be used.

In this chapter, based on an explanation of the need and required quantities of energy storage at large scale, the technical background for hydrogen underground storage and a thorough economic analysis are provided.[1]

7.2 Markets for hydrogen

Although newly invented, the concept of PtG dates back to early ideas of using hydrogen as a "universal energy vector" next to electricity, which concept has been prominently assessed by several international industry-backed studies such as (Bahbout et al., 2012) for Europe/Canada and (WE-NET, 2001) for Japan. Many other international distribution pathways for hydrogen as an energy carrier are listed in the EU-funded ENCOURAGED project (Landinger, 2005). It is the versatility of hydrogen that makes it unique with regard to its use for a range of end-use applications. It can be applied in re-electrification as a "classical" electricity storage option, but it can also be used in other sectors, such as to power fuel cell electric vehicles (FCEVs), as a feedstock and heating fuel in industry (chemical, petrochemical, and steel) and for the "greening" of natural gas by admixing hydrogen into the natural gas grid. A number of industries are involved:

- Electricity industry: Operates a large electricity transport and distribution grid, which when accepted by the public is poised to grow in regions with large renewable electricity development potentials and with nodes to connect large energy imports to the grid (e.g., offshore wind-energy) (ENTSO-E, 2011). Hydrogen large-scale storage will be required in the long term, but could also reduce the strain on grid extension in the short term by ancillary services.
- Mobility: A hydrogen refueling infrastructure for refueling hydrogen-powered fuel cell cars and buses may evolve across Europe starting in 2015, with initial user concentration in populated areas such as megacities or city clusters and slower roll-out into the countryside. In Germany alone 400 hydrogen refueling stations have been earmarked for 2023 and about 1000 for 2030 (H_2-Mobility, 2013).
- Chemical industry: In some parts of Europe, the industry has developed hydrogen pipelines or pipeline grids (Perrin et al., 2008), which could be directly linked to a large-scale storage facility. The pipelines connect single production or end-use sites by point-to-point connections (e.g., northern Germany: Brunsbüttel chemical cluster with refinery in Heide) or multiple production and user locations by complex multi-kilometer grids (e.g., Northrhine-Westfalia and Leuna-Bitterfeld). Eventually the hydrogen grid could be merged with the natural gas grid.
- Natural gas industry: This industry is operating a dense pipeline grid covering large areas of Europe and providing options to add synthetic methane gas with no limitations and admix hydrogen in limited quantities depending on natural gas flow variations and allowed maximum admixture rates (Müller-Syring, 2014). This service could support the electricity industry in meeting the flexibility needs.

[1] In all analysis used in this chapter the time horizons "short to medium term" (2025+) and "long term" (2050+) have been applied.

To understand the order of magnitude of potentially emerging markets for green hydrogen in Europe, a ballpark calculation reveals the following market sizes:

- Fuel for transport: 17,925 kt/year (239 Mill. light duty vehicles (ICCT, 2013), out of which 50% FCEVs, 15,000 km/car/year, 1 kg_{H2}/100 km).
- Chemical raw material: 86,700 kt/year (total hydrogen demand by EU industry in 2009 (Suresh et al., 2010), interpreted as potential to be supplied by renewable hydrogen, if consumption does not change until 2050).
- Natural gas industry: 831–4153 kt/year (4620 Bm^3/year estimated NG consumption 2013 in Europe (eurogas, 2014), allowable admixture rate 2–10 vol%), and 41,534 kt/year if all natural gas were substituted by hydrogen.

All the hydrogen demand from these calculations, if supplied from renewable electricity through electrolysis, is additional hydrogen demand, replacing some form of fossil-based primary energy, with the exception of the hydrogen demand by industry, which substitutes for today's hydrogen production from fossil primary energy. To gain more understanding of the order of magnitude of surplus electricity stored as hydrogen, and future hydrogen demand, the HyUnder project has undertaken benchmarking exercises. For example, in the case of Germany it was found that, given the storage requirement for surplus electricity of up to 300 kt/year by 2025 and of 1600 kt/year by 2050 were fully converted to hydrogen from electrolysis (HyUnder, 2014), then just about all of the transport fuel demand for light duty FCEVs could be supplied by 2050.[2]

As a consequence, and seen from a quantitative perspective, hydrogen produced from residual renewable electricity and hydrogen used as a fuel for transport could become an ideal synergy, starting development in the 2020s.

7.3 Technology for large-scale hydrogen storage

7.3.1 Overview

Hydrogen storage at a large scale is an intrinsic part of complete energy chains, from energy provision, that is electricity generation from wind energy, to end use. Due to the relevance of recent developments in the energy markets, this chapter focuses on the use of large-scale hydrogen storage for PtG schemes being used to store residual renewable electricity.

At a large scale, hydrogen can either be stored underground in deep geological formations, preferably in rock salt formations, or alternatively, where geology does not offer this option, in large-scale aboveground containments. Comparing these options, regional differentiation will be an issue in renewable electricity-based energy systems.

Interest in storing hydrogen in underground gas storages has been considerably boosted in recent years by two principally different developments. In the USA in particular the demand for hydrogen for the production of high-quality fuels has been growing continuously: hydrogen caverns have been successfully used by the petrochemical

[2] The figures are based on the assumptions of the German "Leitstudie," anticipating an electricity residual of 15 TWh_{el} by 2025 and 75 TWh_{el} by 2050 (BMU, 2012) and 50% of all German light duty vehicles converted to FCEVs.

industry in Texas for many years to ensure a continuous supply of hydrogen to the refineries; a third hydrogen cavern is scheduled to be completed shortly. In Europe, and in Germany in particular, there is considerable interest in the underground storage of hydrogen to satisfy the demand for electrical energy storage expected in the medium to long term.

For large-scale hydrogen storage the following technology options, all for compressed hydrogen, may play a role in the future, all displayed in Figure 7.1:

- hydrogen underground storage (favorably in salt formations),
- hydrogen storage in buried pipes (made of steel), and
- hydrogen stored in aboveground spherical or cylindrical storage tanks (made of steel).

All three concepts can be characterized by technical, economic, operational, and siting parameters, which make them applicable for different settings. Their key technology parameters are displayed in Table 7.1. The concepts are principally different and have either been applied before (large cylinders or spheres) or have been proposed for realization based on experience gathered from operation with other gases, mainly natural gas (caverns, buried pipes).

Underground storage requires much lower land use than aboveground tanks. The aboveground installations of underground storage need only a very small surface area. However, the use of the area above the caverns and, even more, between caverns or the land on top of a buried tank is limited to agricultural use for safety reasons. The key advantage compared to surface tanks is the much lower impairment of the overall appearance of the landscape.

Using the annual electricity storage demand figures for Germany from the last chapter and translating them to hydrogen, the required number of large underground salt storage caverns would be 15 typically sized caverns by 2030 and about 60 caverns by 2050.[3] These figures are based on a theoretic approach, but the real number will probably be lower, as other flexibility measures will reduce the large-scale storage needs. Using aboveground storage in cylindrical tanks of, for example, 110 m^3 with a hydrogen content of about 270 kg each, then the volume of about 545 of these tanks is equivalent to one salt cavern, if typical equivalent annual full load cycles (EAC) are taken into consideration for cavern (6) and tanks (150).

Figure 7.1 Storage concepts for large-scale compressed hydrogen.
Source: KBB UT, Erdgas Zürich AG, HyOP.

[3] A geometric cavern volume of 500,000 m^3, which is equivalent to a useable hydrogen content of 3733 tons, is assumed in this study (also see next chapter).

Table 7.1 Technical, economic, operational, and siting specific parameters for large-scale hydrogen storage options (all in approximate numbers)

Parameter	Dimension	Underground storage (salt cavern)	Buried pipes	Cylindrical tank aboveground
Siting	–	Central location	Regional location	Onsite (e.g., a refueling station)
Assumed geometric volume	m^3	500,000	6800	110
Length/width	m/m			19/3
Pressure range	MPa	6–18	2–7	2–5
Min. H_2 energy contents	MWh	80,000	370	6.5
Max. H_2 energy contents	MWh	200,000	1200	16
Net usable H_2 energy contents	MWh	125,000	850	9.0
Energy density	Wh/l	250	125	85
Net capacity	t	3700	25	0.3
Investment costs	M€	107 (incl. aboveground plant and cushion gas)	12	0.08–0.12
Life expectancy	a	30	50	20
Interest rate	%/year	5.5	5.5	5.5
Capital costs	€/year	7,500,000	680,000	10,000
Spec. static storage costs[a]	€/kg	2	27	25–37
Equivalent annual full load cycles[b]	Cycles/year	6	100	150
Spec. dynamic storage costs[c]	€/t	330	270	165–250
Source		HyUnder (2014)	Bünger et al. (2014)	Bünger et al. (2014)

[a] Specific in relation to the maximum usable hydrogen content in the cavern/pipe/tank.
[b] Number of typical equivalent annual full-to-empty-to-full cycles (EAC), strongly depending on hydrogen end-use sector. A fueling station storage tank for low-pressure hydrogen with an average capacity of 2 days has a theoretical EAC of 365/2 = 182, being reduced if the tank were also used for ancillary services in the electricity grid.
[c] Specific in relation to the annual hydrogen throughput through the cavern/pipe/tank.

This number compares to a maximum number of hydrogen refueling stations in Germany of about 1000 by 2030 and, for example, 5000 in 2050. Hence, if all these refueling stations were equipped with three of the aboveground tanks each, then about 37–46% of all surplus hydrogen energy from fluctuating renewable energy could be stored at hydrogen fueling stations.

Hydrogen storage at a large scale and for PtG is intrinsically connected with several other processes along complete hydrogen to end use process chains. They are:

- Electrolysis
- Compression
- Drying and purification (depending on the hydrogen grade required by the specific end use)
- Balance of plant

Electrolysis of water is the key hydrogen production technology, also and optionally for the $PtCH_4$ pathway comprising a methanation reactor to produce synthetic methane gas by adding CO_2. Both technologies are presented in another chapter in this book. As the task of large-scale storage in renewable electricity systems is typically to level out production and demand gaps, the dynamics of all components, hydrogen storage and release, electrolysis and methanation process dynamics are often of high relevance. Therefore, flow rates for a large hydrogen salt cavern need to be tailored to the specific application. Table 7.2 contains further technical characterisation of the processes along the hydrogen to end-use process chains.

The process flow for a large hydrogen storage facility for the purpose of PtG is depicted in Figure 7.2. Even though the flow sheet shows a salt cavern storage facility, the elements for hydrogen surface storage systems are similar. Differences are that surface storage will typically not require gas drying and purification, unless a local refueling station for fuel cell vehicles needs to be supplied with 5.0 grade hydrogen. For hydrogen storage in caverns, the effort required for intercooling of hydrogen during compression can be reduced by cooling against air, as storage caverns can accept large quantities of heat of compression dissipating into the cavern walls. On the other hand, intercoolers for the compression step are needed in addition.

In addition, end-use specific infrastructure technology may have to be placed at the storage site, such as high-pressure compression equipment for trailer filling, fuel cells, or combined cycle gas turbines for re-electrification or admixture equipment to mix hydrogen into an adjacent natural gas pipeline.

For underground storage, one must distinguish between below-ground and aboveground components. Whereas the aboveground components are shown in Figure 7.2 below-ground installations comprise the cavern well head, the cemented casings (pipes), and the gas cavern completion.

7.3.2 Geological hydrogen storage concepts

Suitable deep-lying geological structures allow the storage of large quantities of gas. The main reasons are: the availability of natural reservoirs or the ability to artificially construct large underground storage cavities; the ability to maintain very high operating pressures because of the thick layers of up to several hundred meters of overlying rock; the very small footprint for the surface facilities; low specific costs; and, especially, superior operating safety.

Underground gas storage is usually filled by compressing and injecting the gas, and emptied by releasing the gas pressure. The usable mass of gas between the allowable minimum and maximum operating pressures is termed the *working gas*, while the mass

Table 7.2 Dimensions and performance data for aboveground storage process equipment

Component	Typical power rating	Efficiency	Description
Dimension	MW_{el}	%	
Electrolyzer	1–1000 (from onsite to regional to central)	60–70	**Criteria:** footprint, output pressure, dynamics, minimum partload, cold start capability, efficiency **Type:** AEL, PEMEL, HT-SOEL Water consumption: 0.85 l/Nm³ H_2 (9.4 l/kg) **Cooler:** 890 l/kg cooling water **Impurities:** O_2, H_2O, KOH (alkaline electrolysis) Tanks for chemical storage (for de-ionification, KOH for alkaline electrolysis) **Offgas:** O_2 as byproduct or waste
Dryer/purifyer[a]		~95	Adsorption dryer with cyclic adsorption also used for H_2 purification, typically by pressure swing adsorption (PSA) Photo: StyLAir
Compressor[b]	0.1–10	~80	Two-stage piston compressor with intercooler, oil, and water separators for cavern feed P_{in}: 2.5–5.0 MPa, P_{out}: 18–20 MPa (cavern) Multi-stage compressors for providing H_2 to end users such as trailer loading (e.g., 55 MPa) Electrical drive

Balance of plant			H_2-resistant materials such as austenitic steels or Teflon sealings, standard control systems, measurement by ultrasonic sensing, safety equipment for flame detection and fire extinction avoiding explosions, electrical discharges, damage by striking lightning, individual equipment shut-off valves, safety vents, and vent stacks
Floor space			**Depending on plant size:** Cavern and process equipment sites, separate buildings for electrolyzers and compressors, water/air and brine/water coolers, space for end-use specific H_2-handling infrastructure, parking lots, workshop, maintenance, and spare part building

AEL—alkaline electrolysis, PEMEL—PEM electrolysis, HT-SOEL—high temperature solid oxide electrolysis, KOH.

[a] Purification and drying for cavern feed and optionally for further H_2 use, such as trailer filling.

[b] Compressors can be avoided for buried pipe and aboveground tank storage, if the electrolyzers provide enough feed pressure, i.e., up and around 7 MPa.

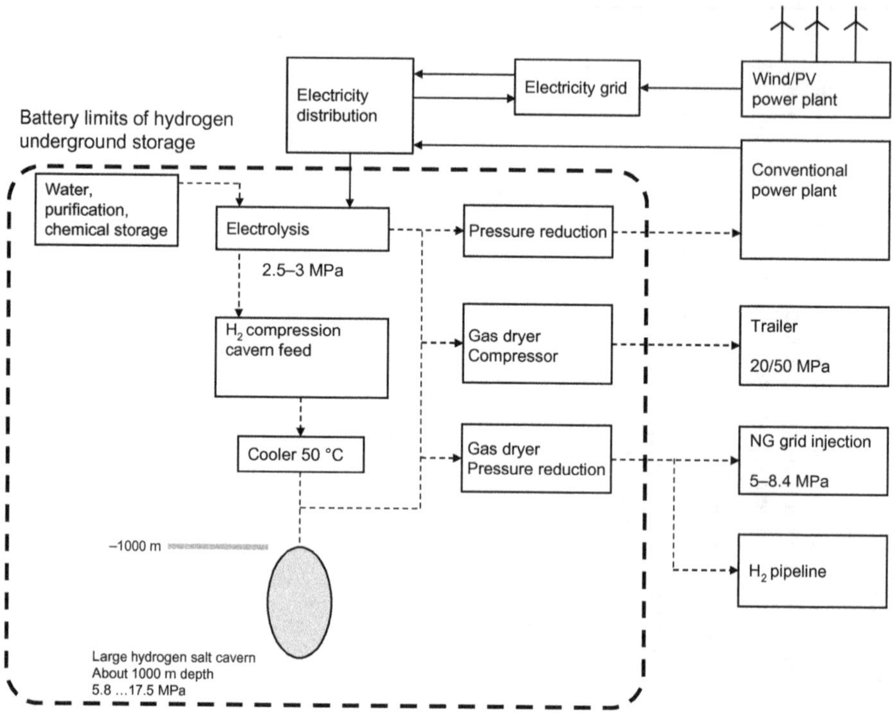

Figure 7.2 Process flow sheet for the aboveground equipment for a large hydrogen underground storage facility (within dotted frame).

of gas remaining in the cavern below the minimum pressure, which is not available for operations, is called the *cushion gas*. The ratio between the cushion gas and the working gas therefore is an important economic storage metric because the cushion gas represents dead capital.

7.3.2.1 Storage in depleted hydrocarbon reservoirs or in natural aquifer formations

Natural gas in particular has been successfully stored underground around the world for many decades. In terms of volume, the most significant storage areas are those in natural porous reservoirs, mostly in depleted natural gas (preferred) or oil reservoirs; see Figure 7.3. The tightness of these reservoirs has already been verified by the existence of the reservoirs over long geological time periods before production of the gas or oil has begun. The reservoir parameters are also already known in detail from the exploration and production phases.

If no depleted reservoirs are available for storage, gas can also be stored in natural porous water-bearing reservoirs known as aquifer formations. The main prerequisites here are the presence of a reservoir with a dome shape or structural fault to enable the gas to be trapped at the top of the structure, and the presence of a seal overlying the reservoir consisting of an impermeable formation; see Figure 7.4.

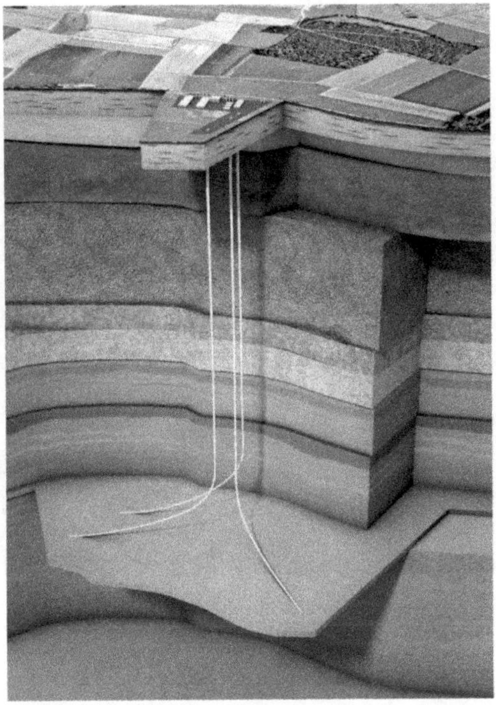

Figure 7.3 Underground gas storage in a porous reservoir.
Source: KBB UT.

Unlike depleted oil and gas fields, aquifer storage usually requires a comprehensive geophysical exploration program before its suitability for gas storage can be verified, and before the capacity and flow properties can be defined with the necessary level of detail.

In both depleted hydrocarbon reservoirs as well as aquifer formations, the gas flows through the small-scale matrix of pores between the boreholes with corresponding flow resistance. Porous storages are therefore more suitable for continuous flow injection and withdrawal rates and for operations involving less frequent cycles. In the past, porous storages have therefore mainly been used for balancing out seasonal fluctuations in demand and less for short-term balancing of gas production and need. Also, the proportion of cushion gas in pore storages is typically 40–50%, much higher as compared to salt caverns.

Assessing the general suitability of porous reservoirs for the storage of hydrogen can draw from experience already gained from the storage of town gas in the past, consisting of up to or above 50 vol% hydrogen. However, this storage option is technically more challenging than salt caverns (see the following section), because possible reactions of hydrogen with minerals within the reservoir rock may occur, e.g., with sulfur as well as with in situ fluids, and any microorganisms if present. These reactions could lead to hydrogen depletion or even the blockage of the very fine pore spaces by reaction products. Several large projects are currently (2014) underway to gain a better understanding

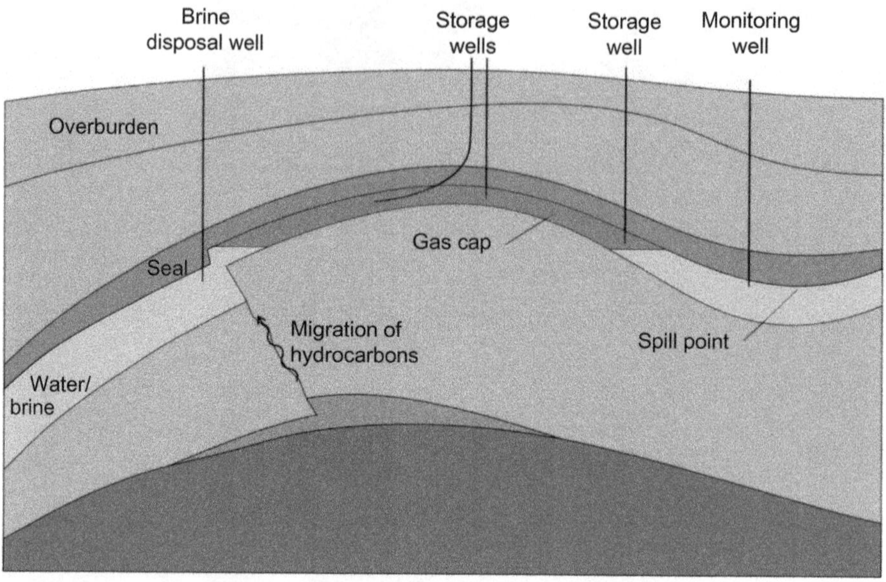

Figure 7.4 Schematic representation of an aquifer gas storage.
Source: KBB UT.

of these potential reactions and their consequences (Perez, 2014; Underground.Sun. Storage, 2014; H2Store, 2012).

7.3.2.2 Storage in man-made salt caverns

Salt caverns are alternatives to porous storages; see Figure 7.5. The caverns first have to be constructed in the salt formation by injecting water through an access well and dissolving the salt. This so-called solution mining process generates large volumes of brine that must be disposed of in an environmentally compatible way. Because of its visco-plastic properties, rock salt is extremely tight to gases like natural gas or hydrogen—even under high pressure. The enormous open cavities which these caverns represent, with volumes from a few 10,000s to more than 1,000,000 m^3 at operating pressures of up to 20 MPa and more, are particularly suitable for flexible gas operations with high production and injection rates and frequent gas cycles. The proportion of cushion gas is typically 30%. Because of these properties salt caverns are most suitable for the future storage of hydrogen from renewable energy. Rock salt does not react with hydrogen, one key advantage compared to porous reservoirs. However, water from the cavern sump will increase the water vapor content of the stored gas.

Salt caverns allow high injection and withdrawal rates; they are, however, limited by the allowable pressure–time gradient $\Delta p/\Delta t$ (common maximum value: 1 MPa/day). Overstepping this value may damage the integrity of the surrounding cavern walls due to thermo-mechanical stress. Numerical simulations for a 500,000 m^3 cavern show mass flow rates of up to 11,000 kg/h.

Figure 7.5 Underground gas storage in a manmade salt cavern.
Source: KBB UT.

Even though hydrogen has been stored in salt caverns successfully in the UK and the USA for many years, to enable this technology to be used in Europe in the future, the technical components of the access wells below and above surface still need to be adapted to present national safety standards.

The completion consists mainly of a retrievable gas production string, subsurface safety valve, and permanent packer (see Figure 7.6).

When comparing the key options for underground gas storage, it becomes obvious for many reasons that salt caverns are the first choice for storing strongly fluctuating wind and solar power. Salt caverns are best suited for flexible operations with high gas injection/withdrawal gradients and frequent turnovers. Furthermore, the share for cushion gas is moderate compared to reservoir storage. Most important, there are no mineralogical or microbiological reactions to be expected. A typical 500,000 m^3 hydrogen cavern can store approx. 3733 tons (working gas). This corresponds to an energy content of around 124 GWh with a maximum power input or output of approx. 0.4 GW.

The choice between the common underground storage options of natural reservoirs or man-made salt caverns ultimately depends not only on technical issues but also on the natural geological conditions, which typically vary from region to region.

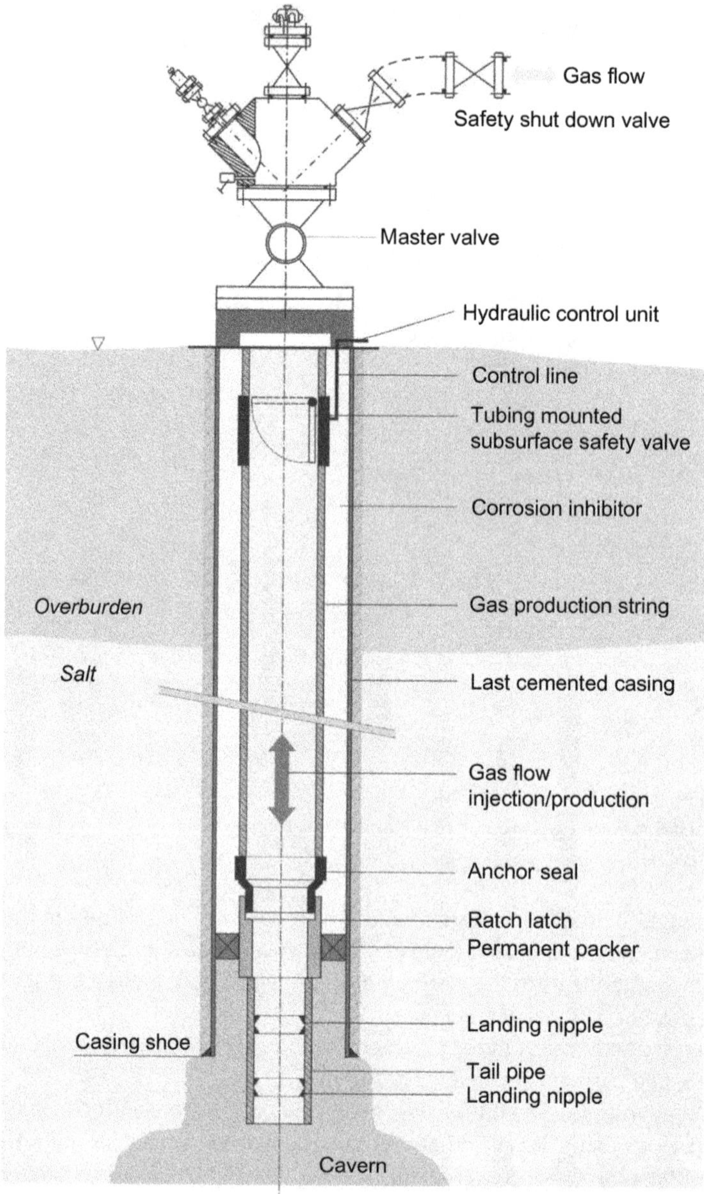

Figure 7.6 Below-ground installation of a large hydrogen salt cavern.
Source: KBB UT.

7.3.2.3 Other storage options

Another option which could be used under certain circumstances is the conversion of abandoned mines into high-pressure gas storage facilities. However, in practice this option plays virtually no role for the storage of hydrogen, the main reason being the difficult proof of tightness.

Another option is lined hard rock caverns constructed by conventional mining technologies. These caverns are lined with stainless steel and sealed off from the surrounding rock walls. Only one project of this kind—the Swedish LRC demonstration project (Tengborg et al., 2014) for natural gas storage—currently exists. The specific costs of rock caverns will remain to be much higher than for solution mined salt caverns. The main challenge is securing the long-term integrity, which still requires overcoming some fundamental challenges.

7.4 Potential for hydrogen underground storage

7.4.1 Salt caverns

Hydrogen could be stored underground primarily in artificially constructed salt caverns, and, subject to certain provisions, also in depleted gas fields or in aquifer formations. The choice of storage option does not only depend on a technical assessment but also on the availability of suitable geological formations.

The map in Figure 7.7 shows the geographical distribution of salt formations across Europe, including realized cavern projects. Although it is a gross simplification, most salt formations suitable for the construction of storage are already being utilized. Another gross simplification is that this means most of the salt formations which have not been used so far are either generally unsuitable or only suitable subject to major reservations—e.g., require previously unimplemented technologies, or are probably associated with very high costs for the disposal of the brine generated during solution mining.

The map clearly depicts the rather regionally limited distribution of salt formations across Europe. The Netherlands and Germany are well supplied, at least in the northern parts of these countries. France on the other hand almost exclusively has suitable salt formations in the extreme southeast of the country; Spain in the north, and further to the south along the Mediterranean coast. The first scientifically based investigation to estimate the potential for future hydrogen storage caverns in relevant countries within the European Union has been finished in 2014 as part of the HyUnder research project funded by the EU (HyUnder, 2014).

Table 7.3 qualitatively describes the geotechnical potential of future hydrogen caverns in selected European industrial countries. This assessment is based on the present technical rules and presently accepted costs for constructing natural gas caverns. However, should the demand for hydrogen storage grow significantly in the future, salt deposits currently considered less suitable may be reconsidered for the construction of further salt storage caverns.

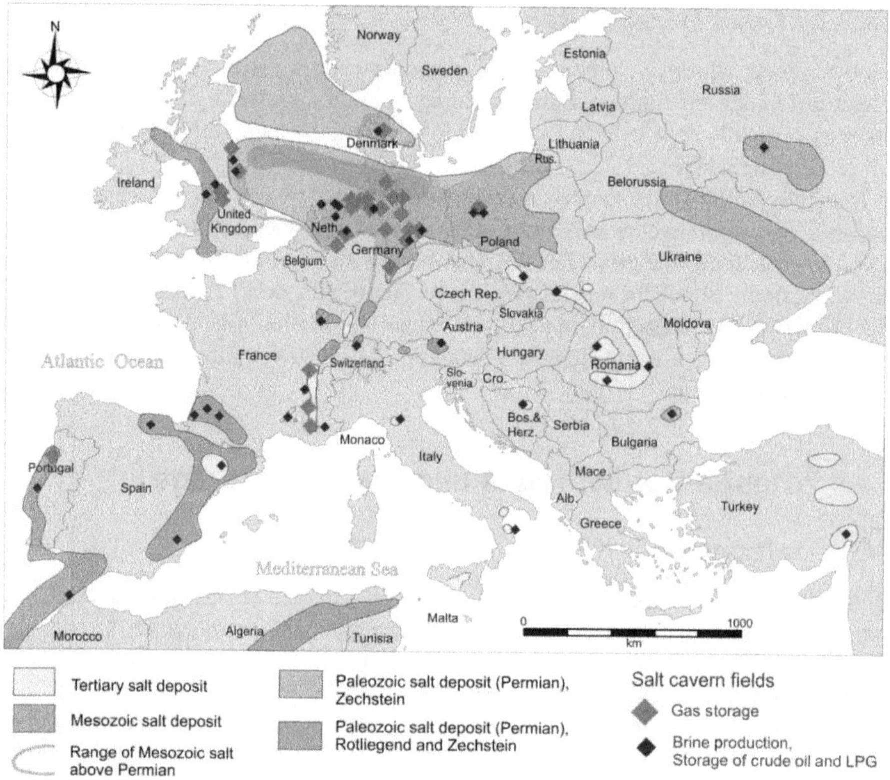

Figure 7.7 Salt formations and salt caverns in Europe.
Source: KBB UT.

In addition to the more detailed European map, the map in Figure 7.8 allows a global overview of the worldwide distribution of salt formations. So far gas storage caverns are operated or under development, e.g., in China and Turkey and particularly in the USA, predominantly along the Gulf Coast of Mexico, as well as in Canada.

7.4.2 Storage in porous reservoirs

As already discussed, depleted natural gas fields, as well as aquifer formations, may be suitable for the storage of gas and possibly also for the storage of hydrogen provided certain geotechnical requirements are fulfilled. Competition for the use of these formations is mainly between natural gas and hydrogen, and to a certain extent also compressed air and carbon dioxide. A comprehensive gas infrastructure has been developed in Europe in recent decades consisting of pipeline grids and underground storage. The strategy to achieve storage capacities of around 20% of annual consumption has so far only been realized in France and Germany. This enormous demand and the profits generated by storage in the past mean that, in general, in most industrial countries most of the geological formations which are basically suitable for this purpose have already been explored.

Table 7.3 Qualitative assessment of potential for future hydrogen caverns of selected European countries

Denmark	Major potential in western Denmark (Jutland)
Germany	Considerable potential for additional caverns in the northwest and the middle of the country
France	Potential in the southeast and possibly also in the extreme southwest
Italy	There are no storage caverns so far in Italy. Any potential for storage caverns, if at all, is likely to be very limited
Spain	Despite a number of salt formations, there are currently no gas caverns in Spain. Problems associated with the salt formations in the center of the country are the provision of water, and even more the disposal of brine. The Torrelavega location on the Atlantic coast is interesting because of salt caverns created during brine production
Portugal	Portugal has salt deposits along its Atlantic coast, and natural gas storage caverns have been constructed at one location for some years now. There should be potential for future hydrogen caverns
The Netherlands	Considerable potential in the northeast of the country and other regions near to the German border
United Kingdom	The salt formations in Cheshire on the west coast and in Yorkshire on the east coast have potential in the UK. The potential for constructing additional gas caverns in Cheshire is limited, but there is still a certain potential available in Yorkshire

Figure 7.8 Occurrence of salt formations worldwide.
Source: KBB UT.

This raises the question of whether at least those gas fields currently being produced could be suitable for the storage of hydrogen in the future. In the early years of oil and gas exploration, the focus tended to be on shallow reservoirs, which also have high permeabilities and porosities. As the potential of these shallow reservoirs is being widely utilized today, the oil and gas industry was forced to develop deeper and less permeable reservoirs. Great depths of several thousand meters and low permeabilities are, however, unfavorable for the requirements of future flexible hydrogen storage. Great depths mean high operating pressures and therefore high investment costs for compressors and the access wells, as well as high operating costs for the compression. It is therefore most likely that neither the presently producing nor future new gas reservoirs will add to the potential for future hydrogen storage.

It can be concluded that two options with smaller reservoir capacities are feasible during the transition phase to satisfy the initial demand for underground hydrogen storage: the use of fields that have not been developed so far for economic reasons, and the conversion of smaller natural gas storage areas which are no longer profitable. At a later stage, as the proportion of renewables increases, also at the expense of the share of natural gas, this could then justify the conversion of larger existing natural gas storages. However, some countries with a well-developed natural gas industry such as Spain and the United Kingdom, own only relatively small storage capacities and therefore also have a corresponding limited potential for future conversion of natural gas storage into hydrogen storage. The unsolved issue of "compatibility of porous storages for the storage of hydrogen" is one of the most urgent issues yet to be tackled.

7.4.3 Other regional siting criteria

Whereas aboveground storage of hydrogen at large scale is only limited by local factors (e.g., safety), below-ground storage is only possible in regions with relevant salt geology (Section 7.4.1). On the other hand, not all storage geologies are suitable for storing hydrogen at large scale for reasons of lacking aboveground infrastructures. In Europe, this is the case for regions in, for example, southwestern Spain, which are poorly linked to existing energy infrastructures.

For this reason, the assessment of energy infrastructures typically accompanies the potentiality assessment, based on an in-depth analysis of existing infrastructures for electricity (transmission grid to connect to renewable electricity supply at shortest distance), for natural gas (pipeline grid for admixing hydrogen or synthetic methane gas), for hydrogen (chemical industry pipeline grid for admixing pure hydrogen) and proximity to large hydrogen end-use (=population) centers with a demand for transport fuel. But also, aboveground large-scale storage sites will require energy infrastructure planning to identify the most suitable locations. A typical region's analysis also considers further decision criteria. A criteria list has been developed and applied to select the most relevant sites or regions, specifically for an early market introduction phase in Germany. The criteria list is presented in Table 7.4 pointing at the relevance of the storage site from a geological, an infrastructural and a political perspective. Depending on the country or the region other criteria may be applied.

Table 7.4 Selection criteria for regions analysis to identify early cavern sites for Germany (Bünger et al., 2011)

Storage cavern geology	
Site existing	To minimize cavern development costs
Geology	For optimum cavern design and operational parameters
Brine disposal	To minimize cavern development costs
Integration in energy environment	
Onshore wind availability	To profit from lowest electricity costs
Offshore wind availability	To profit from lowest electricity costs
El-grid integration	To avoid grid transport (costs)
El-grid bottlenecks	To allow relevant business cases in electricity markets
Fueling station density	To minimize hydrogen transport (costs) to fueling stations
Proximity to industrial users	To profit from existing hydrogen (pipeline) infrastructures
Combined or gas power plants	To allow relevant business cases in electricity markets
Proximity to natural gas pipelines	To use hydrogen in natural gas grid (admixture, methanation)
Politics	

For Europe, the analysis began by preselecting a number of relevant existing salt cavern sites for natural gas. The criteria analysis was then applied to this set of caverns, and a number of regions were identified which are indicated by dotted circles in Figure 7.9.

7.5 Hydrogen storage economics in energy systems with increasing share of intermittent renewable energy

The major objective of this section is to evaluate the potentials of PtG technology for integration of renewable energy supply into the energy system from an economic perspective, here based on detailed analysis for Germany. As described in previous sections, the intermittent electricity might be utilized for electrolytic production of hydrogen, which can be either stored for re-electrification or used for other applications. In this context, this section presents numerical results for a case study carried out in the context of the German energy system with an increasing renewable energy penetration. The analysis provides a comparison of the economic value of different hydrogen applications and detailed insights into the hydrogen plant management. This section is subdivided into four subsections, including a description of the underlying approach; explanation of the major input data and selected scenarios; and presentation of major quantitative economic results as well as outcomes of the sensitivity analysis.

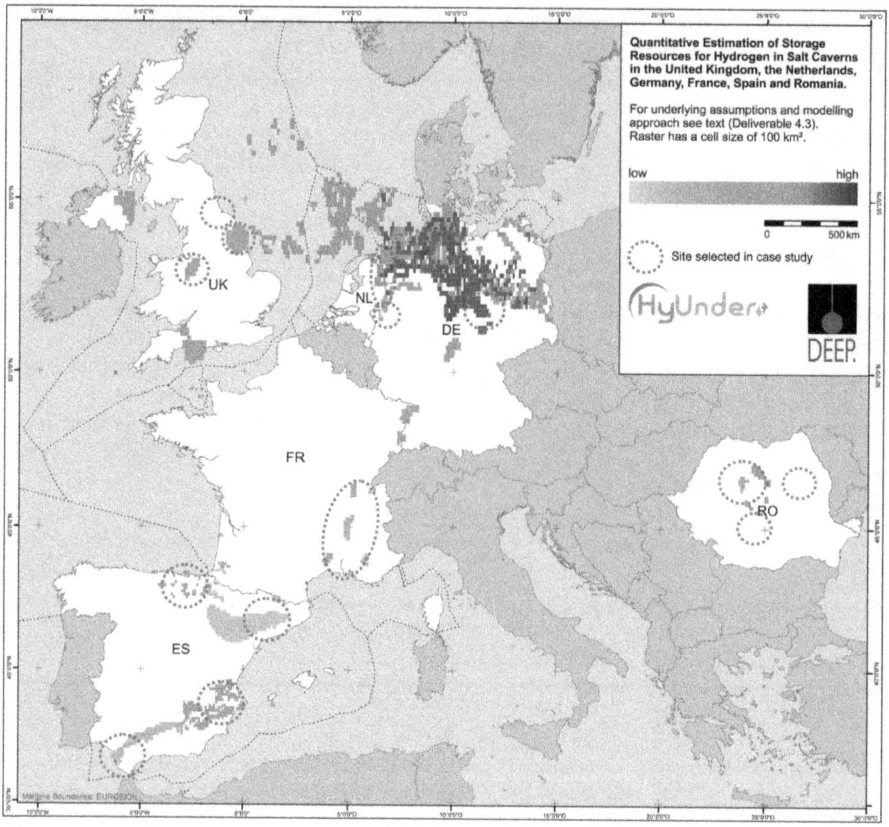

Figure 7.9 Locations resulting from a regions analysis for hydrogen salt caverns in Europe (dotted circles denote regions identified with geological and hydrogen market-specific potential).

7.5.1 Approach, methodology, and major assumptions

In order to provide numerical results for the economic assessment of the PtG technology, a linear programming model is proposed for sizing and operation optimization of a facility consisting of alkaline electrolysis, adequate compression prior to salt cavern storage as underground hydrogen storage, as well as topside equipment after the cavern typical for each application (i.e., hydrogen drying, purification, compression for trailer filling, re-electrification unit, and natural gas grid injection unit). Further elements of the hydrogen infrastructure such as truck delivery or hydrogen refueling station in the mobility case are not included in the analysis. The PtG facility is located in a selected grid node, purchasing electricity from a given intermittent power plant at a forecasted market price.

The underlying optimization problem represents net present value (NPV) maximization of the entire hydrogen PtG facility from a perspective of a single price taking firm without any market power. The investment outlays required to build up the facility are compensated for by a stream of constant annual cash flows from facility operations in a prototypical year over a given number of periods within a detailed planning horizon. At the end of the planning horizon, we assume additional cash inflow from the sales of each facility component at the corresponding residual value under linear depreciation. Equation (7.1) corresponds to the general formulation of the objective function:

$$\max \frac{(1+r)^T - 1}{r(1+r)^T} \times \left[\left(\sum_{h \in H} \mathrm{TR}_h - \mathrm{TEC}_h - \mathrm{SC}_h - \mathrm{TOC}_h \right) - \mathrm{TFC} \right] - \mathrm{TI} + \frac{1}{(1+r)^T} \times \mathrm{TRV} \quad (7.1)$$

The first part of the objective function in Equation (7.1) represents the constant annual cash flow from operations multiplied by an annuity factor for a given number of periods T of the planning horizon and the interest rate r. In this way, we account for the time value of money of prototypical annual cash flows, which are assumed to be identical for the entire planning horizon. The constant annual cash flow consists of total revenues TR_h from sales in all markets in each hour h of the prototypical year and takes into account the corresponding variable costs in each hour h consisting of total electricity costs TEC_h, standby costs of the electrolysis SC_h (i.e., electricity costs in hours when electrolysis is not used) and total other variable costs TOC_h as well as the annual total fixed costs TFC. The second and third parts of the objective function stand for the total investment outlays TI and discounted total residual value TRV of each component of the facility, respectively.

The corresponding decision variables include the quantity of electricity purchased from the intermittent power plant in each hour h, hydrogen provided to every market in each hour h, investment capacity in each component of the facility as well as the initial level of the hydrogen storage. The technical constraint of the maximization problem can be subdivided into market constraints on hydrogen production and sales (Equation 7.2), capacity constraints (Equation 7.3), investment constraints for each component of the facility, storage operation constraints as well as nonnegativity constraints for all decision variables.

The market constraints on hydrogen production and sales are of the type

$$\underline{x}_h \le x_h \le \overline{x}_h, \forall h \quad (7.2)$$

where x_h is the electricity purchase or hydrogen sales decision variable in hour h and $\underline{x}_h$ and $\overline{x}_h$ are the lower and upper bound, respectively, defining the hourly profile either for the intermittent electricity production or hydrogen demand in the selected market.

In general the capacity constraints are represented as

$$x_h \le K + I, \forall h \quad (7.3)$$

expressing that the hourly production decision variable must be lower than the overall capacity of the relevant facility component consisting of the initial capacity K and additional investment I. Further, for the investment decision variable an upper bound is defined in order to assure that the maximization problem has a deterministic solution when investing in new capacities is economically favorable. Otherwise solving the underlying problem might provide under some circumstances an infinite size of the facility.

The storage level in each hour *h* of the prototypical year is an additional auxiliary variable based on the storage level in the previous hour and all storage inflows and outflows in the given hour and the initial storage level at the beginning of the year. The storage inflows come from the hydrogen production via electrolysis multiplied by the efficiencies of the electrolysis and input compressor. Outflows from the storage are caused by the sales to each hydrogen market. The sales decision variables are divided by the efficiencies of corresponding processing steps between storage and delivery to the market to consider the efficiency losses. In this way the storage outflow is higher than the sum of hydrogen sales to all markets.

In this context, the major underlying assumption of the model can be summarized as follows:

- The facility operator corresponds to a sufficiently small firm such that its bidding decisions have no influence on the electricity prices on the market being considered as an exogenous input variable.
- The only source of electricity for hydrogen production via electrolysis in the given grid node is represented by a power plant with an intermittent feed-in.
- Spot-market prices are considered as a fair estimate for the value of renewable electricity in a given point of time and selected grid node.
- Any other ancillary services in the electricity market (e.g., balancing power) are not taken into account for further analysis.

7.5.2 Scenarios and input data

In order to allow for a comprehensive economic assessment of the PtG technology, the case study includes in total eight different scenarios distinguishing between the different hydrogen applications (i.e., mobility, industry, re-electrification, and NG grid) as well as between two time horizons (early hydrogen market, i.e., in 2025, and established hydrogen market, i.e., in 2050). The scenario differences between different hydrogen markets include

- Hydrogen demand profiles (e.g., refueling station demand profile for the mobility case, etc.).
- Necessary topside equipment after the cavern.
- Available intermittent electricity (for some applications such as re-electrification the roundtrip efficiency is lower than for other areas, resulting in a higher overall electricity consumption in order to provide the same energy output).
- Hydrogen sales prices.

The differences between the time horizons stem from the variation of some input parameters for selected facility components (e.g., decrease of specific investment costs for electrolysis according to global economies of scale) and energy prices. It is also worth mentioning that the overall energy amount sold to each market, the constellation of the facility elements prior to the underground storage (i.e., electrolysis and input compressor) and the source of intermittent electricity remain unchanged for all markets. Table 7.5 summarizes the scenarios selected for further analysis.

The major technology input parameters on all elements of the facility are presented in Tables 7.6 and 7.7.

Table 7.5 **Scenario definition for further analysis**

Criteria	Mobility 2025/2050	Industry 2025/2050	Electricity 2025/2050	NG grid 2025/2050
Demand profile	Refueling station	Constant	Electricity demand	Constant
H_2 demand quantity	24,000 t H_2	24,000 t H_2	24,000 t H_2	24,000 t H_2
Available intermittent electricity	1500 GWh	1500 GWh	2600 GWh	1500 GWh
Topside equipment prior to storage	Compressor + other processes	Compressor + other processes	Compressor + other processes	Compressor + other processes
Topside equipment after storage	PSA + compressor	Compressor + drying	CCGT plant + drying	NG grid injection + draying

Table 7.6 **Techno-economic parameters for alkaline electrolysis**

	Investment costs	Efficiency	Stand by	Water consumption	Lifetime	O&M costs
Year	*€/MW$_{el}$*	*%*	*% capacity*	*kg/kg H_2*	*Years*	*% Invest/ year*
2025	991,562	66	1	9	20	4
2050	612,749	66	1	9	20	4

Table 7.7 **Techno-economic parameters for all elements of the topside equipment**

	Electricity consumption		O&M costs		Investment costs		Lifetime	
	2025	**2050**	**2025**	**2050**	**2025**	**2050**	**2025**	**2050**
Technology	*kWh/ kg H_2*	*kWh/ kg H_2*	*% Invest*	*% Invest*	*€/kW H_2*	*€/kW H_2*	*Years*	*Years*
Compressor in	1.00	1.00	4	4	300	240	15	15
Compressor out	1.00	1.00	4	4	300	240	15	15
PSA	0.20	0.20	4	4	84	84	15	15
H_2 drying	0.02	0.02	4	4	6	6	15	15
NG grid injection	0.00	0.00	4	4	0	0	15	15
CCGT	0.00	0.00	0	0	700	700	25	25

Further parameters are:

- Duration of the planning horizon: 10 years.
- Interest rate: 8%.
- Electrolysis water consumption: 9 kg/kg$_{H2}$.
- Price for process water for both time horizons: 0.0015 €/kg.

For further calculations the case study assumes a large cavern of ca. 500,000 Sm^3 of net storage volume (or ca. 133 GWh$_{H2}$ net capacity). The corresponding specific investment outlays are ca. 300 €/MWh$_{H2}$, fixed O&M costs are 4% of the investment outlays and lifetime is 30 years. The statistical characteristics on energy prices considered in this case study are summarized in Table 7.8.

Whether a potential hydrogen market is relevant for hydrogen from PtG also depends on whether the hydrogen can be produced for costs below the price of hydrogen from the most relevant benchmarking technology. This "allowable hydrogen price" takes the efficiency of the relevant end use technologies into consideration. For example, for mobility, the costs of using gasoline in an internal combustion engine are translated to the allowable hydrogen price by comparing its efficiency with that of a fuel cell drive.

The cost comparison depends on a variety of relevant investment and operational parameters and the given time horizon. Table 7.9 depicts relevant cost/price data for various cases of hydrogen production and end use for the short- and medium-term horizon. The hydrogen prices in Table 7.9 are to be interpreted as the maximum allowable price for hydrogen for each individual market.

The case study in this chapter includes also a sensitivity analysis of major parameters in order to better understand the impact of the input data on the economic feasibility of the PtG technology. The sensitivity analysis is based on the results from the scenario on the mobility sector in the established market (2050) and varies the following parameters:

- Relaxation of the intermittent electricity supply constraint: In this case the electricity for hydrogen production can be purchased from the market regardless of the intermittent pattern of the interconnected power plant. In this way the value of the dedicated renewable hydrogen production can be estimated.
- Specific investment outlays for electrolysis: expensive electrolysis technology (ca. 1,200,000 €/MW) and cheap electrolysis technology (ca. 300,000 €/MW).
- Average electricity price: expensive electricity (ca. 200 €/MWh) and cheap electricity (ca. 50 €/MWh).
- Volatility (standard deviation) of electricity price: high volatility (200% of the reference) and low volatility (50% of the reference).

Table 7.8 Statistical characterization of energy prices

	Electricity prices		NG prices	
	2025	**2050**	**2025**	**2050**
Min (€/MWh)	0	0	12	33
Max (€/MWh)	185	535	27	48
Mean (€/MWh)	49	101	38	59
Volatility (€/MWh)	25	72	2	2

Table 7.9 **Cost-specific benchmarking assumptions and resulting costs for hydrogen production and allowable prices for hydrogen for different applications (HyUnder, 2014)**

Market segment	Energy benchmark	Major assumptions (2025/2050)	Allowable prices (€/kg_{H2})	
			2025	**2050**
Mobility	Diesel fuel gasoline	Diesel price: 1.70/2.21 €/l Gasoline price: 1.80/2.25 €/l Diesel consumption: 2.95 l/100 km Gasoline consumption: 3.50 l/100 km H_2 consumption: 0.54 kg/100 km Infrastructure costs: 3 €/kg_{H2} VAT: 19% Fuel tax for H_2: 13.90/57.15 €/MWh H_2-equivalent uses average from gasoline and diesel costs	4.38–6.38 Average 5.38	5.31–7.43 Average 6.37
Industry	NG (incl. SMR plant)	NG price: 28.28/39.82 €/MWh Plant power: 844 MW NG Efficiency: 76%, Life time: 25 years Interest rate: 8% Annual full load equivalent operating hours: 7000 h Capital investment: 262 M€ CO_2 price: 29.13/59.13 €/t_{CO2}	1.64 in 2025 and 2.41 in 2050 Average ca. 2	
Natural gas	NG price from modeling	Average prices are: NG price 2025: 29 €/MWh NG price 2050: 40 €/MWh Modeling results shown	0.99	1.36
Electricity	Electricity as modeled for this study	Assumptions on electricity market as described above, using same assumptions as for hydrogen production from electricity (EEX price). Modeling results are shown	1.66	4.11

7.5.3 Major quantitative economic results

The major quantitative economic results for the optimized sizing and operation of the PtG facility in each market and time horizon are presented by a comparison of the specific hydrogen costs with allowable hydrogen prices. The specific hydrogen costs are considered as per unit costs including all operating costs and annualized capital expenditures (i.e., overall costs divided by the total hydrogen quantity sold to the market). As illustrated in Figure 7.10 the specific hydrogen costs range between 4 and 6 €/kg_{H2} for most applications in both time steps. The only exception is represented by the electricity market with specific hydrogen costs between 8 and 10 €/kg_{H2} due to additional investment outlays for the re-electrification unit and low round-trip efficiency of the system (i.e., converting hydrogen into electricity causes additional efficiency losses and requires additional costly electricity purchases). The major cost components are represented by annualized capital expenditures (with major share of electrolysis investments and minor share of investment outlays in underground storage and other topside equipment) and electricity costs from electrolysis operation. All other costs such as fixed O&M costs have only limited impact on the economic results. This cost structure is similar for all hydrogen applications. Thus, the costs associated with the build-up and operation of the electrolysis have a large impact on the overall economics of the PtG facility results and should be carefully analyzed for any investment decisions in this technology.

In general, the optimal electrolysis size, utilization, and operation mode is a trade-off between the capital expenditures in additional electrolysis capacities and corresponding electricity costs. On the one hand, for a given hydrogen output decreasing specific investment outlays allow for the build-up of larger capacities, which are then utilized in a

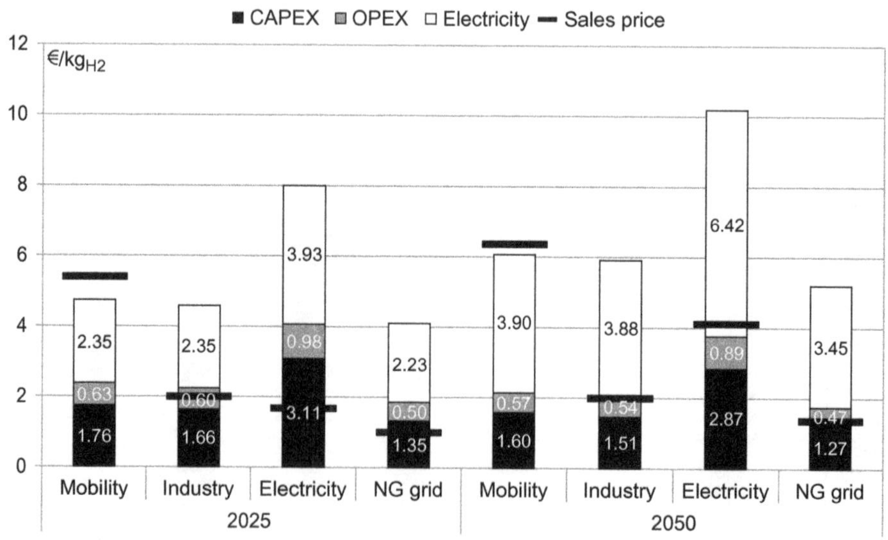

Figure 7.10 Comparison of hydrogen production costs with allowable hydrogen prices for hydrogen for (a) transport, (b) industry, (c) re-electrification, and (d) use in the natural gas grid.

smaller number of hours, lowering the overall electricity costs. On the other hand, decreasing electricity prices (or increasing price volatility) allow for a better utilization of the electrolysis and lower capacity investments. However, in the context of integration of renewable energy this trade-off is additionally constrained by the intermittent pattern of renewable electricity generation and the effect of decreasing specific outlays or average electricity prices cannot be fully utilized.

In fact, in the underlying case study the intermittent pattern of the wind onshore power plants is mainly responsible for the increase of specific hydrogen costs between both time steps. On the one hand, the specific costs related to annualized capital expenditures slightly decrease despite higher electrolysis capacity (increase from 440 MW_{el} in 2025 to 600 MW_{el} in 2050 for electricity market application and from ca. 250 MW_{el} in 2025 to ca. 280–600 MW_{el} in 2050 for all other applications) and lower electrolysis utilization (ca. 4,800 full load hours in 2025 and ca. 3,200 full load hours in 2050). This is due to the expected decrease of specific investment outlays for electrolysis capacities from 2025 to 2050. On the other hand, however, the specific electricity costs increase substantially by more than 50% from 2025 to 2050, overcompensating the former effect. This is mainly due to the fact that electricity purchases are constrained by the wind feed-in pattern and therefore the operator of the facility is not able to fully take advantage of low electricity prices. The average electricity purchase price jumps from ca. 40 €/MWh in 2025 to more than 65 €/MWh in 2050. Nevertheless, the underground storage has seasonal character and plays an increasingly important role as its utilization, represented by the number of full cycle equivalents, goes up between 2025 and 2050.

In comparison to the allowable hydrogen prices the only application with a positive margin (i.e., hydrogen sales prices higher than the specific hydrogen costs) is represented by the mobility sector. This margin decreases only slightly between 2025 and 2050 due to higher hydrogen production costs based on higher electricity prices and constrained by intermittent wind feed-in. In all other hydrogen markets the sales prices are considered to be insufficient to cover the specific costs. Hence, the NPV is positive only in the mobility case and negative in all other scenarios. However, it is important to mention that the hydrogen produced under the feed-in constraint potentially improves the system security as in this case electricity demand is in line with the electricity generation. As this additional benefit is not rewarded explicitly in the model, the results of this case study might underestimate the true value of the PtG facility directly converting renewable electricity into hydrogen.

7.5.4 Sensitivity analysis

The results of the sensitivity analysis for the mobility scenario in 2050 as a reference are illustrated in Figure 7.11. The overall level of the electricity prices represented by the price average has the largest impact on the specific hydrogen costs. The lower the electricity prices, the larger the electrolysis capacity (and the lower the corresponding utilization). However, the substantial decrease in the overall electricity costs overcompensates for the slight increase of the specific capital expenditures.

A Similar behavior can be observed also for the change of the price volatility (represented by the standard deviation of the price time series). Higher volatility leads to

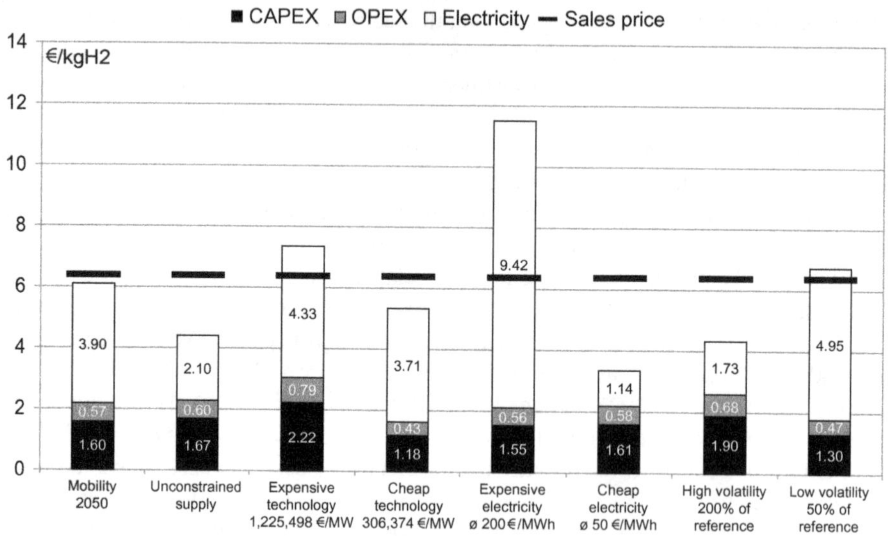

Figure 7.11 Results of the sensitivity analysis for the mobility scenario in 2050 as a reference.

lower electricity purchase prices decreasing the overall specific hydrogen costs whereas for low volatility the opposite is true. This is due to the fact that high volatility in general increases the number of hours within a year with free or very cheap electricity.

As expected the specific investment outlays for the electrolysis capacity has also a large impact on the economic value of the PtG facility. Cheap technology not only lowers the annualized capital expenditures of a given investment but also allows for enhanced operation of the facility in hours with very cheap electricity, simultaneously decreasing the actual electricity costs.

Finally, relaxing the constraint on the intermittent electricity supply allows a further decrease in the specific hydrogen costs. Thanks to increased flexibility of the electrolysis operations, hydrogen can be preferably produced in hours with cheap electricity which otherwise was not available from the intermittent power plant. This also leads to slightly higher electrolysis capacity, lower electrolysis utilization, and thus higher capital expenditures. Interestingly, the underground hydrogen storage becomes more important as its utilization represented by the full cycle equivalents also increases. For the mobility scenario in the established market of 2050 the value of green hydrogen, i.e., the difference between the constrained and unconstrained case, is in the order of 2 €/kg_{H2}.

7.6 State-of-discussion and development perspectives

The electricity industry has motored the first activities to analyze the large-scale storage options needed for balancing the electricity excess and shortages from introducing fluctuating renewable electricity at large scale. Analysis on the need for hydrogen storage at large scale is characterized by wide bandwidths ranging from a few to three-digit Terawatthours for Europe by 2050. Nevertheless, common sense has developed that large-scale storage will be needed for extended periods of time once significant levels of renewable energy beyond 30–40% will have been reached and two-digit Terawatthours beyond renewable electricity shares of 60–80%. Furthermore, it is believed that only chemical energy can supply sufficiently high storage quantities across Europe, below or aboveground.

Hydrogen underground storage, and in regions where there is no geological potential also aboveground storage, is a viable, flexible means to support the electricity sector in its endeavor to integrate renewable electricity at large scale. The storage of gases in salt caverns has a safe operational record and is a well-established technology for use in natural gas markets and also to store hydrogen in an industrial context.

Potential business cases for hydrogen underground are very difficult to justify in the short term and in the existing energy system structure. Important ingredients to be analyzed and demonstrated for a future business case development are the assessment of further services to the electricity industry (ancillary services) in addition to large-scale storage on the one hand and the identification of synergies by utilizing this infrastructure for several end users simultaneously (transport fuel, chemical feedstock) on the other. Without these being developed, future business cases are believed to be challenging. Both of these important ingredients are further described in the following list:

- For ancillary services: In addition to large-scale storage, the storage needs of the power sector also span all ancillary services from momentary reserve, primary and secondary control power, minute reserve, redispatch, and black-start capability to reserve for seasonal balancing. Hydrogen storage could provide one single solution for several of the ancillary services at the same time.
- Synergies with other hydrogen markets: The benchmarking of hydrogen-powered fuel cell vehicles against today's cars has shown that the most viable and early business cases could open up for hydrogen as a transport fuel. Other assessments have given evidence that also the chemical or petrochemical industries have an interest in hydrogen technology, once the economic conditions become suitable.

Finally, also the natural gas industry has offered their widely established gas transport and distribution infrastructure and services to support the electricity sector by applying hydrogen in their grid as flexibility service through the concept of PtH_2 (admixture) or $PtCH_4$ (methanation of hydrogen with CO_2).

The quantitative results of recent analysis in various projects have demonstrated that in the short term (>2020–2025) realistic and competitive cases for hydrogen from PtG can be identified only for mobility (HyUnder, 2014; Bünger et al., 2014). The interpretation of these findings gives clear evidence that PtG concepts still lack a thorough understanding of the currently emerging energy and electricity market structures. Sensitivity analysis shows that one important impact on a commercial development of large-scale hydrogen storage concepts could be imposed by a strongly growing volatility of future electricity prices. The volatility is most likely strongly underestimated at present.

It is anticipated that further research and analysis will be required for the identification of potential businesses with hydrogen storage at large scale. These are believed to emerge with regional differences, early and aboveground in tanks or pipes, and later below ground in caverns in relevant regions that combine the suitable geology (salt caverns) with relevant infrastructures for hydrogen, electricity, and end-use infrastructures.

Furthermore, consequential and continuous policy support will be needed to develop the business cases, energy storage being understood as an important means to accompany the introduction of renewable electricity into the grid. In preparing for utilizing the potential of hydrogen storage at large scale, it is important to understand the long development lead times, at least in the order of 10 years. These will be needed to identify further geologic potentials beyond salt cavern storage, to pave the way towards a wide market introduction by introduction of the necessary regulations and to observe and better understand the emergence of the energy/electricity markets.

References

Bahbout, A., Kluyskens, D., Wurster, R., 2012. Status report on the demonstration phases of the Euro-Québec Hydro-Hydrogen Pilot Project [EQHHPP]. In: Proc. World Hydrogen Energy Conference, Toronto.

BMU, 2012. Deutsches Zentrum für Luft- und Raumfahrt (DLR), Fraunhofer Institut IWES, Ingenieurbüro für neue Energien IFNE, 2012a. Langfristszenarien und Strategien für den Ausbau der erneuerbaren Energien in Deutschland bei Berücksichtigung der Entwicklung in Europa und global Schlussbericht ("Leitstudie 2011"), Stuttgart, Kassel, Teltow, March 2012.

Bünger, U., Crotogino, F., Donadei, S., Kruck, O., Michalski, J., Stiller, Ch., Weindorf, W., Zündel, F., 2011. Wasserstoff aus erneuerbaren Energien: Produktion, Kavernenspeicherung, und Nutzung—detaillierte techno-ökonomische Betrachtung anhand realitätsnaher Annahmen. Study for Volkswagen AG, February 2011.

Bünger, U., Landinger, H., Pschorr-Schoberer, E., Schmidt, P., Weindorf, W., Jöhrens, J., Lambrecht, U., Naumann, K., Lischke, A., 2014. Power-to-Gas (PtG) for Mobility—Status and Development Perspectives. Study for the German Ministry for Transport and Digital Infrastructure (BMVI) in the framework of the scientific consultancy as part of the Mobility and Fuel Strategy, April 2014.

Dena, 2010. German Energy Agency (dena): power grid study II: Integration Erneuerbarer Energien in die deutsche Stromversorgung im Zeitraum 2015–2020 mit Ausblick 2025, Report, Berlin, 2010.

ENTSO-E, 2011. Offshore Grid Development in the North Seas—ENTSO-E views. Secretariat of ENTSO-E AISBL, Brussels, February 2011. http://www.google.de/url?sa=t&rct=j&q=&esrc=s&source=web&cd=1&ved=0CDQQFjAA&url=http%3A%2F%2Fwww.eurelectric.org%2FDownload%2FDownload.aspx%3FDocumentFileID%3D71840&ei=QXljU_PnKIfJ4wS73YCgBQ&usg=AFQjCNGxG-zT07RJapCg8yIIkDQvFqJT3A&bvm=bv.65788261,d.bGE (accessed 02.05.15).

ER2050, 2011. Energy Roadmap 2050 (ER2050). European Commission, COM(2011) 885/2, 2011. http://ec.europa.eu/energy/energy2020/roadmap/doc/com_2011_8852_en.pdf.

ES2020, 2010. Energy 2020. A strategy for competitive, sustainable and secure energy (ES2020). European Commission, COM(2010) 639 final, 2010. http://eur-lex.europa.eu/LexUriServ/LexUriServ.do?uri=COM:2010:0639:FIN:EN:PDF.

Eurogas, 2014. Drop in 2013 EU gas demand emphasizes need for swift change. Press release 2014. http://www.eurogas.org/uploads/media/Eurogas_Press_Release_-_Drop_in_2013_EU_gas_demand_emphasises_need_for_swift_change.pdf (accessed 19.07.14).

H2-Mobility, 2013. H2-Mobility initiative agrees on $474M plan for hydrogen refueling network in Germany; 400 stations by 2023. Announcement by Green Car Congress on

30 September 2013, following a press conference at the F-Cell conference, Stuttgart, http://www.greencarcongress.com/2013/09/20130930-h2mob.html.
H2STORE, 2012. The H2STORE Project—Experimental and numerical simulation approach to investigate processes in underground hydrogen reservoir storage. Ongoing research, funded by the German Ministry for Industry. http://www.h2store.eu/ (accessed 14.07.14).
HyUnder, 2014. Assessment of the Potential, the Actors and Relevant Business Cases for Large Scale and Long Term Storage of Renewable Electricity by Hydrogen Underground Storage in Europe, Deliverable No. 6.2a, Representative Case Studies with a focus on salt cavern storage, Case Study Germany, Executive Summary, 17 February 2014.
ICCT, 2013. European Vehicle Market Statistics—Pocketbook 2013. The International Council on Clean Transportation, 2013.
Landinger, H., September 2009. Overview of existing hydrogen corridor studies and their main results. Study for the EU funded project ENCOURAGED-SSP6-CT2004-006588.
Müller-Syring, G., 2014. Establishing a Pan-European understanding of admissible hydrogen concentration in the natural gas grid, DBI Gas und Umwelttechnik. In: 14th Global Energy Village Conference, Copenhagen, 23–25 April.
Pape, C., Gerhardt, N., Härtel, Ph., Scholz, A., Schwinn, R., Drees, T., Maaz, A., Sprey, J., Breuer, Ch., Moser, A., Sailer, F., Reuter, S., Müller, Th., 2014. Roadmap Speicher—Speicherbedarf für erneuerbare Energien—Speicheralternativen—Speicheranreiz—Überwindung rechtlicher Hemmnisse. Study for the German Ministry for the Economy by Fraunhofer IWES, IAEW, Stiftung Umweltenergierecht, June 2014.
Perez, R.A., 2014. Hychico project: results and perspectives from Patagonia. In: Pilot project to test hydrogen admixture to a depleted gas field in Patagonia, Argentina, Presentation at the HyUnder Final Project Workshop, Brussels, 26 June 2014.
Perrin, J., Steinberger-Wilckens, R., Trümper, S.-Ch., 2007. European hydrogen infrastructure atlas and industrial excess hydrogen analysis. Deliverable 2.1 and 2.1a, part III: industrial distribution infrastructure, Roads2HyCOM project, funded by EC, Document R2H2007PU.1, 3 July 2007.
Sterner, M., Schmidt, J. (Eds.), 2009. Bioenergy and Renewable Power Methane in Integrated 100% Renewable Energy Systems. Limiting Global Warming by Transforming Energy Systems. Kassel University Press, Kassel.
Suresh, B., Schlag, St., Kumamoto, T., Ping, Y., 2010. CEH Marketing Research Report HYDROGEN, Industrial Gases—Chemical Economics Handbook, SRI Consulting, July 2010.
SUSPLAN, 2011. PLANning for SUStainability, EU-project 2008-2011: Efficient integration of renewable energy into future energy systems, Development of European energy infrastructures in the period 2030 to 2050; www.susplan.eu.
Tengborg, P., Johansson, J., Durup, G., 2014. Storage of highly compressed gases in underground Lined Rock Caverns—More than 10 years of experience. In: Proceedings of the World Tunnel Congress 2014—Tunnels for a Better Life. Foz do Iguaçu, Brazil.
TYNDP, 2012. Ten-Year Network Development Plan (TYNDP), 2012. European Network of Transmission System Operators for Electricity, 2012 https://www.entsoe.eu/major-projects/ten-year-network-development-plan/tyndp-2012/.
Underground.Sun.Storage, 2014. Website of the research and Power-to-Gas demonstration project on underground gas storage. http://www.underground-sun-storage.at/ (accessed 10.09.14).
VDE, 2008. VDE study, Energy storage in supply systems with large shares of REN electricity. Task Force ETG—Energy Storage, 2008.
WE-NET, 2001. Fukuda, K., Ogata, K., Kobayashi, O. A hydrogen introduction scenario in Japan. First Canadian Hydrogen Conference. Victoria, Canada, 2001, 17–20.

Hydrogen admixture to the natural gas grid

8

R. Judd, D. Pinchbeck
On Behalf of GERG, European Gas Research Group, Brussels, Belgium

Abbreviations

Bcm	billion cubic meters
Bn	billion (10^9)
CE	mandatory conformity marking for products sold within the European Economic Area
CEN	European Committee for Standardization
CHP	combined heat and power
CNG	compressed natural gas
CO_2	carbon dioxide
DIAL	differential absorption laser spectroscopy
DIN	Deutsches Institut für Normung e.V.
DLN	dry low NO_x
EASEE-gas	European Association for the Streamlining of Energy Exchange-gas
EN 437	European standard for test gases, test pressure and appliance categories (May 2003)
FID	flame ionization detection
G20	a reference test-gas (pure methane) for all GAD appliances
G222	a reference test-gas (a mixture of 23% H_2/77% CH_4) for all GAD appliances
GAD	Gas Appliances Directive (Directive 2009/142/EC on appliances burning gaseous fuels)
GERG	European Gas Research Group
H_2	hydrogen
H gas	high calorific gas (consist 87–99% methane)
HIPS	hydrogen in pipeline systems
ISO	International Organization for Standardization
kW	kilowatt
kWh/m³	kilowatthour per cubic meter
LNG	liquefied natural gas
LPG	liquefied petroleum gas
Micro-CHP	micro-combined heat and power
Mini-CHP	mini-combined heat and power
M€	million euro
MJ/m³	megajoule per cubic meter

Compendium of Hydrogen Energy. http://dx.doi.org/10.1016/B978-1-78242-364-5.00008-7

MN	methane number
MW	megawatt
MWe	megawatt electrical
MWh	megawatt hour
MPa	MegaPascal
NO*x*	generic term for NO and NO_2 (nitric oxide and nitrogen dioxide)
P2G	power-to-gas
PEM	proton exchange membrane
PGCs	process gas chromatographs
SNG	substitute natural gas
Syngas	a fuel gas mixture consisting of hydrogen, carbon monoxide, and some carbon dioxide
TWh	Terawatt hour
UNECE	United Nations Economic Commission for Europe
vol%	percentage by volume
Wobbe Index	Wobbe index (W) is an indicator of the interchangeability of different fuel gases.
W/m×K	Watt per meter Kelvin

8.1 Introduction

Given recent developments, particularly in the field of wind energy, the well-known problem of electricity storage has gained a new dimension. Pumped storage power stations have been used for decades to store electricity on a larger scale, but the number of power stations and their potential are limited in many countries across Europe.

An objective currently being pursued to enable the storage of surplus "renewable" electricity involves production of hydrogen, by electrolysis, for injection into the natural gas network with its relatively huge storage capacity. If hydrogen from surplus renewable electricity was injected into the natural gas network, the enormous transportation capacity and the huge storage capacity of the existing natural gas infrastructure, including underground storage facilities, could be used directly. Germany, for example, has approximately 500,000 km of pipelines and more than 20 billion m^3 of gas storage facilities. This, and similar situations elsewhere, could make an important contribution to the transportation and storage of surplus or nontransportable renewable electricity, a solution which would be particularly attractive if it helps to avoid construction of new electricity transmission lines.

The following two examples make it clear that injection of a hydrogen volume into the natural gas network, seemingly as low as 10 vol%, would significantly contribute to solving the problem of transporting and storing surplus electricity generated from renewable resources in the natural gas network:

- In Germany almost 1000 TWh of energy are transported annually in the form of natural gas. This is almost twice as much as the electricity consumed; 10% of hydrogen admixed to natural gas would correspond to an energy quantity of approximately 30 TWh. For comparison, the total capacity of the pumped storage power plants in Germany is 0.04 TWh per cycle (40,000 MWh).

- A medium-sized natural gas pipeline has a transport capacity of around 1 million m^3/h. Injection of 10% (100,000 m^3/h) of hydrogen would require an electrical input of more than 400 MW for the electrolysis reaction, which corresponds to the maximum output of several large wind farms taken together.

8.2 Reasons for adding hydrogen to the natural gas grid

It is becoming more widely accepted that hydrogen could become an important energy carrier in the energy mix in the quest for sustainability, because it offers major benefits related particularly to its huge potential for energy storage. Indeed it's possible that, with the existing infrastructure, hydrogen/natural gas mixtures could be transported, stored, and converted into electricity where required.

The preferred approach is power-to-gas (P2G) technology, which is motivated by the volatility of the renewable energy sources that are the primary route to decarbonizing Europe's energy system. However, the wind does not always blow and the sun does not always shine and this intermittent behavior means that there is a growing need for long-term, high-capacity energy storage. Unfortunately, not all of today's electricity grids have sufficient capacity to carry all of the renewable energy produced during periods of strong wind and bright sunshine. This is particularly so in Germany, but also increasingly so in the UK and elsewhere. The result is that, quite often, wind and solar installations have to be shut down to avoid overloading the electricity grid while, at the same time, significant compensation (M€) is currently being paid to wind generators for curtailment[1]; clearly this is an unnecessary waste of both energy and money, and while it can be partly addressed through improved approaches to an integrated energy market, a means of providing grid services to improve flexibility of renewables integration can aid the development of new market rules.

Given the huge storage capacity of the European natural gas grid, a much smarter approach would be to adopt P2G, which uses the renewably generated (green) electricity to power electrolyzers that can split water into hydrogen and oxygen. An alternative P2G solution, which would require a further processing phase, would be methanation, which would combine carbon dioxide (CO_2) with the hydrogen to produce a synthetic and renewable natural gas. Whichever option is chosen, the P2G approach would enable the huge capacity of the existing natural gas grid to be used to store and transport renewably produced gas, wherever it was needed.

It's worth noting that the German government strongly believes that the development of storage technologies is one of the main challenges for its Energiewende[2]

[1] Renewable generators often have preferential access to grids and are therefore compensated if electricity is not accepted.

[2] http://energytransition.de/.

(energy transition) if the integration of wind and solar power is to succeed. Such an approach will, necessarily, cause natural gas and electricity networks to become even more interdependent as shown in Figure 8.1 and, of course, R&D investment will be necessary to achieve robust solutions based on the existing natural gas grid and its various constituent components.

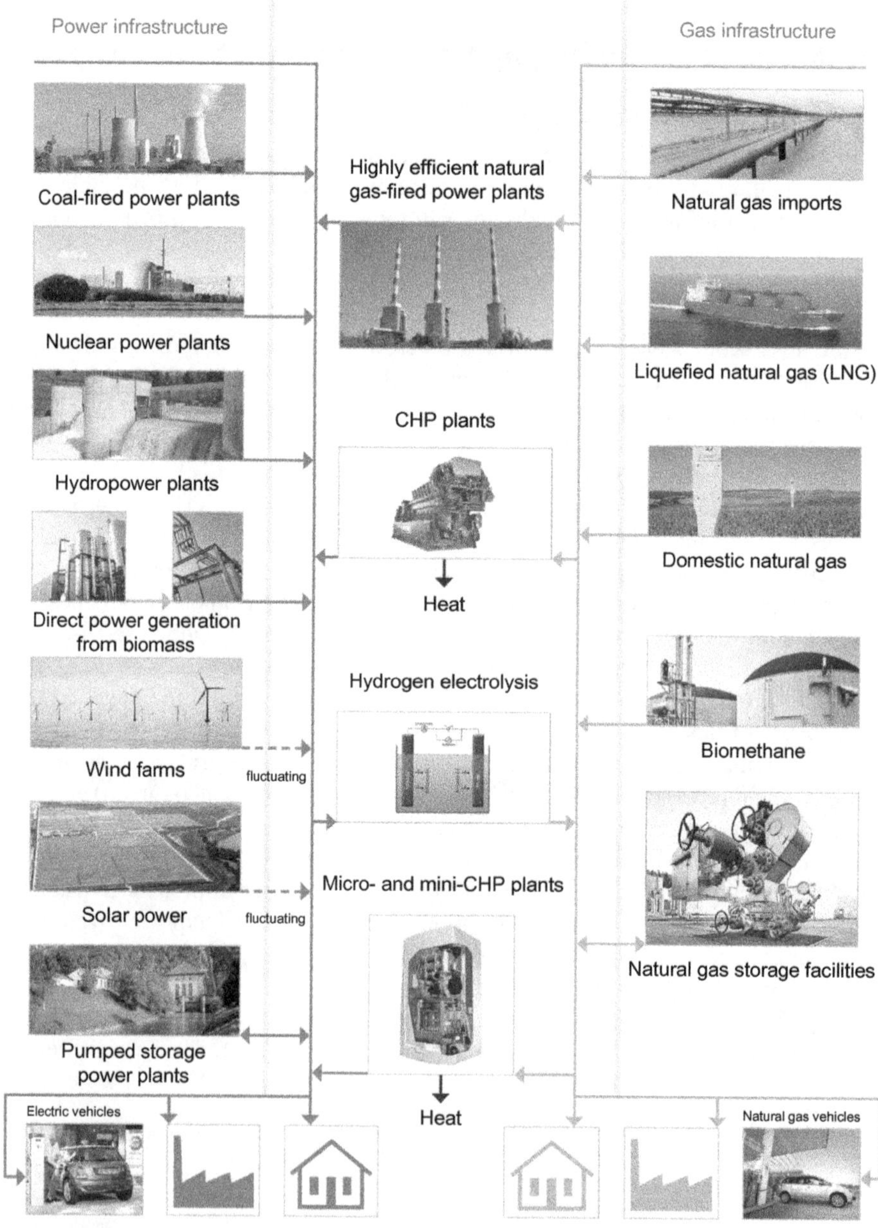

Figure 8.1 Converging electricity and gas infrastructures.

8.3 Potential benefits and problems associated with adding hydrogen to the natural gas grid

In certain parts of Europe, we have the situation already where the generation of "renewable" electricity from wind and solar energy has led, from time to time, to production plants being shut down because the electricity generated exceeds local requirements and the transportation or storage capacities are inadequate. It's a problem that will become even more severe in the future because construction of new electricity lines and high-capacity, pumped storage power plants is a costly and very lengthy process. Investment costs for electricity grid upgrades and interconnections have been estimated at up to €200 Bn and, for example, Germany's transmission system operators have estimated that €20 Bn will be needed to implement the grid upgrade required to phase in the renewable energy necessary to replace nuclear.

It's clear that the European natural gas pipeline network, with its huge storage capacity, has the potential to offer a solution. However, if the addition of small quantities of hydrogen, up to 10% (by volume), to natural gas pipelines is to be accepted, it must guarantee a technically feasible, economically viable and, crucially, safe system of storage, transportation, and use. Several important studies, including the EC-supported NaturalHy project (Florisson et al., 2010) and the GERG "HIPS" project (Altfeld and Pinchbeck, 2013), have examined the feasibility of using it as a means of widespread hydrogen storage and transportation, while, at the same time, conducting a detailed review of some of the potential bottlenecks related to the interaction between hydrogen and the wider European natural gas network, including aspects of end use.

The volume of hydrogen that may be added to natural gas is limited. There are already some very low "ad hoc" limits in place, but earlier studies (Florisson et al., n.d.) have shown that, with certain restrictions, admixture up to 10 vol% is not critical in most cases. However, there are bottlenecks, and attempts have recently been made (Altfeld and Pinchbeck, 2013) both to identify them and, where possible, propose solutions so that the natural gas infrastructure can be developed sufficiently to support the storage and transport of hydrogen–natural gas mixtures in a move toward a low carbon economy.

8.4 State of the art

Studies have shown that most parts of the natural gas system can cope well with hydrogen addition of up to 10 vol%, with no adverse effects. Analyses of the whole natural gas system (Florisson et al., n.d.; Altfeld and Pinchbeck, 2013) have concluded that there are particular components or "noncritical aspects" which should cause no problems and they are listed here:

- Natural gas transmission pipelines and compressors, despite concerns about hydrogen embrittlement.
- Gas distribution pipework systems, including metering and billing equipment, seals, etc., where leakage has been shown to be negligible.
- In-house pipework systems, with no problems reported at all.

- Industrial applications, where no specific problems are anticipated if the Wobbe[3] index of the gas mixture is well within the specified range.
- Safety parameters (e.g., flammability limits, ignition energy, flame speed) are affected only marginally; the increase of risk is very small. However, special attention must be given to gas detection devices. A reassessment of the ATEX zoning may be required in some environments, depending on the methods used.
- Some standards recommend a maximum content of hydrogen and other components of 5 vol% (ISO 6976).

A more detailed analysis is available in Altfeld and Pinchbeck (2013).[4]

8.5 The bottlenecks—Considering a 10 vol% admixture

The volume of hydrogen that may be added to natural gas is limited and studies (Florisson et al., n.d.) have shown that, with certain restrictions, admixture of approx. 10–15 vol% is not critical in most cases. There are, however, several limiting factors, or "bottlenecks," which could delay the introduction of hydrogen into the natural gas system and which require some investigation if they are to be understood and resolved. It's useful here to define what is meant by "sensitive components" and it's simply those elements of the gas system that are affected, or in the longer term, could deteriorate or could cause adverse effects in the presence of hydrogen admixtures up to 10 vol% in natural gas (Altfeld and Pinchbeck, 2013). All of the components known to be sensitive have been considered in detail and are outlined below.

8.5.1 Underground storage

Worldwide there were 688 underground gas storage facilities in operation at the beginning of 2013,[5] representing a working gas capacity of 377 bcm, or 10.3% of 2012 world gas consumption; Figure 8.2 is expected to grow to between 557 and 631 bcm by 2030. Working gas capacity has increased significantly since 2010 (+35 bcm), mainly under the impetus of Europe which added almost 14 bcm of capacity in the past 3 years. Decisions on a large share of these storage facilities were taken in the mid-2000s, before the economic crisis and fall in European gas consumption.

Underground gas storage has been developed in four regions: North America, Europe, the Commonwealth of Independent States (CIS), and Asia-Oceania. More than two-thirds of the storage sites are concentrated in North America, with 414 in the United States, and 59 in Canada with a combined capacity of 152 bcm (40% of the global total). Europe is ranked second in terms of number of facilities at 144 (99 bcm), followed by the CIS with 51 facilities. Asia-Oceania has 18 sites (9.3 bcm working capacity) and Argentina and Iran have one site each.

[3] Wobbe index (W) is an indicator of the interchangeability of different fuel gases. Regardless of calorific value, gases with the same W produce the same heat load in a gas burner.

[4] Copies are available from GERG, the European Gas Research Group. See www.gerg.eu.

[5] "Underground gas storage in the world—2013," CEDIGAS.

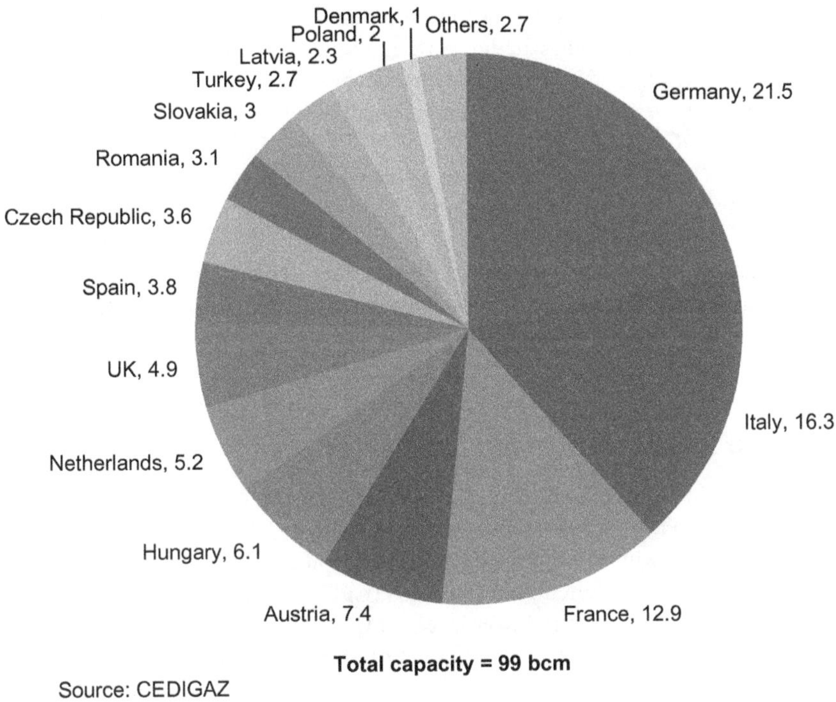

Figure 8.2 European storage capacity (bcm)—2013.

Depleted field storage predominates because it enables storage of large volumes of gas which are mainly used to balance seasonal swings in demand. With 509 facilities in the world, depleted fields represent 74% of the total number of sites. However, market liberalization has created a need for shorter-term storage and this has seen the importance of salt cavern storage in North America and Europe grow. As of January 2013, 94 salt cavern facilities were in operation in the world (up from 76 in 2010), representing 14% of the total number of sites.

A closer look at the type of facilities in operation in the world reveals important disparities from one region to another. Porous reservoirs (depleted fields and aquifers) largely dominate the number of storage facilities in North America (90%) and the CIS (96%); however, their share falls to 71% in Europe, where salt caverns represent a higher proportion than in other regions.

Clearly gas storage is a key component in the natural gas chain and, in a future scenario, its various facilities could come into contact with natural gas/hydrogen mixtures. Little consideration has been given to this prospect until very recently, and experts have been reluctant to suggest a limit value for hydrogen addition because of the difficulty of identifying and quantifying the relevant processes among all possible reactions in underground storage facilities.

Approximately 20 reservoir phenomena have been identified, all of which could impact reservoir exploitation, of which the most serious potential issue identified, particularly in aquifers and oil/gas depleted fields, is the potential for bacterial growth

(DGMK, 2014). The associated issues are principally loss of gas volume and disappearance of injected hydrogen, whether partial or total. There is also potential for damage to the cavity itself, and production of hydrogen sulfide (H_2S). It's important also to note the potential for leaks from steel-lined rock caverns, as detailed in the DGMK[6] literature study (DGMK, 2014).

It's clear then that the effect of bacteria is the main concern for underground storage of hydrogen and natural gas mixtures, specifically in aquifers and depleted fields, as the interaction phenomena are not well understood, nor is it easy to identify specific bacterial species and to know, or measure, their quantities *in situ*.

In summary, it's not possible at the moment to define a limit value for the maximum acceptable hydrogen admixture for natural gas stored underground, as there has been very little research in this area to date.

8.5.2 CNG steel tanks, metallic and elastomer seals

There are some 15 million natural gas vehicles worldwide, mainly outside Europe in, for example, Iran, Pakistan, Argentina, Brazil, and India; they all run on compressed natural gas (CNG), which is preferred to conventional vehicle fuels such as petrol (gasoline), diesel, and liquefied petroleum gas (LPG) because of its cleaner combustion and, therefore, significantly lower levels of CO_2.

In Europe, CNG represents a small market of around 1 million vehicles, but this is set to grow, especially since final approval was granted by the European Parliament in 2014 to new rules to ensure the build-up of infrastructure for alternative fuels across Europe, including CNG refueling points.

Unfortunately, national and international regulations limit hydrogen in natural gas as a motor fuel to 2 vol% and addition of even small quantities of hydrogen to natural gas networks is currently a show-stopper with regard to steel CNG vehicle tanks. According to UNECE[7] Regulation 110 for CNG vehicles, the H_2 content in CNG is limited to 2 vol%, if the tank cylinders are manufactured from steel with an ultimate tensile strength exceeding 950 MPa. This limit stems from the perceived risk of hydrogen embrittlement, which is known to cause accelerated crack propagation in steel and is, therefore, a critical safety issue. It's worth mentioning that the same 2% limit is echoed in the corresponding ISO standard 11439 (ISO 11439) and under DIN 51624, the German national standard for natural gas as a motor fuel.

In Europe, quenched and tempered steel 34CrMo4 is employed exclusively for CNG tanks and is compatible with hydrogen, provided that the tensile strength of the steel is less than 950 MPa, and that the inner surfaces of the cylinder have been inspected for allowable defects. However, existing steel CNG tanks are made predominantly of steel grades with a tensile strength greater than 950 MPa, because these materials allow smaller wall thicknesses and thus reduced weight of the cylinders, which is preferred for vehicle use. In addition, CNG tanks are not inspected for surface defects (simply because there is no need for it).

[6] Deutsche Wissenschaftliche Gesellschaft für Erdöl, Erdgas und Kohle e.V.

[7] United Nations Economic Commission for Europe.

A key aspect here is that, under UNECE rules, car manufacturers are held responsible for the suitability of car components, including CNG tanks. This inevitably means that CNG vehicles will only be fueled with natural gases containing more than 2 vol% hydrogen when substantial tests have proved that it's safe and the existing regulations have been amended accordingly.

A complete screening of the existing tank population will be complex, as tests must prove that the storage tanks are safe under daily conditions, from −40 to +85 °C, including exposure to frequent cyclic loads induced by the fueling process and consumption and, finally, a lifetime of 20 years. So, clearly this will be a long-term process where, at the moment, final success cannot be guaranteed.

In addition to the well-known embrittlement difficulties with hydrogen, there are concerns regarding leak tightness of seals, both metallic and polymer. All gas carrying components inside the vehicle are currently designed and tested for a maximum 2% H_2. As a result, all such components are potentially critical and their ability to cope with higher H_2 fractions remains to be tested.

8.5.3 Gas engines

Gas engines are used traditionally in CHP installations where power (usually electricity) and heat are produced simultaneously. This allows for very high efficiency when compared to separate production of electricity and heat. A more recent and growing area of application is the transportation sector as CNG- or LNG-fueled prime movers in passenger cars, trucks, and ships.

The total number of gas engines and their output in Europe is not known exactly but, for example, the Netherlands has about 4000 MWe of gas engines in CHP installations, most of which are used for peak-shaving at times of high demand; the average engine power is around 1 MW. In some other countries, kW range micro-CHP is popular and it is understood that some 24,000 micro-CHP units have been installed in Germany alone.

The installed base of stationary gas engines consists almost entirely of four-stroke engines, which can be divided into two basic types, lean-burn and rich-burn. In CHP applications, most of the bigger gas engines, generally in the range 100–150 kW, are of the lean-burn type while smaller gas engines are often of the rich-burn type. In transport applications, both rich-burn and lean-burn engines are used, depending on the manufacturer's preferred emission control strategy. Both types are designed for optimal performance, i.e., high output, high efficiency and NO_x emissions within the legislated limits.

Despite the limited availability of published information in this area, the physics of combustion, supported by experimental evidence from real engines, shows that the increase in flame speed and reactivity caused by hydrogen addition to natural gas typically increases in-cylinder peak pressures. It is also known that the methane number (MN) decreases if the proportion of hydrogen (or higher hydrocarbons) is increased. This can result in:

- increased combustion and end-gas temperature, which leads directly to enhanced sensitivity for engine knock and increased NO_x emissions;
- improved engine efficiency, but with increased engine wear and increased (noncompliant) NO_x emissions;

- reduced power output or tripping, for engines with knock control;
- an adverse effect on lambda sensors which can cause an inaccurate (low) measurement of oxygen in the exhaust gas. (This will cause the control system to change the air:fuel ratio, resulting in a leaner mixture than intended, thus influencing performance and increasing both the possibility of misfiring and emission levels, especially NO_x.)

That even low fractions of hydrogen can precipitate engine knock, compared to the natural gas to which it has been added, directly implies one limitation on hydrogen fraction: if the knock resistance of the fuel is at the lowest value acceptable for an engine or population of engines and no adaption of engine operation is possible, then no hydrogen can be added to this gas.

For natural gases with a relatively high knock resistance, such that the engines that use it have a substantial knock "margin," the question of maximum hydrogen addition is complicated by other performance issues, partially related to the large diversity of engine types and field adjustments of the installed base. At the least, installed base engines are not expected to have controls to adapt engine conditions for (fluctuating fractions of) hydrogen addition.

One of these performance issues regards NO_x emissions; many gas engines that are not capable of adapting their operating conditions for hydrogen addition (air:fuel ratio, timing), and are at the permitting limit for NO_x, can also admit no hydrogen. A more complex issue regards the consequences of the higher cylinder pressure for engine/component lifetime, reliability and maintenance requirements, which are more difficult to quantify. However, these issues are not hydrogen specific; for example, the admixture of LPG leads to similar effects.

There are recommendations that 2–5% hydrogen addition should be acceptable for engines, depending on the source of the gas. However, given the large and unknown variation in operating conditions of the installed base of engines, and the dependence of knock and NO_x emissions on the gas composition supplied to any given engine, it is strongly recommended that a case-by-case approach be used to determine the maximum allowable hydrogen fraction.

Most of this information has been derived from experience with stationary engines. Of course, the physical effects are the same for engines used in the transportation sector and, hence, the conclusions made here remain valid.

8.5.4 Gas turbines

When considering the addition of hydrogen to distributed natural gas the gas turbine market can be divided into the power generation market and "others." The power generation market requires the highest efficiency and relies on pipeline natural gas while the chemical industry also requires high efficiency, but may require a process gas to be burned. So the development of gas turbines for this second market demands more attention be paid to fuel diversity than for the development of gas turbines for power generation.

Modern gas turbines use dry low NO_x (DLN) in order to meet emission regulations. These DLN burners are generally less robust to changes in gas composition than diffusion burners. The earlier generations of DLN burners were designed and tested for pipeline gas with the narrow bandwidth that was common at that time. More recent testing has either revealed that these DLN burners tolerate wider bandwidths or has led to burner redesign work to cope with the new gas composition situation. Gas turbines equipped with older DLN burners therefore might not be fit for a wider bandwidth of the gas.

Older (mainly smaller) gas turbines may be equipped with diffusion burners that are relatively robust to changes in the natural gas composition. However, the effect of changes in the natural gas composition on the performance of most gas turbines is not known.

It is widely understood that there are strict limitations to the degree to which hydrogen may be added to gas turbine fuel. It is normal for customers to specify a particular fuel, often depending on what is available locally, sometimes even process gas, so that the gas turbine combustion system could be carefully specified and tuned for optimum operation.

Current fuel specifications for many gas turbines place a limit on hydrogen volume fraction in natural gas below 5%. Exceptions are dedicated (syngas) gas turbines that can accept very high hydrogen fractions (>50%) and some specific gas turbines that are capable of burning natural gas containing 10% hydrogen and even more.

A large amount of literature exists on new gas turbine developments for gases containing high and mostly fixed fractions of hydrogen. However, literature relevant to hydrogen admixture in natural gas for the installed gas turbines is very rare.

From an end-user point of view, Abbott et al. (2012) conclude that fuel composition variation can have an adverse impact on gas turbine operation, despite being within the range allowed in the grid and manufacturers' specifications. The indication is, therefore, that for some gas turbines there is little or no margin for additional variations in fuel quality, which reinforces the view that addition of even very low fractions of hydrogen to natural gas is likely to increase such issues for the installed gas turbine fleet.

It seems clear that, for the installed base of gas turbines, 1% must be considered as the general limit for hydrogen admixture to natural gas in a first step. Again, a case-by-case approach is required with special attention given to early or highly optimized DLN[8] burners. After tuning and/or modifications much of the installed base may be capable of tolerating 5–10% volume hydrogen admixture. Clearly, further work will be necessary to modify this situation.

As with gas engines, admixture of hydrogen to natural gas fuels that are toward the extremes of current acceptability will prove more problematic than admixture to "mid-range" fuels. The development of criteria that limit the amount of hydrogen addition to extreme fuels while allowing more addition to mid-range fuels may aid the introduction of hydrogen to the natural gas network.

8.5.5 Specific gas burners in the domestic sector

The risk when mixing 10% H_2 with natural gas depends on the combination of two factors: the primary air excess and the initial Wobbe Index. Therefore, atmospheric burners used with low-Wobbe gas are more sensitive to H_2 if they have been adjusted with G20 (methane).

[8] Dry low NO_x.

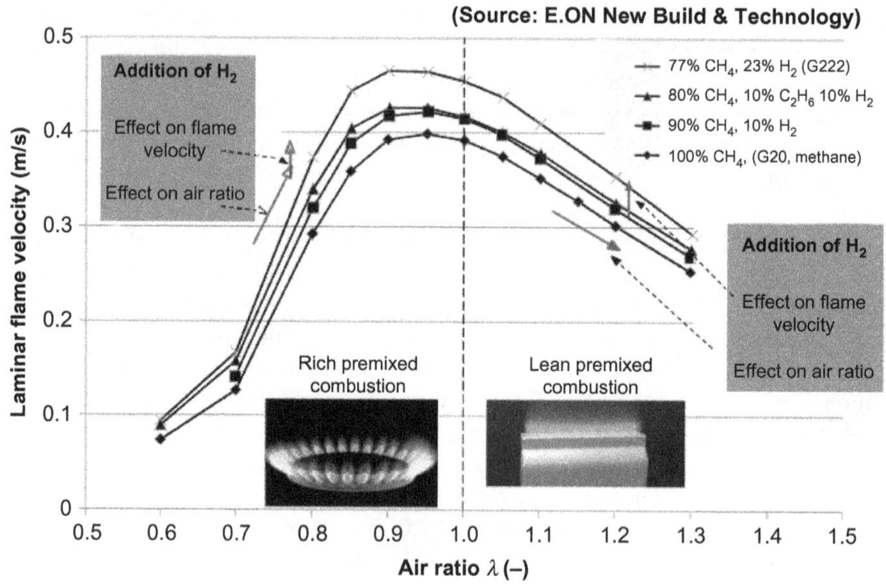

Figure 8.3 Flame speed and air ratio for different gases.

As can be seen in Figure 8.3, the addition of H_2 in natural gas has a direct and indirect effect on the flame speed in burners used in domestic appliances:

- it slightly increases the flame speed;
- it increases the air ratio if rich premix burners are considered (unless there is a system that controls it) and so indirectly changes the flame speed.

It's well known that, for rich premixed combustion (atmospheric burners, partially premixed burners) the addition of H_2 will result in both direct (due to H_2) and indirect (due to air ratio) increase of the flame velocity, making these burners more sensitive to H_2. Cookers, water heaters, and space heaters exclusively use such burners.

For lean premix combustion (fully premixed burners), the flame velocity will increase because of the H_2 content and decrease because of the changed air ratio, indicating a burner technology that is less sensitive. Appliances that use such burners are primarily condensing boilers.

All GAD[9] appliances (i.e., since the beginning of the 1990s) have been routinely tested with test gas G222,[10] which is a strong indication that hydrogen concentrations as high as 23% in natural gas might be acceptable, at least in the short term. There are, however, some limitations to that conclusion, among which are:

- the possible long-term impacts, which are not known;
- the fact that some countries may have gases in which the Wobbe Index can be lower than that of the G222 test gas.

[9] Gas Appliances Directive (Directive 2009/142/EC on appliances burning gaseous fuels).
[10] G222 is a mixture of 23% H_2/77% CH_4.

Table 8.1 Overall sensitivity from different sources by appliance type for 10% hydrogen in natural gas

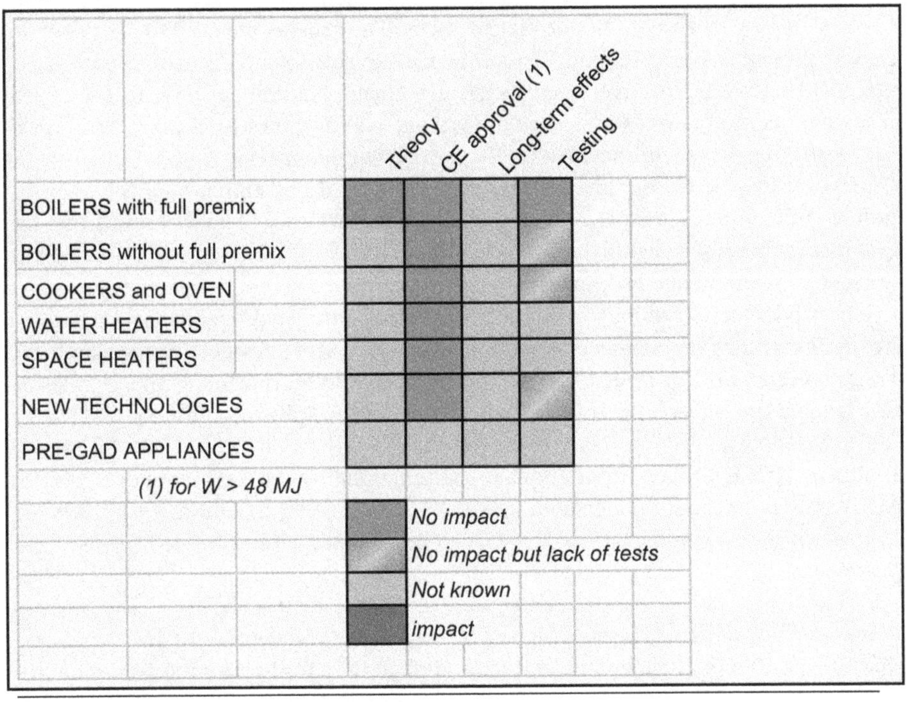

Table 8.1 summarizes the current overall situation and it's clear that some uncertainties remain. (N.B. The results are valid for gas grids with a Wobbe Index exceeding 14 kWh/m^3 (equivalent to 48 MJ/m^3 at a reference temperature of 15 °C/15 °C), 1.01325 bar.)

On this basis, injection of 10% of H_2 into natural gas grids (H gas) seems to be a reasonable future prospect for the domestic and commercial appliances considered. A "safety margin" should be taken into account. However, the uncertainties need to be clarified, and additional testing will be necessary to acquire more data.

8.5.6 Gas chromatographs

Gas composition and pipeline gas quality are generally determined for fiscal and safety purposes using certified gas chromatographs. However, there is a problem with the current generation of process gas chromatographs (PGCs) which use helium as the carrier gas and, as a result, are unable to detect hydrogen because of the relative proximity of their thermal conductivities (helium = 151 W/m*K; hydrogen = 180 W/m*K).

It's possible to solve this by retrofitting an additional separating column of argon as a carrier gas for hydrogen detection or by using new PGCs licensed for the metering of hydrogen. Another possibility might be to use PGCs with two single separating columns and two types of carrier gas. Some manufacturers have already developed new gas chromatographs capable of working with a hydrogen content of up to 10%.

8.5.7 Leak detection

Leak detection is essential for the identification of potentially hazardous gas leaks and is carried out by a variety of appropriate sensors. The sensors are used to alert people, often as part of a gas detection system, and to detect potentially dangerous gases, crucially, before an explosive situation has developed. Natural gas detectors are used for a wide range of applications such as: leakage surveys, industrial plant, refineries, wastewater treatment facilities, but rarely in domestic premises.

However, gas detection devices designed for natural gas may not be accurate for mixtures of natural gas and H_2. Some gas detection devices will be more sensitive for H_2 than for natural gas, while others are not sensitive at all to H_2 and will only react to the fact that the methane is diluted.

Semiconductor technology is suitable for detection of hydrogen/natural gas mixtures as it can identify methane as well as hydrogen. Most devices for measurement of the lower explosion limit of a gas mixture are configured for methane. Alarms are triggered upon reaching 10% or 20% of the lower explosion limit, i.e., 0.44% or 0.88% methane in air. The lower explosion limit declines slightly (4.36%) with admixture of 10% hydrogen. Limitations to the functionality are therefore not expected. Manufacturers generally understand how the addition of H_2 to natural gas affects the accuracy of their equipment and devices can be adjusted and calibrated for use with hydrogen.

FID devices (flame ionization detection based on a hydrogen flame) and thermal conduction sensors have been designed for the specific detection of hydrocarbons, which means that these technologies can be applied only for low admixtures of hydrogen. The usual safety and screening methods with gas detector devices and detectors currently used for pipeline grid inspections (by foot, by vehicle, by helicopter) are typically FID or, in the case of helicopters, DIAL (differential infrared laser absorption spectroscopy); neither of these technologies is capable of detecting hydrogen but would be acceptable, in terms of accuracy, in situations with hydrogen admixtures up to 5% in natural gas, as the main component of the gas remains methane.

We must conclude therefore that the addition of H_2 to natural gas changes the accuracy of gas detectors. Some will react on the safe side and others won't. It is essential, therefore, to recalibrate gas detection devices when H_2 could be present in natural gas to ensure that they will produce a result which will not increase the level of risk.

8.5.8 Safety—ATEX zoning

Safety is paramount in the gas industry and ATEX[11] is the name commonly given to the two European Directives used for determination and control of explosive atmospheres in the workplace:

1) Directive 99/92/EC (also known as ATEX 137 or the ATEX Workplace Directive) on minimum requirements for improving the health and safety protection of workers potentially at risk from explosive atmospheres.

[11] See http://www.hse.gov.uk/fireandexplosion/atex.htm#whatatex.

2) Directive 94/9/EC (also known as ATEX 95 or the ATEX Equipment Directive) on the approximation of the laws of (European) Member States concerning equipment and protective systems intended for use in potentially explosive atmospheres.

On the basis of the two Directives, employers must classify areas where hazardous explosive atmospheres may occur into zones; the classification given to a particular zone, and its size and location, depends on the likelihood of an explosive atmosphere occurring and its persistence.

There is a suggestion that addition of hydrogen to what was previously classified as an industrial methane (natural gas) zone (i.e., IIA) would lower the zone rating. However, the IIA classification zoning (EN 60079–20-1, 2010) is predicated on the assumption that the natural gas "does not contain more than 25% (v/v) of hydrogen" which indicates that zone IIA is appropriate for natural gas with admixture of up to 10% hydrogen.

8.6 R&D necessary to overcome the bottlenecks

8.6.1 Underground storage

As bacteria are considered to cause the most severe problems, there have been attempts at eradication with disinfectants, but trials have been inconclusive to date (froth/foam formation had caused problems). Clearly further investigation is needed to overcome this problem.

An alternative solution may be to separate the hydrogen from the natural gas before injection into storage and to store the hydrogen separately, and then to mix hydrogen and natural gas before injection into the gas network. (N.B. This separation of hydrogen from the natural gas/hydrogen mixture, using specially developed membranes, was investigated in some detail in the NaturalHy (Florisson et al., n.d.) project and was found to be both problematic and expensive.)

To date, there has been no specific research into the effects of hydrogen on the storage capacity employed by natural gas infrastructure—the underground storage facilities. However, this issue has recently been taken up by a consortium led by RAG[12] who will lead a project[13] to investigate underground storage of a mixture of hydrogen and synthetic methane.

In particular, the project will try to determine the effects of hydrogen on:

- the geochemical structure of reservoir rock, fluids, and transport mechanisms;
- microbiological metabolic activities in porous reservoirs;
- corrosion in wet gas environments;
- cement properties.

The research project should be completed in 2016.

In addition, there appears to be potential for estimating the impact of hydrogen addition following earlier attempts at numerical simulation by Maurer (1992) and Bonnaud (2012).

[12] Rohöl-Aufsuchungs Aktiengesellschaft.

[13] www.underground-sun-storage.at.

It may also be instructive to define a model for each kind of reservoir, to run simulations with various scenarios and to compare results with current projects such as HyUnder[14] or Underground Sun Storage.[15]

8.6.2 CNG tanks, metallic and elastomer seals

For the existing fleet of CNG vehicles, with steel tanks (type 1), the hydrogen limit for admixture with natural gas is set in accordance with ECE-R110 and DIN 51624 and CNG customers must be able to rely on the availability of compatible fuel.

A dedicated research program, which may help to determine a higher limit for the existing fleet would be of limited suitability as it cannot replace the certification procedure covering the complete set of relevant specifications. So, backdated fleet approval for higher hydrogen contents on the basis of a test program cannot be expected.

Amendment of the existing regulations would imply that all CNG vehicle components in the field must be thoroughly investigated under all operation conditions, including all relevant parameters of durability such as hydrogen partial pressure, assembly temperature between −40 and +85 °C, relative humidity, etc., in order to be approved for higher hydrogen contents. The enormity of this task suggests that it will probably never happen. However, it may be useful to analyze the theory and assumptions that led to the apparently *ad hoc* 2% limit and it is almost certain that it will be challenged by those wishing to encourage the introduction of hydrogen into the natural gas grid. Such a challenge will, however, require an extensive and rigorous R&D effort to produce sufficiently authoritative results.

Alternatively, it may be possible in future years to introduce CNG vehicles that are specifically designed to tolerate higher hydrogen contents. As this is linked to higher expenditure, indications from automobile manufacturers are that it will be implemented only when the introduction of higher hydrogen fractions in natural gas becomes a realistic prospect.

If we assume increasing market penetration of such advanced CNG cars and the normal, progressive phase-out of older models, a gradual transition to the acceptance of higher hydrogen levels in CNG would appear feasible. It is worth mentioning that this will be a long-term process as, for example, CNG vehicle tanks have a lifetime expectation of 20 years.

8.6.3 Gas engines

To allow higher fractions of hydrogen, engine configuration/adjustment/controls must be adapted to remove the physical cause of the issues:

- For engine knock, NO_x and engine wear, the peak pressures and temperatures must be lowered to those that will negate the effects of hydrogen. However, it's clear that unless the hydrogen fraction, and the natural gas composition to which it is added, can be constant, adaptations will have to accommodate fluctuating amounts of hydrogen.

[14] www.hyunder.eu.

[15] www.undergroundsunstorage.at (operational October 2013).

- Modern controls for air–fuel ratio and/or ignition timing that make use of exhaust NO_x sensors, temperature sensors, or pressure sensors in the combustion chamber may compensate for fluctuating hydrogen fractions in the fuel. However, the adequacy of such solutions and the adaptability to the various engine types must be examined.

So, to enable the fluctuating hydrogen content in natural gas to be increased, further work is required to resolve a number of issues:

- Effect of hydrogen on knock resistance and pre-ignition—a better method of accounting for the effects of hydrogen addition to the knock resistance of natural gas is required.
- The way in which engine control systems handle the effect of hydrogen on NO_x emissions and combustion pressures must be examined. The installed base has a diversity of control systems with various components. Their response to, and the ways they handle, hydrogen admixture vary. Different engine types with different control hardware and software may require different adaptations.
- Risk of occurrence of explosions in intake, crankcase, and exhaust will increase with hydrogen admixture. The effects and measures that should be taken to avoid this must be identified.

8.6.4 Gas turbines

Tests on gas turbines are required with specific attention being paid to starting, flame stability (pulsations and flashback) and emission issues. Two possible approaches are to build experience:

- with new turbines and expand conclusions to the installed base;
- from pilot hydrogen projects that inject hydrogen into the natural gas grid.

For turbines and many other industrial processes, it's important also to know the range and rate of change of the hydrogen content in natural gas. Consequently, this question has to be addressed before considering widespread injection of hydrogen in the gas network.

The development of criteria that limit the amount of hydrogen addition to extreme fuels while allowing more addition to mid-range fuels may be beneficial. Consideration should be given as to whether common criteria may be applicable to gas turbines, gas engines, and other combustion processes.

8.6.5 Specific gas burners

With regard to domestic applications, uncertainties remain that need to be clarified so it would be useful to acquire more data from additional series of tests on:

- atmospheric burners, as the majority of the technologies on the market use these burners and the available test results do not sufficiently cover all segments of appliances;
- new technologies, especially those with features not previously present in the tests;
- pre-GAD appliances, for which we do not have a documented safety margin, such as CE approval.

Further investigations are recommended:

- on potential long-term effects of hydrogen addition to natural gas, such as overheating of burners and heat exchangers;
- to clarify the impact for Wobbe numbers below 14 kWh/m^3 (under which CE approval results cannot be used).

It's also clear that for poorly adjusted and/or unfavorable conditions of natural gas quality, no hydrogen admixture is allowed.

Finally, it is strongly recommended that, in EN 437,[16] there should be a definition of new test gases and test procedures for approval of appliances that operate with hydrogen/natural gas mixtures; this would seem to be a fundamental requirement for future-proofing appliance manufacturing for an evolving market.

8.6.6 Leak detection

Special attention must be given to gas detection devices because some are not sensitive to hydrogen. As a result, they see only the diluting effect of addition of H_2 to natural gas and will therefore give an inaccurate response.

For measuring systems that are not able to detect hydrogen explicitly, such as FID and DIAL, modification or replacement is recommended for admixtures of hydrogen of more than 10 vol%. This statement is a first recommendation and certainly needs further investigation.

8.7 Additional requirements

In addition to removing the bottlenecks that are currently preventing this approach to energy storage and transportation from working for both the energy industry and the consumers in Europe, it's important to consider other factors that will come into play when placing P2G in the broader context of our transforming energy system. Indeed, there are a number of overarching R&D and policy needs:

- a Europe-wide energy system model that incorporates the natural gas infrastructure as a key element;
- SMART grid concepts that incorporate gas generation, transport, storage, and use;
- work to reduce the cost and improve the efficiency of peak and flexible power provision;
- support for injection of renewable gases—standards and low-cost technology (supported by positive regulatory messages);
- options assessments and demonstrations of P2G—hydrogen and methanation, electrolysis;
- future-proofing the natural gas network by making it H_2 ready;
- developing advanced end-use technologies and hybrid systems for end-use efficiency gains;
- EU and International Government support.

[16] European standard for: Test gases, Test pressure and Appliance categories (May 2003).

8.8 Key technologies

In addition to the network impacts, it will be important to develop the enabling technologies for P2G which will allow the re-injection of renewable gas into the grid and this section outlines two specific, practical examples, one in Europe and one in North America.

When hydrogen production from wind or solar energy is considered, the clear front-runner is electrolysis, and this is where most effort is currently concentrated. However, other innovative hydrogen production technologies should not be discounted, and could include photocatalytic water splitting, plasma waste gasification, and plasma-based pyrolysis.

Electrolyzers are key to P2G, for their ability to convert excess renewable electricity to hydrogen and oxygen by electrolysis of water. However, it's clear that in order to make this approach economically viable on a large scale, technical breakthroughs will be necessary. Electrolyzers are fairly straightforward technology but they are expensive. Developments must be targeted toward significant reductions in investment cost and increasing efficiency and durability (lifetime) of catalysts.

There are developments taking place (Hydrogenics) and hydrogen and utility-scale next-generation electrolysis can take gas–electricity convergence into new territory (National Centre for Hydrogen). However, to provide useful grid service, electrolyzers will have to be scaled up significantly, probably to the 10 MW range (in distribution grids), and even higher for transmission grids. They will need also to be capable of higher pressure operation with greater flexibility and turn down and reliability.

Developments in PEM[17] electrolyzers may provide a way forward, as this approach can offer short start-up times and flexibility across a wide range of loads. A recent example is the PEM electrolyzer, an example of which is shown in Figure 8.4, which was

Figure 8.4 Self-pressurizing 80 bar PEM stack.
Courtesy of ITM Power plc.

[17] PEM = proton exchange membrane.

delivered to the Thüga Group's P2G plant in Frankfurt am Main, Germany, at the end of 2013; it was used to inject hydrogen into the Frankfurt gas distribution network and, as such, became the first plant to inject electrolytically generated hydrogen into the German gas distribution network. The modular, turn-key system[18] was commissioned and finally accepted in the first quarter of 2014.

One of the key challenges for electrolyzer developers is in making their technologies economically viable with variable utilization, both in terms of maintaining efficiencies, and also in meeting capital and operational cost targets at often low utilization. This challenge is of course not dissimilar to the one faced by the nascent wind energy industry, and investment in developments leading to cost reductions through economies of scale will help, as will innovative solutions for modular integration of electrolyzers.

In North America, the first commercial P2G project is advancing in Ontario, Canada (Teichroeb) and, in July 2014, Ontario's independent electricity system operator (IESO) selected 34 megawatts (MW) of energy storage in a competitive RFP (request for proposals). Included in the selection process was a 2 MW P2G energy storage project that will provide regulation services (frequency control) to the power grid. Hydrogenics' PEM electrolyzer technology will be incorporated into the project, which will be co-developed by Hydrogenics and Enbridge. While the project's primary service will be the provision of regulation services to the power grid, additional phases of the project include the development of hydrogen blending standards to support the future injection of this renewable energy into the local natural gas network. The development of blending standards is expected to help advance the commercialization of P2G in various North American markets.

8.9 Future trends: The methanation option

Although much of the focus so far has been on hydrogen injection directly into the pipeline, hydrogen from an electrolyzer could also be used to produce methane, the main constituent of natural gas. Although the process would involve further capital expenditure and energy losses, it has real potential advantages. Not least of these is that the natural gas network has been designed and built specifically for natural gas, and many of the technical issues, the "bottlenecks" described earlier, and by extension the costs of mitigation, would disappear. The other advantage is that methanation of hydrogen removes and re-uses CO_2 from the ecosystem. This approach is of considerable interest and worthy of research, and there are promising and novel technology options which, with appropriate support, could yield real results in the coming years.

Methanation is a process that is used for the generation of SNG that can be fed directly into the gas grid. It can use either a classical chemical process or more recently developed approaches using biological catalysts which, for example, feed CO_2 and

[18] See: http://www.itm-power.com/.

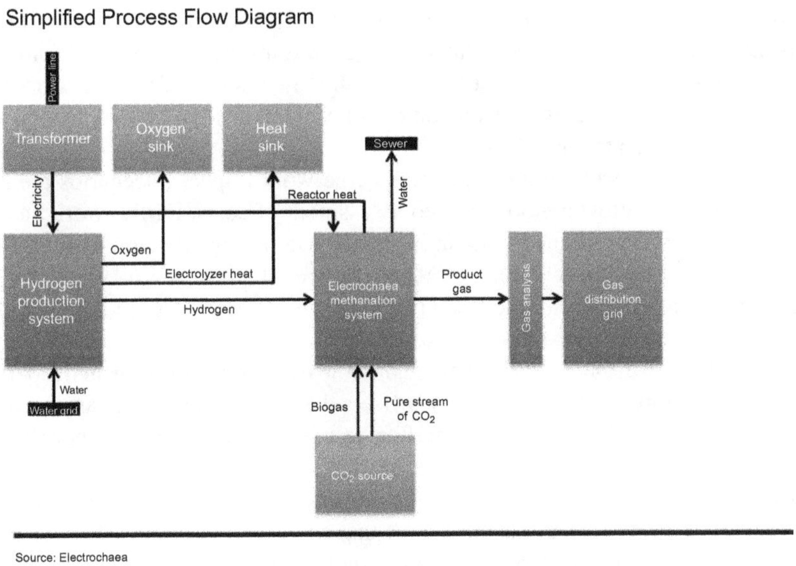

Figure 8.5 Proof of concept demonstration (Foulum, Denmark).

hydrogen to methane-producing microbacteria (archaea) in a biological methanation process called methanogenesis. The necessary reaction is:

$$CO_2 + 4H_2 \rightarrow CH_4$$

A typical system for $CO_2 + 2H_2O$ methanation, with options for storage of both CO_2 and renewable energy, uses well-established and reliable technologies (electrolysis of water, methanation reactors, fuel cells, etc.). The technology is not yet completely mature and improvements in efficiency need to be made.

However, a breakthrough methanation-based application is being built by the Audi Group, where CO_2 from biomethane is combined with renewable hydrogen to produce SNG or “e-gas.” This will be injected in the grid in sufficient quantities to fuel a new range of commercially available Audi TCNG A3 vehicles (http://www.volkswagenag.com/content/vwcorp/info_center/en/themes/2012/12/audi_e_gas_plant.html).

The biological catalyst (Electrochaea) approach has shown promise in the laboratory and is of considerable interest. A small scale, proof-of-concept demonstration is in progress in Foulum, Denmark (Figure 8.5), with plans for a purpose-designed and scaled-up system using a more realistically sized 4 MW electrolyzer.

8.10 Economic considerations

Solving technical challenges is of little use if the solution cannot be implemented economically. The full business case for P2G is yet to be written, as it will depend to a certain extent on the costs of mitigation of the network issues related to hydrogen injection, as well as the capital and operating cost of plant which will be used to enable its introduction. However, a number of organizations are already developing P2G demonstrations at a scale of several hundred cubic meters of hydrogen per hour into local gas distribution systems. The outcome of these demonstrations will help to build a business as well as a technical case for P2G.

Based on current levels of understanding of network impacts, scenarios are being considered that will allow a more detailed assessment of costs. Earlier work has indicated that using P2G combined with an increased use of cogeneration from gas-fired plant could enable carbon reduction targets to be met at a cost of €6–30 per tonne of CO_2 abatement, as opposed to the €120 per tonne that would be required using reinforcement of electricity transmission systems.

It is also the case that, although methanation has higher upfront and operating costs, requiring more energy input, it also offsets the costs of mitigating any hydrogen impacts on the network and, through carbon re-use, would potentially benefit from higher emission credits.

A significant barrier that remains in achieving economically viable P2G implementation is regulation. At present the only value that can be gained from P2G technology is the production of a gas with a resale value, such as hydrogen or SNG. Much of the real value to the energy system rests elsewhere—in the avoided capital cost of the extra infrastructure, in the enabling of maximum utilization of renewable electricity and in the increase in renewable content of the gas networks. All these need to be assigned a value through appropriate regulation so that P2G developers have the confidence to implement market-ready solutions.

8.11 Regulatory issues

Feeding hydrogen into the natural gas infrastructure will, inevitably, result in cross-border transport of gas mixtures with a knock-on effect on gas quality parameters. To date, gas quality requirements for use in the natural gas grid have been regulated neither by international nor European standardization bodies. ISO 13686 "Natural gas—Quality designation" contains a general description of different parameters rather than concrete parameters in the sense of thresholds, limits, or ranges.

The ISO 13686 standard refers gas quality parameters, on an informal basis, to the Common Business Practice (CBP 2005-001/02), released by EASEE-gas[19] in 2005.[20] The CBP has the title "Harmonisation of Gas Qualities" and recommends natural gas quality specifications, parameters, and parameter ranges to streamline interoperability at cross-border points in Europe. Hydrogen concentration limits are not specified in particular, but only "insignificant levels of hydrogen" are tolerated.

As the CBP has not been issued by an authorized body or organization, it is not legally binding for the transportation of natural gas, unless agreed upon under

[19] European Association for the Streamlining of Energy Exchange-gas.

[20] http://easee-gas.eu/cbps.

private law agreements. It is therefore necessary to pay attention to the national standards and legislation for gas quality. An overview of applicable hydrogen concentration limits in Europe may be derived from research conducted by AFNOR[21] on behalf of the European Committee for Standardization (CEN) in 2011 (CEN European Standards Committee, 2011) or an investigation by the GASQUAL project (2010).

National standards for hydrogen concentration in the transmission network seem to exist in only a few European countries. This means that feed-in of hydrogen is not prohibited although the concentration may be limited by other gas parameters, such as Wobbe index or MN.

Work is in hand by the "HIPS-NET" network to collate information on applicable national standards or national regulations and this will be available during 2014.

At the market level, the current situation where Distribution Grid Operators are required to pay curtailment penalties to renewable generators who have preferential access to grids can be changed by new approaches to regulation of the integrated energy market. However, this requires the technical tools that allow balancing and flexibility mechanisms, and other grid services to be tested and available. P2G is one of these tools, and the business case starts to emerge more clearly when the avoided costs of grid reinforcement are incorporated into appropriate market codes, but will require favorable regulation to take off.

8.12 Practical recommendations for hydrogen injection

1. Work to date has shown that a case-by-case analysis is necessary before injecting hydrogen into the natural gas network.
2. For the time being, porous rock underground gas storage is a "show stopper."
3. Most gas chromatographs will require modification.
4. It is recommended that manufacturers' specifications should always be followed, particularly when gas turbines or gas engines are connected to the network.
5. On the basis that much of the natural gas system can tolerate admixture of up to 10% by volume of hydrogen, depending on the specific local situation, the following maximum hydrogen concentrations are recommended:
   - 2%—if a CNG filling station is connected;
   - 5%—if no filling station, no gas turbines and no gas engines with a hydrogen specification <5% are connected;
   - 10%—if no filling station, no gas turbines and no gas engines with a hydrogen specification <10% are connected.

   N.B. For both 5% and 10%, care should be taken to ensure that the Wobbe index and MN of the natural gas/hydrogen mixture are not close to the existing limit values for the network ("safety margins" for Wobbe index and MN).
6. Injection of hydrogen should be carefully controlled to avoid sudden increases of the hydrogen concentration in the natural gas (e.g., speed of change <2%/min).

[21] http://www.afnor.org.

8.13 Conclusions

It's clear that P2G, in both its forms, hydrogen production and methanation, can help to develop workable solutions to some of the problems of renewables integration which have resulted from a sometimes haphazard and less than integrated approach to planning future energy networks.

The proposed P2G solutions clearly fall into the category of "win–win":

- for regulators who need to develop market mechanisms that discourage curtailment of renewable generation and a level playing field.
- for the public who will benefit from a greener and more secure supply of energy, from reduced levels of CO_2 in the environment and from reduced need for additional electricity infrastructure;
- for gas companies who can earn revenue from transporting and storing the gas while, at the same time, "greening" their operations and image;
- for policy makers who will be able to use P2G to help meet their own demanding targets for CO_2 reduction, security of supply and competitiveness.

P2G is not being developed in isolation from other advances in the energy system. The energy landscape is being transformed as we move to a low-carbon future. Electrolyzers are not only being developed for P2G; a future incorporating renewable hydrogen and fuel cells depends on their effectiveness.

Renewable gas is already being injected into gas networks from a variety of sources. In the context of a roadmap to the future, P2G could be seen as occupying a pivotal position. It both enables and benefits from the energy transition (Energiewende) to provide a future based on an increasing renewable content in the energy supply. It's clear that whichever P2G approach is adopted, Europe can only benefit from using the enormous storage capacity of the European natural gas system.

A considerable amount of R&D will be necessary to make this work. The business case is far from being clear, and will emerge from close cooperation between regulators and industry and the R&D community. In particular, the utilization and turn-down capabilities of electrolyzers need to be addressed for a development angle. However, it is clear that many actors believe that P2G can provide a significant storage solution and have already begun to act.

8.14 Sources of further information

Several technology platforms/programs have been set up to monitor/encourage/promote the notion of P2G:

1. "HIPS-NET" led by DBI-GUT.
2. "North Sea Power to Gas platform" led by DNV-GL (http://www.northseapowertogas.com/).
3. The HYREADY Joint Industry Programme: "Recommended practices for preparing natural gas networks for H_2 injection" led by DNV-GL.

Recommended further reading

1. Power-to-gas
   - Linke, G., Steiner, K., Hoppe, M., Mlaker, H., E.ON Ruhrgas, May 2011. Power Storage in Smart Natural Gas Grids: Fiction or Fact? EGATEC, Copenhagen.
   - Altfeld, K., Schley, P., E.ON Ruhrgas, 2011. Development of natural gas qualities in Europe. GWF International.
   - Judd, R., Pinchbeck, D., 2013. Power to gas research roadmap. Offering a solution to the energy storage problem? Gas for Energy (2).
2. Underground storage
   - Schulze, K., 2013. Hydrogen and bacteria—the big unknown. In: 4th RWE Dea Research and Development Day, Hamburg 16th May 2013, Wietze Lab.
   - Pichler, M., 2013. Assessment of hydrogen-rock interactions during geological storage of CH_4–H_2 mixtures. Master thesis, Department Mineral Resources & Petroleum Engineering Chair of Reservoir Engineering.
   - Garcia, J.-L., 1998. Les bactéries méthanogénèses—The Methanogenic Archaea. C. R. Acad. Agric. Fr. 84, 23–33.
   - Wagner, M., Ballerstedt, H., 2013. Influence of bio-methane and hydrogen on the microbiology of underground gas storage/Einfluss von Biogas und Wasserstoff auf die Mikrobiologie in Untertagegasspeichern. Literature Study/Literaturstudie (DGMK-Project 756)/DGMK-Projekt 756, ISBN 978-3-941721-36-4.[22]
3. CNG tanks
   - UN ECE Regulation No. 110: Uniform provisions concerning the approval of:
     (i) Specific components of motor vehicles using CNG in their propulsion system.
     (ii) Vehicles with regard to the installation of specific components of an approved type for the use of CNG in their propulsion system.
   - Barthélémy, H., 2007. Compatibility of metallic materials with hydrogen. Review of the present knowledge. In: International Conference on Hydrogen Safety, San Sebastián, Spain, September.
   - EU-Project, "Integrated Gas Powertrain," Final report, 2012. http://www.ingas-eu.org.
4. Gas engines
   - Gillingham, K. Hydrogen Internal Combustion Engine Vehicles: A Prudent Intermediate Step or a Step in the Wrong Direction? Stanford University Department of Management Science & Engineering.
   - Orhan Akansua, S., Dulgerb, Z., Kahramana, N., Nejat Vezirogluc, T., 2004. Internal combustion engines fuelled by natural gas-hydrogen mixtures. International Journal of Hydrogen Energy 29.
   - Klell, M., Eichsleder, H., Sartory, M., 2012. Mixtures of hydrogen and methane in the internal combustion engine—synergies, potential and regulations. International Journal of Hydrogen Energy 37.
   - DIN 51624; Kraftstoffe für Kraftfahrzeuge—Erdgas—Anforderungen und Prüfverfahren Automotive fuels—Compressed natural gas—Requirements and test methods.
5. Gas turbines
   - Andersson, M., Larfeldt, J., Larsson, A., 2013. Co-firing with hydrogen in industrial gas turbines. SGC Rapport.

[22] Available from: www.dgmk.de/upstream/agber_untertagespeichert.html.

- Marchmont, C., Florjancic, S., 2012. Alstom gas turbine technology trends. In: Proceedings of ASME Turbo Expo, 2012 GT2012 June 11–15, Copenhagen, Denmark; Switzerland.
- Wu, S.J., Brown, P., Diakunchak, I., Gulati, A., 2007. Siemens Power Generation, Inc., Orlando, USA; Lenze, M., Koestlin, B., Siemens Power Generation Inc., Muelheim, Germany; Advanced gas turbine combustion system development for high hydrogen fuels. In: Proceedings of GT2007 ASME Turbo Expo 2007: Power for Land, Sea and Air; May 4–17, 2007, Montreal, Canada.

6. Gas burners
- Nitschke-Kowsky, P., Wessing, W.E., E.ON Ruhrgas. 2012. Impact of hydrogen admixture on installed gas appliances. World Gas Conference (WGC), Kuala Lumpur.
- EN 437 Test gases. Test pressures. Appliance categories.
- Huang, Z., et al., 2006. Measurements of laminar burning velocities for natural gas-hydrogen-air mixtures. Combustion and Flame 146.
- Di Sarli, V., Di Benedetto, A., 2007. Laminar burning velocity of hydrogen-methane-air/ premixed flames. International Journal of Hydrogen Energy 32.
- Ayoub, M., et al., 2012. An experimental study of mild flameless combustion of methane/ hydrogen mixtures. International Journal of Hydrogen Energy, 37.
- Levinsky, H.B., 2004. Consequences of "new" gases for the behaviour of gas utilization equipment. International Gas Research Conference, Vancouver.
- Hermanns, R.T.E., 2007. Laminar burning velocities of methane–hydrogen–air mixtures. Technische Universität Eindhoven, Diss.
- Hu, E., Huang, Z., He, J., Jin, C., Zheng, J., 2009. Experimental and numerical study on laminar burning characteristics of premixed methane–hydrogen–air flames. International Journal of Hydrogen Energy 34 (11), 4876–4888.
- Huang, Z., Zhang, Y., Zeng, K., Liu, B., Wang, Q., Jiang, D., 2006. Measurements of laminar burning velocities for natural gas–hydrogen–air mixtures. Combustion and Flame 146 (1–2), 302–311.
- Miao, H., Jiao, Q., Huang, Z., Jiang, D., 2008. Effect of initial pressure on laminar combustion characteristics of hydrogen-enriched natural gas. International Journal of Hydrogen Energy 33 (14), 3876–3885.
- Brower, M., Petersen, E., Metcalfe, W., Curran, H.J., Füri, M., Bourque, G., Aluri, N., Güthe, F., 2012. Ignition delay time and laminar flame speed calculations for natural gas/ hydrogen blends at elevated pressures. In: ASME Paper GT2012-69310, Proceedings of ASME Turbo Expo 2012, June 11–15, Copenhagen, Denmark.

7. Safety
- De Vries, H., Florisson, O., Tiekstra, G.C., Safe operation of natural gas appliances fuelled with hydrogen/natural gas mixtures (progress obtained in the NaturalHy-project)., International Conference on Hydrogen Safety (ICHS 2007), San Sebastian, Spain, September.
- Jurdik, E., Darmeveil, H., Levinsky, H.B., 2004. Flashback in mixtures of molecular hydrogen and natural gas. International Gas Research Conference, Vancouver.
- Cammarota, F., Benedetto, A.D., Sarli, V.D., Salzano, E., Russo, G., 2009. Combined effects of initial pressure and turbulence on explosions of hydrogen-enriched methane/ air mixtures. Journal of Loss Prevention in the Process Industries 22 (5), 607–613.
- Lowesmith, B.J., Hankinson, G., 2012. Large scale high pressure jet fires involving natural gas and natural gas/hydrogen mixtures. Process Safety and Environmental Protection 90 (2).

See also the following:

- DVGW project G1-02-12 "Information and data collection for hydrogen tolerance of the gas grid."
- DVGW project G7-02-11 "Analysis of the hydrogen tolerance of CNG vehicle tanks."
- DVGW project G106-10: "Gas quality changes: consequences for industrial combustion processes."
- DVGW project G3-02-12: Influence of hydrogen on the energy metering and billing.
- International Energy Agency (IEA)—Technology Roadmap on Hydrogen—in preparation 2013–14.
- Iskov, H., Rasmussen, N.B., 2013. Global screening of projects and technologies for Power-to-Gas (P2G) and Bio-SNG, Danish Gas Technology Centre (DGC), November.

Acknowledgments

The authors would like to thank GERG,[23] the HIPS (Altfeld and Pinchbeck, 2013) project partners, the "HIPS-NET" consortium, DGMK, Germany, DBI-GUT, Germany, and Enbridge, Inc., Canada for their assistance and support.

References

Abbott, D.J., Bowers, J.P., James, S.R., 2012. The impact of natural gas composition variations on the operation of gas turbines for power generation. In: 6th International Gas Turbine Conference, 17–18 October 2012, Brussels, Belgium.

Altfeld, K., Pinchbeck, D., 2013. Admissible Hydrogen Concentrations in Natural Gas Systems. Gas for Energy 03/2013.

Bonnaud, E., 2012. Hétérogénéités compositionnelles dans les réservoirs de gaz acides. Compréhension et modélisation du rôle d'un aquifère actif. Ph.D. thesis, École nationale supérieure des mines de Paris—Spécialité "Hydrologie et Hydrogéologie quantitatives, June 29.

CEN European Standards Committee 2011. Gases from non-conventional sources—Injection into natural gas grids—Requirements and recommendations (CEN/TC 234 WG 9N 54:201) ns1.

DGMK Research Report 752—Influence of hydrogen on underground gas storage. Literature study, 2014.

Electrochaea (www.electrochaea.com).

EN 60079–20-1:2010. Explosive Atmospheres Part 20–1: Material Characteristics for gas and vapour classification—Test methods and data, Section 5.2.6 Methane, Group IIA.

Florisson, O., et al., 2010. NaturalHy—preparing for the hydrogen economy by using the existing natural gas system as a catalyst: an integrated project. Final Publishable Activity Report http://www.naturalhy.net/.

GASQUAL, 2010. Investigations on new acceptable EU limits for gas quality. Influence on the performance of new and installed gas appliances (Phase 1 of Mandate M/400).

[23] The European Gas Research Group (www.gerg.eu).

http://www.volkswagenag.com/content/vwcorp/info_center/en/themes/2012/12/audi_e_gas_plant.html.
Hydrogenics, NEL Hydrogen, ITM-Power, H-Tec Systems, etc.
ISO Standard 11439: Gas cylinders—High pressure cylinders for the on-board storage of natural gas as a fuel for automotive vehicles.
Maurer, O., 1992. Etude de la distribution des espèces soufrées et de la formation de l'hydrogène sulfuré dans les stockages de gaz naturel en aquifère, ENPC thesis, décembre.
National Centre for Hydrogen, Puertollano (Spain).
Private communication with Dr. D. Teichroeb, Enbridge, Inc., Ontario, Canada.

Part Three

Hydrogen safety

Hydrogen safety: An overview

9

R. Wurster
Ludwig-Bölkow-Systemtechnik GmbH, Ottobrunn, Germany

9.1 Introduction

In this chapter, hydrogen properties are described in comparison to other similar energy carriers or fuels. The implications of these hydrogen properties—for example, the interaction of hydrogen with other materials—and the resulting behavior and consequences for the safe use of hydrogen are addressed.

The potential hazards of hydrogen use in storage systems and their periphery are also presented and solutions are outlined as to how the risks can be minimized, mitigated, or avoided. The use of stored hydrogen in vehicles (road vehicles) and potential risks resulting therefrom for the vehicle and its users are outlined, and potential risks of such a vehicle for its surrounding environment are described. Similar descriptions are provided for the use of hydrogen storage systems in other vehicles (nonroad vehicles), trains, aircrafts, ships, and stationary applications.

A discussion follows of the safe design of hydrogen storage systems and their periphery as required for their systems integration. The role of appropriate regulations and standards is highlighted.

Concepts for accident prevention are explained, as well as how these concepts are, or can be, implemented.

Future trends, in particular for hydrogen use in road vehicles, are outlined, as hydrogen use in newer applications, other than the traditional industrial uses, is farthest advanced in this area.

The relevant sources for further information on the subject matter are provided. In the references section all sources used and/or quoted are listed. Sources from which whole passages have been quoted in the chapter are highlighted separately.

9.2 Properties of hydrogen and their implications for safety

In Table 9.1, the main properties of hydrogen relevant to safety are shown in comparison to other fuels like methane, propane, and gasoline.

Hydrogen not only is lightweight, but it also mixes with air so that an ignitable mixture has a short residence time in which it is located near the ground ignition sources nearby. In this aspect, hydrogen is different from gasoline and liquefied petroleum gas (LPG—propane/butane), both of which are heavier than air and linger at the outlet at the bottom. Due to their longer residence time, a higher probability of ignition exists.

Compendium of Hydrogen Energy. http://dx.doi.org/10.1016/B978-1-78242-364-5.00009-9

Table 9.1 **Safety-relevant characteristics of hydrogen in comparison (Alcock et al., 2001)**

Fuel properties	Hydrogen	Methane	Propane	Gasoline
Flammability limits (vol.% in air)				
Lower limit (LFL)	4	5.3	2.1	1
Upper limit (UFL)	75	15	9.5	7.8
Minimum ignition energy (mJ)	0.02	0.29	0.26	0.24
Autoignition temperatures (°C)				
Minimum	585	540	487	228–471
Heated air jet (0.4 cm diameter)	670	1220	885	1040
Nichrome wire	750	1220	1050	
Adiabatic flame temperature in air (K)	2318	2158	2198	2470
% Thermal energy radiated from flame to surroundings*	5–10	10–33	10–50	10–50
Quenching gap at NTP (mm)	0.6	2	2	2
Detonability limits (vol. % in air)				
Lower detonation limit (LDL)	11–18	6.3	3.1	1.1
Upper detonation limit (UDL)	59	13.5	7	3.3
Maximum burning velocity (m/s)	3.46	0.43	0.47	0.42
Concentration at maximum (vol.%)	42.5	10.2	4.3	1.8
Burning velocity at stoichiometric (m/s)	2.37	0.42	0.46	
Concentration at stoichiometric (vol.%)	29.5	9.5	4.1	
Flow parameters				
Diffusion coefficient in air at NTPa (cm^2/s)	0.61	0.61	0.12	
Viscosity at NTP (g/cm-s $\times 10^{-5}$)	89	11.7	80	
Density at NTP (kg/m^3)	0.0838	0.6512	1.870	
Ratio of specific heats, Cp/Cv at NTP	1.308	1.383	1.14	

*The energy radiated to the surrounding to a wide extent depends on the carbon content of the fuel and the carbon black created by combustion of the fuel.
Thus hydrogen radiates only about 1/3–1/5 compared to other fuels and in this aspect is much less dangerous regarding heat impact.

The behavior of hydrogen is different from hydrocarbons:

- wide limits of flammability and detonability;
- ignition and detonation energies low;
- nonluminous flame;
- very buoyant and diffusive—flammable cloud disperses rapidly.

For safety analysis, one needs to be able to estimate the consequences of a release:

- How likely to ignite?
- How likely to detonate vs. deflagrate?

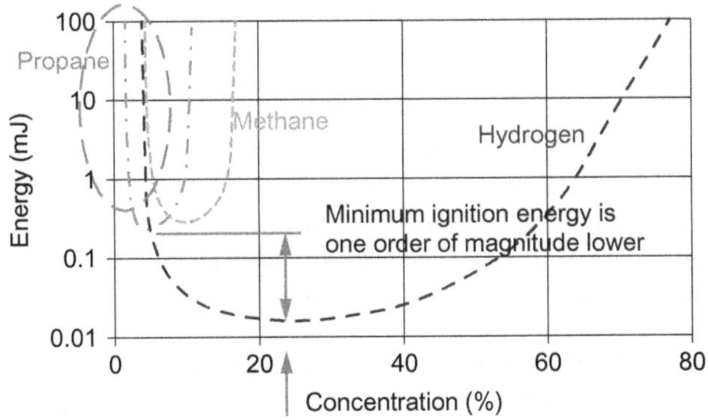

Figure 9.1 Comparison of ignition energy vs. concentration for hydrogen, methane, and propane.

- What overpressures would be generated?
- What level of injury/structural damage/escalation is possible?

Near the lower explosion limits (LEL), the differences in the applicable ignition energies between the three gases compared (propane, methane, and hydrogen) are not significant. The energy of most ignition sources is higher than these limits at any rate. The theoretical physical minimum ignition energy of hydrogen is one order of magnitude lower than that of propane or methane and is reached at comparatively high concentrations of hydrogen in air of above 20% (see Figure 9.1). As the lower flammability limit of hydrogen with 4% hydrogen in air is only slightly lower than that of methane of 5.3%, and at least double that of propane (2.1%) or gasoline (1%), hydrogen air mixtures can be ignited by most ignition sources at this concentration level. Consequently, acceptable ignition or explosion prevention measures are required for both hydrogen and methane and should not be based on minimum ignition energies but rather on limiting concentration levels or avoiding ignition sources as much as possible.

9.3 Hazards of hydrogen

9.3.1 General consideration of hydrogen hazards

9.3.1.1 Explosion

Definition of an explosion (reproduced here from Schmidtchen, 2009):

> *For the start let us define a few terms which are important and are frequently mixed up by people who do not have full understanding of the matter.*
> *An explosion is a self-sustained combustion of a gas mixture (flammable plus oxidizing gas, e.g., oxygen or chlorine) which releases energy as heat and a shock wave. (Dust or solid explosions are not discussed here, only gas explosions.) The process can be further subdivided as follows:*
>
> - *A deflagration is distinguished by the feature that reaction and flame-front velocity are below the velocity of sound.*

  - *The characteristic features of a slow deflagration are laminar flow and a speed of the flame front much lower than the velocity of sound.*
  - *The characteristic features of a fast deflagration are turbulent flow and a speed of the flame front of the order of the velocity of sound; the effects of a fast deflagration may be rather similar to those of a detonation.*
- *A detonation spreads with a velocity much higher than the velocity of sound; the shock wave is much higher than in the case of a deflagration; it is also much sharper (higher rate of pressure rise), and the damage resulting from a detonation is usually more severe than in the case of a deflagration.*

Deflagration and detonation are the two main subtypes of an explosion.

In common language, including that of the media, this distinction is usually unknown. Any type of explosion is called "detonation," and even purely physical processes like the burst of a pressure vessel due to excess pressure, which is not even an explosion, get this label as long as they are accompanied by a loud and sudden noise.

In order to facilitate an explosion several conditions have to be fulfilled:

- A mixture of a flammable gas and an oxidizing gas has to be formed.
- A concentration of the flammable gas in this mixture has to be above the LEL which is 4% for hydrogen in air and below the upper explosion limit (UEL) which is 75% for hydrogen in air.
- Availability of either the ignition energy which can trigger an explosion or a temperature in the gas mixture which is equivalent to the auto ignition temperature which generates sufficient gas molecule movement in order to start the explosion reaction without infusion of an external ignition energy.

In order to facilitate a detonation further additional boundary conditions have to be fulfilled (Schmidtchen, 2009; ISO 15916):

- The detonation range for hydrogen is narrower than its explosion range and is characterized by a lower detonation limit (LDL) of 11% (mainly due to the influence of semi-confined spatial settings; see Alcock et al., 2001), compared to that usually quoted, which is 18% and an upper detonation limit (UDL) of 59%.
- Whether a detonation is initiated not only depends on the mixture of hydrogen in air but also on the pressure, temperature and possible presence of diluents.
- In open, completely unrestricted geometrical settings a gas mixture cloud is very unlikely to detonate, while in semiconfined areas (e.g., a floor and two walls, each at an angle under 90° connected to each other) or confined areas (e.g., a room or a long tube) faster deflagrations under the presence of obstacles might easily develop nonlaminar/turbulent flows and thus result in detonations (DDT=deflagration-detonation-transition); consequently at a given flame acceleration level the blockage ratio also has an influence on the initiation and the magnitude of a detonation.

Hydrogen itself consequently is not "explosive" at all. Any explosion with an involvement of hydrogen requires any of these previously described ingredients or boundary conditions in order to trigger an explosion (in this aspect it is different from "explosives" which do not require an oxidizing agent or from acetylene which is an unstable gas that can decompose into hydrogen and carbon and thus trigger explosions).

As long as no access is permitted of an oxidizing agent (e.g., oxygen) to a flammable gas (except acetylene), this gas can safely be stored under high pressures. Even the presence of ignition energy will not cause any ignition or explosion as no oxidizing agent is available. This is a safety advantage of compressed gases like hydrogen in comparison to liquid fuels like gasoline or kerosene which, as the tank is emptied, are replaced by air and thus an oxidizing agent (due to this characteristic, modern aviation tanks have to be inertized).

9.3.1.2 Invisible flame

Usually the hydrogen flame is almost invisible in bright daylight due to its emission wavelength near the ultraviolet range (311 nm). Detection devices sensitive to this wavelength range can detect hydrogen fires quite reliably and are used by fire brigades or in chemical industry installations. In some cases hydrogen fires are visible when particles of metallic materials of, for example, orifices, etc., are available in the hydrogen flame.

9.3.1.3 Instantaneous ignition

In some cases instantaneous ignition events have been perceived which were thought to be caused by "auto ignition." None of these events could be reproduced reliably in order to determine the real cause or causes of ignition (such as electrostatic charges, diffusion ignition, hot surface ignition, sudden adiabatic compression, inverse Joule–Thomson effect). The most likely explanation lies with electrostatic charges, but further analytical and testing efforts are needed.

9.3.1.4 Embrittlement

Embrittlement is an effect usually observed with metallic materials (metallic lattice) when exposed to hydrogen atoms which can be dissociated from HCl, HCN, H_2S, or H_2. Mainly, the effect occurs with ferritic steels or metallic materials of similar crystallographic structure (body-centered cubic layout) like chromium or molybdenum. Less vulnerable are face-centered cubic lattices, as typical for austenitic steel, aluminum, nickel, or copper. Nonetheless, embrittlement is a scientific and technical issue which is being dealt with by proper choice of suitable materials.

Hydrogen embrittlement usually occurs when a metal surface is exposed to ionized hydrogen and when the hydrogen migrates faster into the construction material than it recombines with molecules still on the surface. In the case of hydrogen migration into the metal lattice, part of the hydrogen is integrated into the lattice. It can either react to hydrides or settle on defective areas or grain boundaries. In both cases, embrittlement of the metal will result. Other preferred locations for such reactions are tips of cracks or other locations of high tension, where hydrogen weakens the cohesion of the metal lattice. Under static or cyclic tensile strengths, cracks can be induced, and for incipient cracks, crack propagation can be induced (accelerated crack propagation). For the possibly elapsing damaging process/effect, the dissociation of molecular hydrogen adsorbed at the metallic surface is of key importance as it requires a chemisorption process that can occur only at a clean surface. Such surfaces can be created by plastic deformation. In practice this means that dissociation of hydrogen on metal surfaces can occur only under cyclic loads leading to plastic deformation at indents or tips of cracks. The probability of such damages is influenced by the type and magnitude of the pulsating stress, frequency, surface roughness, hydrogen pressure, temperature and strength of the steel (Behrens, 1986; Wurster and Schmidtchen, 2011).

When designing safe hydrogen systems, the risk of embrittlement has to be taken into account. This is achieved by the selection of construction materials that are suitable for the expected loads. Furthermore, the operating conditions have to be considered. If certain loads on the material are not to be exceeded, a less strong construction material may be sufficient. Construction materials have to be chosen in compliance with the requirements to be fulfilled.

Organic materials, such as composite materials used for parts of high-pressure compressed storage tanks, are not susceptible to hydrogen interaction and thus embrittlement (Wurster and Schmidtchen, 2011).

9.3.1.5 High permeation rate

Hydrogen over time can migrate through almost all materials. Contrary to what is sometimes stated—that hydrogen cannot be retained in pressurized steel or composite material vessels—this is not correct. Hydrogen has been enclosed safely in pressure steel vessels for almost seven decades (Wurster and Schmidtchen, 2011). The rate of diffusion for metallic vessels or liners for practical conditions of use is insignificant. Even for the materials having higher permeation rates, as applicable to full composite-material pressurized storage containers, hydrogen losses may occur only after many months or years and do not pose any safety issues in everyday use when the safety concept takes this into account. Duroplastic liner materials can achieve diffusion rates of, for example, aluminum liners, which will lead to a complete emptying of the pressure container in about a century—thus irrelevant for automotive or other daily stationary storage uses.

9.3.1.6 Risk

The risk associated with a certain event has been defined by experts as follows:

$$\text{Risk} = \left(\text{probability of an event}\right) \times \left(\begin{array}{l}\text{foreseeable damage caused by event} = \\ \text{consequence}\end{array}\right),$$

The potential magnitude of a risk is determined by the damage caused by an event/incident (the consequence) as well as by the probability of its occurrence. The acceptable risk level depends fundamentally on value judgments that cannot be standardized or quantified (see Table 9.2).

This means that societal conventions and comparative experience with the occurrence of events and their consequences (i.e., perception) play a significant role in determining an acceptable risk level. Also, the question whether a technology can be replaced by another similar or equally beneficial one can determine the level of acceptance of risk. Airplane travel, for example, may not be easily replaceable, whereas in fuel use some flexibility may exist, as well as in fuel storage technologies.

Table 9.2 **Risk matrix and types of risk levels in the risk matrix for acceptance criteria proposed by EIHP (2003)**

Consequence	Improbable	Remote	Occasional	Probable
Likelihood				
Extremely severe damage	H	H	H	H
Severe damage	M	H	H	H
Damage	M	M	H	H
Small damage	L	L	M	H
Minor damage	L	L	L	M
Risk level	***Description***			
H (high)	Risk is not tolerable. Remedial actions should be considered to reduce risk to a tolerable level			
M (medium)	In principle, risk cannot be tolerable. It can be accepted only when risk reduction cannot be achieved by reasonably practical action			
L (low)	Tolerable. Further risk reduction is not necessarily required			

9.3.2 Hydrogen hazards in storage

Hydrogen can be stored in different modes: as compressed gas, as absorbed or adsorbed gas or as cryogenic liquid gas. The widest application in industry is compressed gaseous storage and bulk cryogenic (−253 °C) liquid storage. Compressed hydrogen can be stored in steel containers, as practiced for more than a century, and in fiber composite-material vessels as they are typically used in cars, buses, truck trailer transport or stationary onsite storage at hydrogen refueling stations (HRSs), industrial or commercial sites. Typical pressure levels range from 5 MPa, over 20 MPa, 30 or 45 MPa, up to 95 MPa.

For automotive uses, the industry has committed itself to the use of compressed storage systems, usually in partially or fully composite material storage (typically type III, metallic liners and fully fiber wrapped, and increasingly type IV technology, plastic liners and fully fiber wrapped) at onboard filling pressures of 35 or 70 MPa.

Also, in principle, storage under very low overpressure is possible (0.0022 MPa) in gasometers, as they are known, for the storage of city gas.

Very large quantities of hydrogen have been stored by the chemical industry for decades in underground salt cavern storage facilities of between 70,000 and 600,000 m^3 geometrical volume under pressures of up to 15 or 20 MPa.

Smaller amounts of hydrogen can be stored in metal hydride storage systems in which hydrogen is stored in a chemical compound formed by hydrogen and metals or metal alloys. Decade-long experience exists for this type of hydrogen storage. Further methods, as, for example, the storage of hydrogen in complex hydrides or the physisorption of (or physical adsorption) of hydrogen molecules in porous materials, are limited to niche applications or are still a matter of basic research.

Today, hydrogen is used mainly in industry and stored onsite for commercial/industrial uses in nonpublicly accessible areas. Typical modes of hydrogen storage are pressurized gaseous storage in cylinders or tubes made of steel or aluminum with fiber-wrapped composite materials or full fiber composite materials; liquid storage in cryogenic multilayer super-insulated spherical or cylindrical storages made of stainless steel or aluminum; or in very large quantities in salt underground caverns (e.g., in the UK and USA). Hydrogen is also transported in truck trailers from production sites to customer sites. Modes of transport are steel tube trailers, steel bundle trailers or composite materials bundle trailers, the latter ones recently with pressures of up to 50 MPa.

Of growing interest is the use of hydrogen in refueling stations, onsite storage of remote telecommunication back-up power systems, and in power-to-gas power plants converting renewable electricity via electrolysis into hydrogen for further use as vehicle fuel, for direct injection into the natural gas grid or for methanation and subsequent injection into the natural gas grid. For these uses (use as vehicle fuel, for methanation) intermediate storage of hydrogen before end use is required, usually in the form of compressed gas.

In particular, HRSs are publicly accessible areas and therefore will have more stringent safety requirements than those areas where hydrogen is handled by trained personnel, such as in industrial complexes.

For use in HRSs, hydrogen will be handled in highly pressurized storage systems, compressed by multistage compressors, transported in pressurized pipelines, controlled by pressure valves and dispensed via high-pressure hoses and nozzles to vehicle tank receptacles and thus high-pressure vehicle tanks. All these components and their interfaces are prone to potential ruptures, breakages or leakages, caused either by material defects or by operation errors. The pressure levels involved in the refueling of today's state-of-the-art hydrogen vehicles are nominal onboard storage pressures of 70 MPa (at 15 °C) and pressure levels during precooled fast refueling of 88 MPa from conditioning pressure vessels made of full composite materials (type IV) of up to 95 MPa. The layout described is that of a typical German H_2Mobility station set-up (H2M, 2010).

Sudden pressure releases from high-pressure pipes, hoses, or vessels are already dangerous due to the pressure differential in the case of a malfunctioning or an accident. Due to the direct exposure to such high-pressure components (e.g., a hose connecting the dispenser and the nozzle), vehicle users would be affected immediately. For this reason the manufacturers of these components in the EU must respect the requirements laid down, for example in the pressure equipment directive (97/23/EC). An ignition of a released hydrogen cloud would not be necessary in order for damage to be caused.

9.3.3 Hydrogen hazards in vehicles

9.3.3.1 Vessel disintegration

Failure modes are global failure events, such as ruptures or leakage. Failure mechanisms of composite-material pressurized storage containers are different from those of metallic tanks. Failure mechanisms are failures on the material level, such as fiber failure and matrix cracking (Table 9.3). The tank designer has to address all critical

Table 9.3 Components and their failure mechanisms

Inner liner (made of polymer)	Fracture Yield Change of permeability Environmental degradation
Composite laminate	Fiber failure Matrix cracking Delamination Environmental degradation
Outer liner (made of polymer)	Fracture Yield Change of permeability Environmental degradation
Boss (made of metal)	Fracture Yield Corrosion
Interfaces	Adhesive failure Clamping failure Environmental degradation

failure mechanisms and integrate them into the design concepts for the tanks. Then, the tank has to undergo the legally required testing procedures in order to be qualified for road vehicle use.

Echtermeyer and Lasn (2012) has provided an overview of composite-material components and their failure mechanisms, discussed in the following sections.

9.3.3.2 Pipe rupture

Due to hydrogen in ionized form being the smallest atom, it has a high tendency to diffuse or permeate. Such effects can be avoided by the appropriate selection of suitable construction materials. In case of a leak in a pipe or other component, hydrogen will escape at a high flow rate. Therefore, piping and seals applied need to be suitable for hydrogen over the whole life of the system. Wherever leaks in high-pressure hydrogen pipes must be avoided, welded joints are the connection of choice. In order to reduce risks in case of a rupture of a pipe, the supply pressure from a high-pressure vehicle storage tank to, for example, the fuel cell should already be reduced at the tank to a pressure level of 0.2–0.3 MPa in order to minimize the flow rate, and thus the risks.

9.3.3.3 Valve functioning

Tanks and pipes carrying pressurized hydrogen should be protected against overpressurization by installation of pressure-relief devices like valves and burst disks. Typically the impact of a fire or the failure of a pressure regulator can cause overpressure in the system, which has to be relieved by pressure-relief devices and vented outside the system.

9.3.3.4 Crash safe integration

Depending on the tank integration concept, the piping system must be protected against a crash.

9.3.4 Hydrogen hazards in other applications

9.3.4.1 Trailer transport

Some of the potentially largest storage vessels may be used in bulk transport of hydrogen on roads. In Europe today, according to ADR (European Agreement Concerning the International Carriage of Dangerous Goods by Road) (ADR, 2015), the largest permissible single-vessel sizes are 3000L geometrical volume with no pressure limitation. There are plans to extend these sizes to up to 10,000L for type IV composite-material storage tubes. Such tube sizes are already in use for the transport of compressed natural gas at 25MPa pressure level in other world regions (e.g., the Americas, Japan, and Thailand). In case of an incident or accident, the release of the content of such large vessels can constitute the release of a large volume of explosive gas mixture (geometrical volume × pressure/compressibility factor) with a considerable damage potential. In order to implement such storage systems safely, a risk analysis would have to justify that this volume will reduce the overall risk due to the use of less tubing or fewer valves at, for example, a nominal fill pressure exceeding 50MPa, meeting improved (optimized) design requirements, for instance with regards to the burst pressure ratio, compared to today's accepted multicylinder/tube constructions composed of several single cylinders or tubes of between 50 and 3000L each. For tubes with larger diameters, experience already has shown that the stiffness and stability of these tubes in case of accident or exterior impact is larger than for tubes with smaller diameters, due to the greater amount of fibers in the composite required to balance radial by tangential forces.

9.3.4.2 Refueling stations

In addition, for HRSs the major risk event is mainly determined by the total volume of hydrogen stored onsite. Typically this amount is larger for trucking-in stations where CGH_2 or LH_2 are delivered and stored onsite. Onsite production HRSs, for example with an electrolyzer producing hydrogen onsite and storing more moderate hydrogen volumes onsite, can have a reduced risk potential. Assuming an HRS layout as foreseen by the H_2Mobility station approach, the majority of the hydrogen to be stored onsite will be stored in systems of about 45MPa storage pressure and only a comparatively small amount, intended for differential refueling into the vehicle tank, will be stored in very high-pressure 95MPa type IV composite-material storage cylinders. These very high-pressure cylinder systems will have to be laid out for high cycle rates as they may be charged and discharged several times per day. They will consist of the composite-material vessels with the highest cycle rates considered in any use today. Furthermore, all pipes and connections will have to be laid out and manufactured gas tight for high pressures, which is achieved best with welded joints.

9.3.4.3 Material handling vehicles

Hydrogen-powered material handling vehicles are already widely used in North America in several thousand units. These units operate typically with 35 MPa onboard storage systems and undergo fast refueling with compressed hydrogen at typically 44 MPa performed by the operating personnel at least twice a day. The safety record so far is outstanding.

9.3.4.4 Aviation

The consideration of the use of hydrogen in aviation was investigated, tested and conceived in design studies in the late 1980s to the mid-1990s, much more broadly than is being done at present. The advantages of the use of hydrogen in aviation are obvious when it comes to payload. Setbacks have to be considered, however, when it comes to required storage volumes. Safe solutions were identified in the Cryoplane project, which would enable safe operation of aircraft fueled with LH_2. Recently, hydrogen is being considered as a fuel for the operation of onboard fuel cell APUs, for example to be integrated into commercial Airbus aircrafts well after 2020. A hydrogen/fuel cell-operated APU generator would allow for the complete redesign of the aircraft (bleed air avoided, full electric operation, less water to be taken on board, O_2 reduction for onboard use of inert gas for kerosene fuel tank inertization, zero emission operation on the ground, etc.).

9.3.4.5 Back-up power

Small hydrogen back-up power systems for mobile phone communication towers have become a real business case in several countries (in India, for example) in some of which conventional fuels (e.g., diesel) are subject to theft, whereas hydrogen is not subject to theft. Due to industrial standardized bottled hydrogen supply for back-up power systems, the risk in handling these small volumes of hydrogen is moderate, managed professionally as a daily practice by the gas technology companies.

9.3.4.6 Residential use

Several private houses have been operated with hydrogen as energy storage systems in research and demonstration projects over the last several decades in Europe, the USA, and Asia. The application of state-of-the-art safety requirements have shown that these residential hydrogen uses can be operated safely. Safety design criteria have to be followed and safety equipment and devices must be installed in order to ensure safe use of hydrogen systems in homes.

9.3.4.7 Pipeline transport

Globally more than 2100 km of hydrogen pipelines are in operation by several chemical companies in Belgium, Brazil, Canada, France, Germany, the Netherlands, Thailand, the UK, and the USA, of which in Europe alone about 1600 km of hydrogen pipelines are operational. The safety records of these hydrogen pipeline systems operated with pressures of between 2.5 and 10 MPa are excellent and proven in some cases (of hundreds of kilometers) for more than 70 years. In Frankfurt, Germany, an HRS is supplied with chemical by-product hydrogen via a 100 MPa pipeline of 1.7 km length.

9.3.4.8 Safe design

This section considers vehicle design issues and is largely adapted from Wurster and Schmidtchen, 2011:

Safety is an inherent characteristic which must be built into hydrogen systems. A safe system comprises, among other things, redundant safety features (single failure tolerance principle), back-up of critical components, fail-safe position of valves (in system failure setback to safe position), and other safety devices, sensors, automatic control, and emergency systems (ISO/TR 15916:2004, ISO/DTR 15916 Basic considerations for the safety of hydrogen systems).

For hydrogen to be released in an accident in large quantities requires a massive crash, presumably with total loss. Hydrogen pressure tanks are by design much more resistant than conventional fuel tanks for gasoline or LPG. Hydrogen pressure tanks must withstand very high pressures: at 70 MPa for CGH_2, around 3.5 times as much as in a CNG tank with 20 MPa. In case of damage to valves at the high-pressure tank, damage consequences, among others, can be narrowed such that pressure reduction valves reduce the pressure level from 70 MPa inside the tank to the pressure level outside the tank supplied to the consumer, such as a fuel cell.

The art of vehicle design involves anticipating all possible realistic and catchable accident types and considering them in the construction of the hydrogen system itself and its implementation in the vehicle.

To be well prepared for an accident caused by mechanical deformation, various preventive design criteria are followed. The preferred installation location for pressure tanks in vehicles is the subfloor (e.g., the sandwich floor of the Mercedes A- and B-Class) or the space between the two rear wheel houses (away from the rear bumper and the side protected by the wheels) or city bus roof integration (on the one hand, to ensure low-floor characteristics and on the other hand, also as far as possible away from the collision impact region as well as at the highest point, which is also advantageous in case of any leaks). All tanks are equipped with pressure relief devices such as melting fuses or glass bulb devices that are responsive to pressure or temperature, respectively.

The pressure regulator reduces the gas pressure from the storage pressure level (70 MPa) to the pressure level of the gas delivery conduit to the fuel cell or the motor (e.g., in two stages to a level of from 0.2 to 0.3 MPa). This has significant safety advantages: if during an impact the pressure line is pulled from the tank, then only the significantly reduced pressure in the supply line is pending.

Some manufacturers also envisage a housing-like impact protection around the valve or the valve integrated into the pressure vessel. The manufacturers also try to get along with as few tanks as possible, to keep the number of valves and connections as low as possible and the lengths of the pipes as short as possible. The pipe should be welded. The installation space of the hydrogen tank group is vented and sealed gas-tight from the passenger compartment of the vehicle. Furthermore, the vehicle manufacturers arrange for hydrogen sensors at the highest point of the passenger compartment, in the engine compartment and also on the pressure tank group. These hydrogen sensors, if released, provide a signal to the control system which immediately initiates an orderly shutdown of all vehicle functions. The pressure relief line of the tank relief valve is usually installed underneath the car so that in the event of an external fire it will release the hydrogen bleed-off due to the associated increase in pressure and burn it without allowing the formation of an explosive H2/air mixture (this is a result of fire tests).

On various occasions one hears concerns articulated about the safety of hydrogen vehicles in garages. In the case of single garages that are closed spaces and only have certain natural or artificial predetermined ventilation, this concern is understandable—in particular, since hydrogen, when it is released as a gas, spreads rapidly, due to its high diffusion rate, in all directions and rapidly mixes with air.

The safety concept of the car manufacturers therefore provides in principle that the tank and piping systems and valves must be completely sealed and thus tight on board a vehicle. Therefore, such vehicles being technically leak tight can enter also into a garage and be parked without additional measures.

In the event of a fire, it is ensured that the hydrogen heated by the fire, which leads to a volumetric expansion of the gas and thus causes an increase in pressure in the tank, is vented through a relief valve, for example, downwards under the vehicle and burned in the surrounding fire before it could form an explosive mixture.

The amounts of gaseous hydrogen, which due to the inevitable heat input at normal state into a liquid hydrogen tank or which is being caused by the purging process of a fuel cell, are typically converted catalytically to form water. If this hydrogen, however, is directly introduced into the garage, the legally required air exchange rates would be sufficient to prevent the formation of dangerous or explosive mixtures. However, it is up to each private garage operator to establish his own rules. Insurance aspects may play a role here. However, one has the impression that the often-sighted signs "Banned for vehicles with gas drive" just hang there for decades and no one has taken them off.

Proper design, quality control during manufacturing, supervision and control during the installation of these components as well as their recurring inspection are indispensable.

Taking all these considerations into account, nevertheless the fire brigades in the German-speaking countries do not regard hydrogen to be more dangerous than other fuels. One must understand the peculiarities, recognize dangers in advance and respond and adapt in the treatment thereof.

9.4 Management for accident prevention

The main focus of accident prevention with regard to hydrogen storage systems is to avoid the formation of significant flammable gas clouds. A second focus is on limiting overpressures resulting from any possible explosions.

Both aspects can be addressed by keeping confinement as small as possible, thus enhancing the free flow of any gas/air mixtures and their upward diversion exploiting the buoyancy characteristics of hydrogen. Two walls arranged under an angle of 90° connected to a floor may already constitute sufficient confinement, allowing the formation of explosive clouds. Ceilings should be designed to allow the escape of mixture clouds through vent openings. Vent openings and vent areas will facilitate lower explosion pressures due to lower tendency of turbulent flow propagation and flame acceleration. Any geometrical obstruction of gas cloud distribution causes more turbulent and faster flame propagation and thus higher explosion pressures, which should be avoided.

On the other hand, leaking hydrogen should be diverted as fast as possible in an upward direction. High-momentum leaks in the horizontal direction should be avoided by constructive measures, for example vertical walls close to possible leak sources.

Avoiding walls in general, while also aiming to reduce the level of confinement by installing a wall as a flow diverter, results in contradicting designs. In such a case, the proper selection of the appropriate design will depend on the frequency and consequences of the various incident scenarios to be applied. The evaluation of the optimum approach has to be based on methods that take the complexity of the phenomena into consideration, such as, for example, CFD (computational fluid dynamics) modelling and risk assessment.

Regulations, standards, and established training procedures try to ensure safe manufacturing, installation and operation of components, subsystems and thereof configured systems, like an HRS or a vehicle. All components destined for stationary use with pressurized fluids (in this regulation, PED, gases are considered to be fluids also!) are subject to certification by the PED (European Pressure Equipment Directive). Those destined for transport on roads are subject to approval by the TPED (European Transportable Pressure Equipment Directive) and the ADR (European Agreement Concerning the International Carriage of Dangerous Goods by Road). Transport on rail and inland waterways is regulated by RID (European Agreement on the International Carriage of Dangerous Goods by Rail) and AND (European Agreement Concerning the International Carriage of Dangerous Goods by Inland Waterways), respectively. Hydrogen vehicles have to undergo a whole-vehicle type approval in the European Union according to 79/2009/EC (EC, 2009) and 406/2010/EU (EU, 2010) in order to acquire roadworthiness. In the attempt to harmonize vehicle approval requirements globally, the GTR (Global Technical Regulation) (GTR, 2013) has been adopted in Europe (R134, 2015) and will replace the EC regulations. In order to ensure the safe stationary use of not only the equipment installed but the entire system configured from such certified components, the European Union in its proposed Alternative Fuels Infrastructure Directive (AFID, 2013, 2014) plans to reference existing standards harmonizing the requirements for the safe use of such components, for example in a refueling station. Such a standard for the layout, operation and maintenance of an HRS is under development at ISO (International Standardization Organization) in Geneva. As soon as this ISO 19880-1 standard becomes available, it can be referenced and thus be granted regulatory recognition (i.e., mandatory to be observed). An overview of applicable regulations for hydrogen is provided in HyFACTS (2013).

9.5 Future trends

9.5.1 Fail-safe design

In the design phase of a critical hydrogen component such as, for example, a hydrogen storage system, an independent expert organization performs FMECAs (failure modes, effects, and criticality analyses) and also fault tree analyses in order to minimize potential risks due to design failures.

Complex subsystems or components are analyzed by finite element method simulation in order to determine potential design flaws that might materialize in a weak design when exposed to pressure, thermal stress, dynamic loads, or other critical load regimes. Additionally, also according to existing regulatory requirements like 79/2009/EC and 406/2010/EU and UNECE R134, destructive and nondestructive testing will have to be performed on metallic and nonmetallic components for qualification reasons.

Hydrogen components and systems designed according to these criteria, in the case of accidents or system failures will have to be operated in such a way that control electronics switch the system to a safe state of operation.

9.5.2 Global harmonization of interfaces and requirements

Many hydrogen components are traded internationally, some globally. In particular, hydrogen vehicles will be sold into many markets all over the world. Already the lead markets perceived presently (California/USA, Germany, Japan, Korea, Scandinavia, UK) have very different legal requirements and sometimes national standards, combined with differences in local safety philosophies. Therefore, it is of utmost importance for the vehicle manufacturers to design their vehicles in order to comply with all valid requirements in the target markets. The better the requirements for components, subsystems and their interfaces are harmonized, the easier this can be accomplished. Therefore, the harmonization of, for example, refueling interfaces for compressed hydrogen is already accomplished globally (ISO 17268). The requirements for the approval and homologation of hydrogen road vehicles are on their way to being globally harmonized by GTR (2013) as adopted by all signatory states. In order to come to a harmonized set of minimum requirements for the layout, operation and maintenance of HRSs, an ISO standard (ISO 19880-1) is presently under development that will one day allow cross-referencing from regional and national laws as e.g. the AFID and thus will lead to widely harmonized requirements and easier implementation of similar or identical designs all over the world. This facilitates safer systems as fewer differing subsystems, components or interfaces will have to be taken into account. Also the economics of the systems will improve.

9.5.2.1 Better understanding of safety issues

In order to improve the common understanding of hydrogen safety issues among experts and the scientific community, the US and Europe have established a variety of information systems (HyFACTS, 2013):

- The Hydrogen Incident and Accident Database (Hiad-DB) is a European knowledge base and reporting regime to assist industry and authorities in better understanding the relevance of hydrogen-related incidents and accidents as well as the safety actions taken (https://odin.jrc.ec.europa.eu/engineering-databases.jsp).
- H2Incidents is a database-driven website intended to facilitate the sharing of lessons learned and other relevant information gained from actual experiences using and working with hydrogen (http://h2tools.org/lessons).
- Detonation Database aims to compile, catalogue, and present on gaseous detonations including cell width, critical tube diameter, initiation energy, and minimum tube diameter (http://www.galcit.caltech.edu/detn_db/html/db.html).
- The Hydrogen Safety Bibliographic Database provides references to reports, articles, books, and other resources for information on hydrogen safety as it relates to production, storage, distribution, and use (http://nrelpubs.nrel.gov/Webtop/ws/hsdb/www/hydrogen/SearchForm).
- Technical Reference for Hydrogen Compatibility of Materials, Sandia National Laboratories (US) (http://www.sandia.gov/matlsTechRef/).

These databases will help to develop common understanding and approaches for addressing hydrogen safety issues, and will assist in integrating the experience and knowledge of hydrogen safety into daily research, design and engineering. Also the integration and harmonization of the fragmented research base can be facilitated.

9.5.2.2 Information and training of fire brigades and first responders

HRSs: In Germany, before issuing the construction permit for a HRS, authorities require an operation plan (to be established by the fire brigade) and an emergency response plan (to be established by the municipality and the rescue services). As soon as the construction starts, local fire brigades shall be involved in order to establish a specific rescue plan for the HRS. A meeting in which the applicant and the HRS operator present the project to the approval authorities, the approved body, component manufacturers and the building authority can be very helpful in providing information and collecting requirements.

According to existing legal requirements, for any refueling station the fire brigade requires a clearly defined scope of information and the determination of harmonized layout plans, for example. The retailer operating the HRS has to provide this information in agreement with the local authorities. An early involvement of the fire brigades during the application for the approval process can ease the whole permitting process. All necessary information for firefighting is part of the emergency response plan, which is deposited at HRSs or in service workshops. In Berlin, the information systems of all fire brigades dispose of this information for all operational HRSs. All firefighters have undergone corresponding training courses.

Service workshops: Fire brigades are provided with technical insights into the vehicle manufacturers' rescue manuals for each hydrogen vehicle in operation. Furthermore, for each hydrogen vehicle in operation in Germany, as for any other conventional vehicle, rescue cards exist from which it can be derived how and where vehicles can be cut safely without damaging high-voltage lines or high-pressure pipes or tank systems. Based on the rescue manual's explosion zone plans, specification sheets for equipment to be delivered, a firefighting plan or a fire protection regulation, a plan for emergency escape routes, operation manuals and recurring education and training for the staff will be developed (Wurster et al., 2011).

9.5.2.3 Better training, communication and creation of trust and acceptance

It has been almost 1½ decades since the first demonstration project covering vehicle and refueling station infrastructure handled instructing the fire brigades and first responders on how to act in the case of incidents or accidents (AGBF, 2008).

Several higher education programs on hydrogen and fuel cells are offered in Europe in the form of short courses (mainly in Germany), summer schools and hydrogen-related undergraduate and postgraduate programs, single modules or lectures offered by several universities in Denmark, Germany, the Netherlands, Norway, Sweden, and the UK (HyFACTS, 2013).

In the HyFACTS project, training materials for authorities, manufacturers, and operating companies have been prepared and published (http://hyfacts.eu/2014/training-material/).

The project "hyTRUST—on the road to hydrogen-driven society (2009–2013)" (hyTRUST, 2013) has "looked into the effects on society caused by the introduction of hydrogen and fuel cell technology in mobility. On the focus of the project were questions about the acceptance of the technology and about the trust that the public puts in technology-driving players." In total more than 2500 citizens and relevant stakeholders were involved in the project. "A media analysis of German-language newspapers and magazines reveals how the topic hydrogen mobility is constructed in the mass media discourse, how the concept is linguistically occupied by the different actors and how it is communicated to the public."

Only by taking into consideration better training for, communication about and creation of trust in and acceptance of hydrogen, together with a better understanding of hydrogen safety issues, will the same level of safety, reliability, improved technical culture in handling and comfort of use as we have with today's energy carriers and fuels be achieved for hydrogen.

9.6 Conclusions

Hydrogen is not necessarily more dangerous than other fuels if its safety-relevant characteristics are properly taken into account when designing and operating technical systems handling hydrogen. Hydrogen systems can be designed to be at least as safe as other systems using combustible fuels without compromising the health of persons or the environment.

Accident prevention in design, operation, and maintenance is a key approach for the safe use of hydrogen systems and can be facilitated by appropriate information, application of valid regulations and standards, and by education and training. Global information exchange and harmonization of requirements for layout, operation, and maintenance will further facilitate the evolution of safe and easy-to-operate hydrogen systems.

9.7 Sources of further information

9.7.1 Europe

HySafe: http://www.hysafe.info/.
DWV: http://www.dwv-info.de/publikationen/2011/sicher.pdf.
Safe Hydrogen Infrastructure: http://www.now-gmbh.de/fileadmin/user_upload/RE-Presse_Downloads/Sichere-H2-Infra_NOW-RCS_FR_final_02MAR2012.pdf.
Wasserstoff und dessen Gefahren—Ein Leitfaden für Feuerwehren: http://www.bmvi.de/SharedDocs/DE/Anlage/VerkehrUndMobilitaet/Gefahrgut/wasserstoff-und-dessen-gefahren-ein-leitfaden-fuer-feuerwehren.pdf?__blob=publicationFile.
EMPA: http://www.empa.ch/plugin/template/empa/1064/*/---/l=1.
HSL/HSE: http://www.hsl.gov.uk/hydrogen-safety.
UNIV. ULSTER: http://hysafer.ulster.ac.uk/.
H2TRUST: http://h2trust.eu/news/events/5th-international-conference-on-hydrogen-safety-ichs-2013/.
HyFACTS: http://www.hyfacts.eu/.

9.7.2 United States

http://h2tools.org/lessons
http://www.hydrogensafety.com/
www.fuelcellstandards.com
www.eere.energy.gov/hydrogenandfuelcells/codes
http://www.hydrogen.energy.gov/safety.html
http://www.sandia.gov/matlsTechRef/
http://energy.sandia.gov/?page_id=3725
http://h2tools.org/bestpractices

9.7.3 International Organizations

IEA: http://www.ieahydrogensafety.com/
As from 1 January 2013 and until 30 June 2015: http://www.unece.org/trans/danger/publi/adr/adr2013/13contentse.html
As from 1 January 2015: http://www.unece.org/trans/danger/publi/adr/adr2015/15contentse.html

References

ADR, 2015. United Nations Economic Commission for Europe, Committee on Inland Transport, "ADR—European Agreement Concerning the International Carriage of Dangerous Goods by Road", vols. I and II, 1 January 2015.

AFID, 2013. Interinstitutional File: 2013/0012 (COD), with reference to the proposal for a Directive of the European Parliament and the Council on the deployment of Alternative Fuels Infrastructure, Brussels, 5 December 2013.

AFID, 2014. Directive 2014/94/EU of the European Parliament and of the Council of 22 October 2014 on the deployment of alternative fuels infrastructure, 28 October 2014, http://eur-lex.europa.eu/legal-content/EN/TXT/PDF/?uri=CELEX:32014L0094&from=EN.

AGBF, 2008. Wasserstoff und dessen Gefahren—Ein Leitfaden für Feuerwehren, AGBF-Bund AG-G, October 2008.

Alcock, J.L., Shirvill, L.C., Cracknell, R.F., 2001. Compilation of existing safety data on hydrogen and comparative fuels. Deliverable Report EIHP2, February 2001. Download at: http://www.eihp.org/public/Reports/Final_Report/Sub-Task_Reports/ST5.1/CompilationExistingSafetyData_on_H2_and_ComparativeFuels_S.pdf.

Behrens, D. (Hrsg.): Wasserstofftechnologie—Perspektiven für Forschung und Entwicklung, DECHEMA, Frankfurt, 1986.

EC, 2009. Regulation (EC) No 79/2009 of the European Parliament and of the Council of 14 January 2009 on type-approval of hydrogen-powered motor vehicles, and amending Directive 2007/46/EC, 04.02.2009.

Echtermeyer, A., Lasn, K., 2012. Critical failure modes and load conditions for composite pressure vessels, Deliverable 5.1, DeliverHy project, December 1.

EIHP, 2003. Methodology for Rapid Risk Ranking of H2 Refuelling Station Concepts, Norsk Hydro ASA and DNV, Sept 2003, for WP5.2 Risk Analysis of Refuelling Infrastructure, EIHP2 project.

EU, 2010. Commission Regulation (EU) No 406/2010 of 26 April 2010 implementing Regulation (EC) No 79/2009 of the European Parliament and of the Council on type-approval of hydrogen-powered motor vehicles, 18.05.2010.

GTR, 2013. Proposal for a global technical regulation on hydrogen and fuel cell vehicles, UN Economic Commission for Europe, ECE/TRANS/WP.29/2013/41, 12 April 2013.

H2M, 2010. H2Mobility 70 MPa Hydrogen Refuelling Station Standardization—Functional Description of Station Modules, 07 June 2010, http://www.now-gmbh.de/fileadmin/user_upload/RE_Publikationen_NEU_2013/Publikationen_NIP/H2Mobility_HRS_Functional_Description.pdf

HyFACTS, 2013. Regulations at: http://hyfacts.eu/2014/regulations/.

hyTRUST, 2013. Socio-scientific research for the implementation of hydrogen as fuel, http://www.hytrust.de/.

ISO 15916. ISO/PDTR 15916 "Basic considerations for the safety of hydrogen systems", committee draft standard, 19 February 2014.

R134, 2015: UNECE R134 [ECE/TRANS/505/Rev.2/Add.133] Uniform provisions concerning the approval of motor vehicles and their components with regard to the safety-related performance of hydrogenfuelled vehicles (HFCV), United Nations, Geneva, 25 June 2015

Schmidtchen, U., 2009. Hydrogen safety: facts and myths, March.

Wurster, R., Schmidtchen, U., DWV Wasserstoff-Sicherheits-Kompendium, November 2011.

Wurster, R., Zerta, M.—LBST, Lehan, A.—BASt, Dey, R.—CCS Global, Born, M.—TÜV SÜD, Sichere Wasserstoff-Infrastruktur (Safe Hydrogen Infrastructure). NOW, 9 December 2011.

Hydrogen sensors and detectors

10

G. Manjavacas, B. Nieto
Centro Nacional del Hidrógeno, Puertollano, Spain

Abbreviations

ATEX	explosive atmosphere
CAT	catalytic sensor
CFD	computational fluid dynamics
DOE	U.S. Department of Energy
IEC	International Electrotechnical Commission
ISO	International Organization for Standardization
JRC	Joint Research Centre (Europe)
LEL	lower explosion limit
MOX	metal-oxide sensor
NREL	National Renewable Energy Laboratory (USA)
TC	thermal conductivity sensor
TR	Technical Report from ISO documents
UEL	upper explosion limit
VIM	International Vocabulary of Metrology

Units

Temperature	K, Kelvin
Pressure	Pa, Pascal

10.1 Introduction

The use of detectors is necessary in all installations where hydrogen is present. As is known, hydrogen is a colorless, odorless, and tasteless gas; the only way of detecting it is by using an adequate sensor. Also, hydrogen is the lightest of elements and the smallest molecule, having the greatest propensity to leak.

This chapter uses the nomenclature presented in International Electrotechnical Commission (IEC) 60079-29-2:2007 to distinguish between sensor/detector and sensing element. As explained in Section 10.2, a sensing element is the component (electrochemical, thermal conductivity (TC), etc.) that reacts in the presence of a flammable gas mixture to produce some physical or chemical response that can be used to activate a measuring or alarm function, or both. Typically, the term *sensor* is often used by developers of sensing technology to describe just the sensing element; however, in this chapter, *sensor* or *detector* will refer equally to an instrumented system composed of a sensing element, control circuitry, and a user interface that provides analytically useful information to the end user.

Compendium of Hydrogen Energy. http://dx.doi.org/10.1016/B978-1-78242-364-5.00010-5

Hydrogen sensor installation is generally presented as the most common and most important way to avoid any hazardous event—for example, to prevent and control the risk of explosions—the sensor being a part of a complex system to detect hydrogen, acting to different alarms and exhaust systems, in both mobile and stationary applications. The main concern is that safety is considered a prerequisite where hydrogen is involved and it is expected that hydrogen technologies and infrastructures would be engineered to be safe (Gandía, 2013b: 437–452).

Nowadays, different types of detectors are on the market, based on several technologies depending on the final application. Selecting and using the best hydrogen sensor properly for a particular application is the most important factor in having a safe installation. For this reason, the aim of this chapter is to contribute to clarifying all the concepts related to hydrogen sensing technologies, their importance, applications, and limitations. For this purpose, Section 10.2 provides relevant definitions related to hydrogen detectors prior to Section 10.3, which focuses on requirements for detectors. In Section 10.4, a review of current detectors, explaining the available technologies and operation principles of sensing elements, is given. Complementing the previous section, Section 10.5 provides an overview of the current research and development related to hydrogen detectors on which manufacturers and researchers are focused. Finally, Section 10.6 outlines other important aspects of study, like area coverage, distribution, calibration, and maintenance.

10.2 Terms and definitions

This section provides relevant terms and definitions related to hydrogen sensing:

Accuracy (accuracy of measurement)—Closeness of agreement between a measured quantity value and a true quality value of a measurement (Joint Committee for Guides of Metrology, VIM, 2012).

Calibration—Operation carried out periodically to check and adjust the zero signal and the sensitivity of the sensor with a known calibration gas mixture without any changing of the parameters, type of gas, measuring range or specific application (see term recalibration from IEC 60079-29-2:2007).

Drift—Variation in the apparatus indication with time at any fixed gas volume fraction (including clean air) under constant ambient conditions (IEC 60079-29-1:2007).

Full scale (measurement range)—The full scale range defines the maximum and minimum values of the measured property.

Lifetime—It is the acceptable period of use in service for a device.

Poisons (of sensors)—Substances that lead to temporary or permanent loss of sensitivity of the sensors (IEC 60079-29-2:2007).

Precision (measurement precision)—Closeness of agreement between indications or measured quantity values obtained by replicating measurements on the same or similar objects under specified condition measurements (Joint Committee for Guides of Metrology, VIM, 2012).

Selectivity—Parameter that studies the change of an indication of a measuring system and the corresponding change in a value of a quantity being measured (Joint Committee for Guides of Metrology, VIM, 2012).

Sensing element—Part of a sensor that reacts in the presence of a flammable gas mixture to produce some physical or chemical change that can be used to activate a measuring or alarm function, or both (IEC 60079-29-2:2007).

Sensitivity—The ratio of change produced in the apparatus by a known concentration of gas or vapor (IEC 60079-29-2:2007). Depending on context, this can refer to the minimum change in concentration of gas or vapor that the apparatus will detect. In general, high sensitivity implies that low concentrations can be measured.

Sensor—Assembly in which the sensing element is housed that may also contain associated circuit components (IEC 60079-29-2:2007). In the present chapter, it is assumed that the term *Detector* is equivalent to *Sensor*.

Span—Upscale reading on the normal test gas of the apparatus (IEC 60079-29-2:2007).

Span gas—Calibration gas which consists of a mixture of the target gas balanced in the background environmental air.

Time of response $T(x)$—Time interval, with the apparatus in a warmed-up condition, between the time when an instantaneous change between clean air and the standard test gas, or vice versa, is produced at the apparatus inlet, and the time when the response reaches a stated percentage (x) of the stabilized on the standard test gas (IEC 60079-29-1:2007). This response time is based on gas being injected directly into the sensor head.

T_{90}—It is usually defined as the time that the output of the sensor needs to reach 90% of its final value.

T_{60}—Time that the output of the sensor needs to reach 60% of its final value.

Zero gas—Gas recommended by the manufacturer, which is free of flammable gases, and interfering and contaminating substances, the purpose being calibration (IEC 60079-29-1:2007).

10.3 Requirements of hydrogen sensors and detectors

Because of the increasing use of hydrogen in the so-called hydrogen economy, identification of a few generic requirements for hydrogen detectors is needed. In this context, hydrogen sensors can be used to detect releases, to automatically shut down systems, to activate alarms and ventilation systems and to notify emergency responders.

The application of hydrogen detectors in potentially explosive atmospheres (ATEX) requires compliance with appropriate regulations, codes and standards, in which are established the main requirements for health protection and safety. Also, the use of sensors in those areas involves consideration of three basic aspects for which the requirements are fixed in specific standards:

- The sensor itself must be explosion protected, as it is defined in the series of standards IEC 60079.
- The hydrogen sensor shall fulfill performance requirements to detect hydrogen as a flammable gas, as it is described in IEC 60079-29-1:2007 and ISO 26142:2010.
- Hydrogen sensors should comply with a functional safety as it is generally described in the series of standards IEC 61508 for electrical, electronic, and programmable electronic safety-related systems.

As mentioned previously, hydrogen sensors should fulfill general requirements described in IEC 60079-29-1:2007. Recently, the new standard ISO 26142:2010 defines the performance requirements and test methods of hydrogen sensors that are designed for measuring and monitoring hydrogen concentrations in stationary applications. The provisions in ISO 26142:2010 cover hydrogen detectors used to achieve the single and/or multilevel safety operations, such as nitrogen purging or ventilation and/or system shut-off corresponding to the hydrogen concentration. It sets out only the applicable requirements to a product standard for hydrogen detectors, such as precision, response time, stability, measuring range, selectivity and poisoning, and its purpose is to be used for certification.

Depending on the information source, different requirements are outlined. For example, Buttner et al. (2011) summarize the most important requirements in the specifications shown in Table 10.1, which are the target specifications prescribed by U.S. Department of Energy (DOE) in 2007 (see Section 10.8).

Nevertheless, it is remarkable that target specifications from Table 10.1, prescribed by DOE in 2007, were reviewed in 2011 to refine hydrogen sensor requirements for safety purposes for specific hydrogen applications. The 2007 targets were developed for generalized applications, since, for example, the required response time is a very ambitious target for some applications, and for this reason application-specific requirements were explored.

Summarizing and focusing on the final use of hydrogen (hydrogen production, storage, and transportation), four basic parameters should be taken into account:

1. Performance. Sensors must be wide-ranging in air, nitrogen, and inert environments and have good sensitivity below the lower explosion limit (LEL), 4% H_2, in air. The sensitivity of hydrogen sensors is also a very important parameter and a specification of the lower detection limit should be noted too. Apart from features related to the chemical parts of these devices, physical properties have to be considered as well. In this way, T_{90} is a variable to be noticed, as well as temperature, pressure, and ambient humidity range that are parameters to consider, depending on the type of detector, as explained in Section 10.4.
2. Lifetime. Sensors must have a lifetime according to the intended application. In general, the following range of lifetime can be established depending on the application: less than 10 years for stationary power systems, establishing a lifetime of 3–5 years in many industrial processes and more than 10 years for transportation (although automobile manufacturers would prefer over 15 years for hydrogen safety sensors, which is far beyond current

Table 10.1 Target specifications for hydrogen sensors

Parameter	Value
Measurement range	0.1–10%
Operating temperature	−30 to 80 °C
Response time	<1 s
Accuracy	5% of full scale
Gas environment	Ambient air, 10–98% RH
Lifetime	10 years
Interference	Resistance

capabilities). During this period of time, sensors should be operational without replacing, cleaning or frequent calibration.

3. Reliability. Sensors must provide the security of no false alarms to ensure stable and safe use of hydrogen as well as a response with enough accuracy and sensitivity. Sensors should be capable of surviving high hydrogen concentrations without being affected in its integrity and their measures must be validated and certified according to international standards.
4. Cost. The relationship between cost of purchase, installation, and maintenance with the technology of detectors should be taken into account. It also depends on the final application for which the sensor is needed, so it is necessary to compare all the available technologies in the market. The ratio of cost/technology varies constantly because of the new developments in this technology field, as discussed in Section 10.5.

10.4 Current hydrogen sensors and detectors on the market: Technologies and operation principles

A review of current hydrogen sensors is given in this section, explaining the available technologies and operation principles. Each type of sensor has its operating principles, suited for determined applications. Furthermore, each type of available technology is explained, paying attention to its advantages and disadvantages.

10.4.1 Electrochemical sensors

Electrochemical sensors operate on the principle that an electrical current passes through a sensing electrode produced by an electrochemical reaction, which takes place at the surface of a sensing electrode coated with a catalyst, such as platinum. Basically, an electrochemical sensor is based on a metallic anode and a metallic cathode, submerged in an electrolytic solution (H_2SO_4 for example) to allow ion transport between both electrodes. Generally, electrochemical sensors have two or three different electrode configurations with a membrane for gas transport, the electrical current being proportional to the hydrogen concentration and this current can be measured to determine the gas concentration (see Figure 10.1). Sometimes the electrolyte is a solid polymer, which removes the possibility of leakage that may occur in the use of liquid electrolytes.

Amperometric and potentiometric sensors are the two main configurations of electrochemical sensors. The difference between them is that amperometric sensors work at a constant applied voltage and the sensor signal is a current, and potentiometric sensors operate at zero current and the sensor signal is the potential difference between the sensing electrode and a reference electrode.

10.4.1.1 Advantages

Electrochemical sensors have high sensitivity to hydrogen, consume very little power during operation and are well-established commercially. Also, they have a small size and good price, very good precision, and good selectivity.

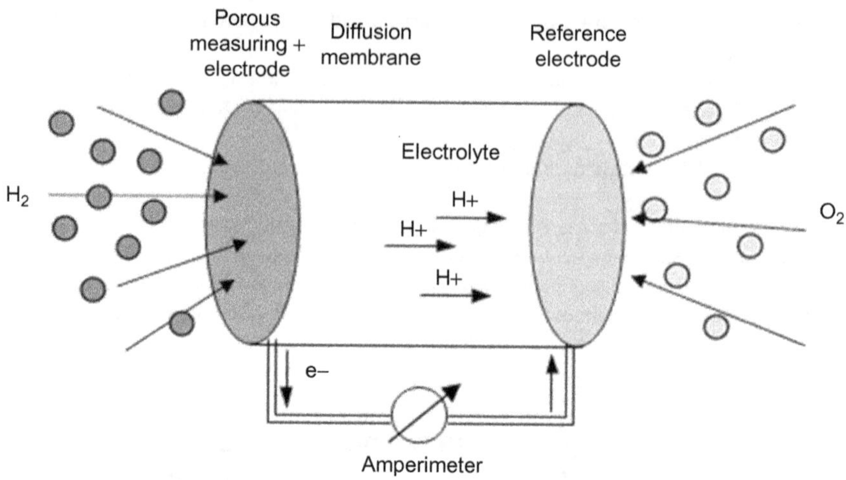

$$H_2 \longrightarrow 2H^+ + 2e^-$$

$$\frac{1}{2} O_2 + 2H^+ + 2e^- \longrightarrow H_2O$$

Figure 10.1 Scheme of an electrochemical sensor measuring principle (amperometric configuration).

10.4.1.2 Disadvantages

Despite the fact that electrochemical sensors have a high sensitivity to hydrogen, it decreases with time due to the degradation of the electrode catalyst, being easily contaminated by process gases in process applications. Also, they work with a restricted temperature range and have moderate selectivity.

10.4.2 Thermal conductivity sensors

The TC sensor operating principle is based on temperature-induced change of an electrically heated sensing element. Thermal conductivity is a property of each gas. Readings are positive for hydrogen, using air as the reference gas. The reason for this is because the thermal conductivity coefficient for hydrogen at normal conditions (273 K and 101,325 Pa) is the greatest of all known gases.

As can be seen in Figure 10.2, a TC sensor measures a concentration of a gas in a binary mixture by measuring the thermal conductivity of a sample gas and comparing Thermal conductivity to the reference gas (Gupta, 2008). Thermistors are used to form the sensing element, one in contact with the sample gas and the other one in contact with the reference gas. The sensing element temperature, that determines the electrical resistance, is conditioned by the heat loss through the surrounding gas, the sensor signal being a change in resistance. This change is proportional to the hydrogen concentration in the gas mixture.

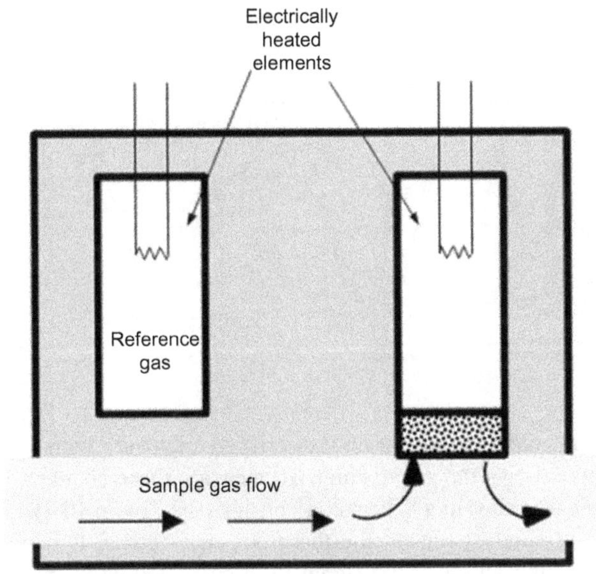

Figure 10.2 Scheme of a thermal conductivity sensor.

10.4.2.1 Advantages

TC sensors are stable devices due to the fact that there is no chemical interaction, so they are less susceptible to contamination. They have a wide detection range (<1–100% H_2) and long operation lifetimes (>5 years). Also they are highly reliable and accurate.

10.4.2.2 Disadvantages

TC sensors have difficulty detecting very low concentrations of hydrogen; for this reason they are usually used in combination with other types of sensors. Besides, they have a low gas selectivity which is a problem in process applications but not when only a gas combustible is present. These disadvantages are being studied currently, with attempts to improve them by miniaturization (see Section 10.5).

10.4.3 Catalytic sensors

Catalytic sensors (CATs) are based on gas oxidation on the surface of a catalytic element electrically heated. This oxidation uses the oxygen of the air and causes a temperature increase on the sensing element, which depends on the gas concentration. The most common type of detector is the "pellistor" type (see Figure 10.3), formed by two ceramic beads with platinum wires embedded, one of them being coated with a catalyst material in which hydrogen oxidation is produced. The gas oxidation produces a temperature increase on the catalyst bead, causing a change in electrical resistance of the platinum wire, also acting as the heater, which is a measure of gas concentration.

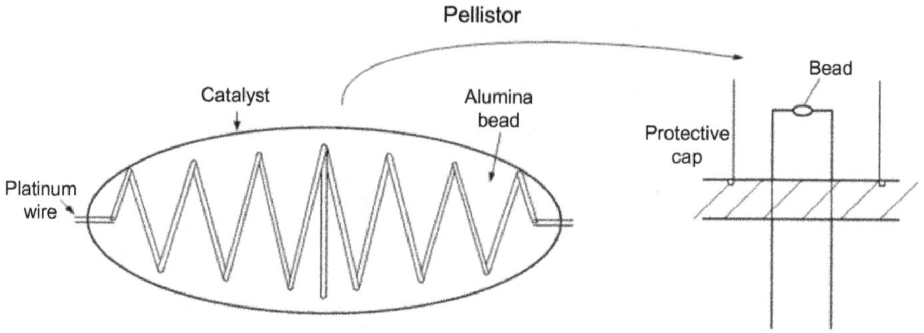

Figure 10.3 Pellistor scheme.

The heated wire is contained within an Ex-certified enclosure with a porous sintered metal inserted that allows the gas to enter. To measure these changes, both pellistors are connected to each other in a Wheatstone bridge (see Figure 10.4).

Another type of catalyst sensor, the thermoelectric sensor, is based on the same principle of generating an electrical signal by a catalyzed exothermic oxidation reaction of hydrogen but, in this case, it uses the thermoelectric effect, which basically consists of a direct conversion of temperature difference to electrical voltage, to generate the electrical signal.

10.4.3.1 Advantages

CATs have a well-developed technology and it can be used to detect any combustible gas. These detectors are small and used for detecting flammable gases from 0% to 100% LEL.

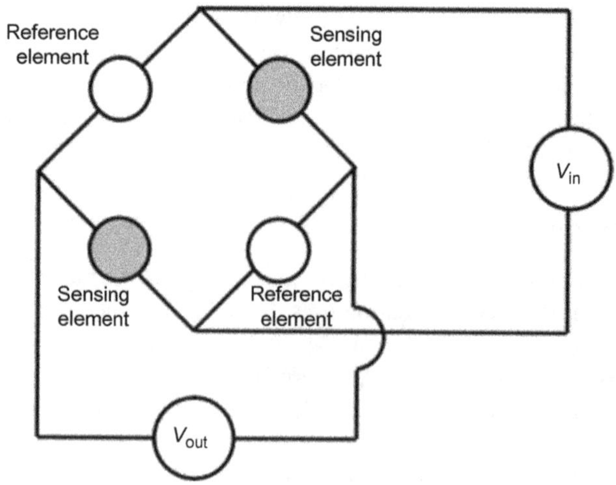

Figure 10.4 Catalytic sensor measure principle.

10.4.3.2 Disadvantages

CATs are not specific to hydrogen, and often cannot differentiate between combustible gases. Oxygen presence is required for their operation and is not recommended above the LEL. Also they can give false readings in gas-rich atmospheres, for example, above the upper explosion limit (UEL). The catalyst can be poisoned by trace gases such as silicones and hydrogen sulfide and it needs regular calibration and replacement.

10.4.4 Semiconductive metal-oxide sensors

The operating principle of metal-oxide sensors (MOX) is that a surface interaction between a reducing gas and a gas-sensitive semiconductor modifies the conductivity of the latter. Basically, a metal-oxide film is applied on a substrate material between two electrodes (Hübert et al., 2011), which shows sensitivity toward hydrogen gas (see Figure 10.5). The change in electrical conductivity of the semiconductor is a measure of the concentration of hydrogen gas.

10.4.4.1 Advantages

This detector can have a fast response and acceptable lifetime. Also, it is a low-cost, small sensor and has tolerable power consumption.

10.4.4.2 Disadvantages

MOX sensors are sensitive to water vapor and many other gases that may produce a false reading and they are not considered selective devices. Also, they have a long and nonlinear response time, being susceptible to contamination as well.

10.4.5 Optical sensors

Optical sensors are based on an optically active material that transforms the hydrogen concentration to an optical signal (Gupta, 2008). They are adequate to operate

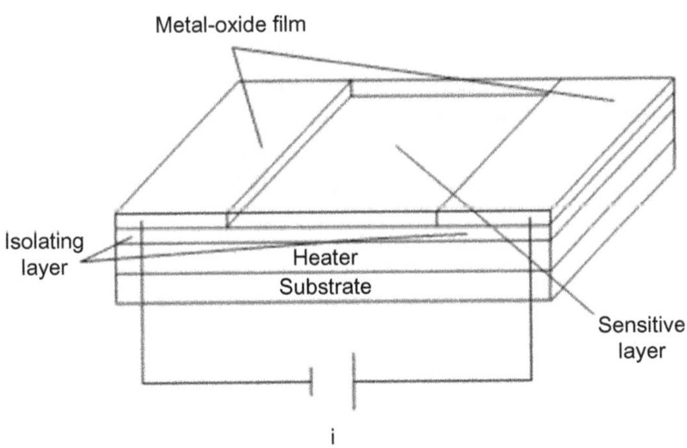

Figure 10.5 Schematic metal-oxide sensor.

in ATEX because they are electrically isolated (Hübert et al., 2014). There are many types of optical sensors, the most referenced being the devices based on optical properties of palladium films. The exposure to hydrogen produces a dimensional change in this metal, causing a modification in its effective optical path, which is proportional to the hydrogen concentration. Thus, various techniques are employed to measure this dimensional change, for example interferometric or reflectivity measurements.

10.4.5.1 Advantages

The optical sensor eliminates the risk of providing a source of ignition in the place of the leak because it is an optical signal rather than electrical and due to its configuration in the field it could cover a wide monitoring area using only one device. Besides, it is less sensitive to electromagnetic noise than others and may operate in the absence of oxygen.

10.4.5.2 Disadvantages

Optical sensors may be sensitive to interference from ambient light and to temperature changes.

10.4.6 Comparison of hydrogen sensor technologies

After the previous sections, typical characteristics of the sensors described are summarized in Table 10.2, which reviews some technical data of different detectors available commercially, showing in some cases more details than previously.

Related to the different parameters that appear in Table 10.2, it is important to take a look at the influence of *temperature* and *relative humidity* on the performance of the detectors. During normal operation, most false alarms are caused by the physical operational environment (see Section 10.3). For instance, electrochemical sensors are influenced by temperature because the electrochemical reactions and the rates of diffusion of gases in order to generate them vary with temperature. In general, and depending on the technology, to compensate for this effect, detectors usually incorporate a temperature sensor.

The other important variable is the relative humidity of the gas or ambient under investigation. It can affect the proper working of the detector, specifically the sensing element. To avoid this, it is common that manufacturers protect the sensors to prevent condensation on the surface. Nevertheless, to obtain a true measure it could be possible to integrate a humidity-sensing element in the hydrogen detection device.

10.5 Current research and development in hydrogen sensors and detectors

According to the previous sections, it can be concluded that the technology most suited to a given application depends on the operating conditions. The most important requirements were summarized in Section 10.3. However, not all sensor performance

Table 10.2 Comparison of sensor technologies

		Sensor type				
		Electrochemical	**Thermal conductivity**	**Catalytic**	**MOX**	**Optical**
	Operating principle	**Electrical current**	**Temperature change**	**Temperature resistance**	**Conductivity change**	**Optically active material**
Performance	Measuring range (vol%)	<4%	<1–100%	<4%	<2%	0.1–100%
	T_{90} (seg)	<30	<15	<20	<30	<60
	Lifetime (years)	2	>5	>5	2–4	<2
	Selectivity	Acceptable	Low	Low	Low	–
	Temperature influenced	Yes	Yes	No	Yes	Yes
	Humidity influenced	Yes	Yes	No	Yes	No
	Power consumption	Low	Low	Tolerable	Tolerable	Tolerable
	Cost	Good	Low	Low	Low	High
Characteristics	Advantages	Small size Good precision Well-stablished commercially	Small size Stable devices	Small size Well-developed technology	High sensitivity Small size	No source of ignition No electromagnetic influenced Wide area operation
	Disadvantages	Easy contamination	Difficult to detect very low concentrations of H_2	Can give false readings Regular calibration Poisoned by trace gases	Can give false readings Nonlinear response time Susceptible to contamination	Sensitive to ambient light interference
	Common applications	Leak detection + process monitoring	Process monitoring	Leak detection	Leak detection	Leak detection

Based on commercial sensors datasheet.
T_{90} and lifetime depends on the provider datasheet. Typical data are shown in this table.

requirements can be fulfilled for a specific application by only one kind of device itself. So, the improvements of the performance of detection devices, together with the development of new hydrogen detection technologies, are very important to reduce the risks associated with hydrogen technologies. Thus, the employment of detectors which are outside of the current uses is considered essential for safety. A practical case can be found in on-board light duty vehicles in which hydrogen sensors around and inside the vehicle will increase safety. The demands of the industry are summarized in response times lower than 30 s (target ones), an adequate strength to withstand environmental impacts (temperature, pressure, relative humidity, …) and issues related to maintenance.

There are several lines of research trying to fulfill these requirements. The approach of this section is to show a general view toward where the gas detection industry is now being oriented. However, not all of the novel technologies suggested for hydrogen detection are treated deeply. Within these technologies, biosensors for electrochemical detection of hydrogen (range: 1–100%) and the performance of detectors under anaerobic conditions can be studied.

Despite the next sections summarizing the current developments and research in hydrogen detection, there are challenges such as the protection of the sensing element from contamination/poisoning (sulfur mainly) or cross-sensitivity (target interferents: CO, CO_2, CH_4, NH_3) and the reduction of the response time (see Section 10.3) that still need to be satisfactorily addressed. Aside from detection technologies and requirements, there are overall concepts which are being studied too, such as network systems suitable for monitoring the spatial distribution of hydrogen concentration.

10.5.1 Advanced sensing technologies

In order to meet all the performance requirements shown so far, one of the research lines consists of combining different sensing technologies in one detection device. At this point, the concept *"Smart or intelligent sensor"* appears. These kinds of systems are based on the combination of two, at least, of the hydrogen sensing platforms viewed in Section 10.4.

For instance, a commercial sensor module can be found that combines an electrochemical cell and a pellistor sensing platform based on catalytic combustion in a single device (Hübert et al., 2014). The combination of these technologies allows a wide measuring range and a better response time.

Other concepts that could be included in this section are "sensors which are based on *work function.*" It is not indeed a new technology, because devices from Section 10.4.4 belong to this category. Nevertheless, Hübert et al. (2014) include them within lines of development because the research in this field is focused on improvements of the hydrogen sensitive materials and sensor design optimization.

10.5.2 Novel sensor materials

In the last few years, developments in the field of nanotechnology have enabled the application of nanostructures for gas detection. Because of their size, nanoparticles require less gas to cause a measurable change in the electrical properties of the sensing element and as a result extremely low hydrogen concentration measurements are possible (Hübert et al., 2014). In short, the large surface area makes it possible to increase the reactivity.

Within this research area, different technologies have been suggested like: carbon nanotubes, metal-oxide nanowires-nanorods, and palladium nanoparticles (Hübert et al., 2011). This new generation of sensors based on nanomaterials is expected to fill some gaps that exist in some critical areas like analytical parameters (lower response times, mainly) or low power consumption and low running costs within operational parameters.

Another promising material for gas sensing is porous materials, due to the high surface to volume ratio and strong adsorption of gases (Hübert et al., 2014).

10.5.3 Innovative fabrication techniques

An aspect for a safe deployment of equipment belonging to hydrogen technologies will be the availability of low-cost and high-performance hydrogen sensors. One of the issues which has to be considered on the deployment of new sensing elements design is the advanced manufacturing techniques.

Within this field, miniaturized versions of conventional hydrogen gas-sensing elements have already been introduced in the market using *micro-machinery techniques*, trying to cover the previous needs. El Matbouly et al. (2014) defines a micromachined, three-dimensional structure device with micrometer-scale dimensions typically manufactured using silicon microfabrication techniques (absence of any mechanical component). In general, microfabrication is the process for the production of devices in the submicron to millimeter range. The ability to fabricate hydrogen sensors at this scale offers advantages in terms of performance and cost.

Thus, the fact of developing these devices has been one strategy employed by sensor developers to improve the response time so as to meet other targets like miniaturize geometric dimensions of the sensing element.

Micro-machined hydrogen-sensing elements for different technologies are available. Nowadays, the most commercial micromachined hydrogen-sensing elements are based on the catalytic, TC sensor, and MOX (El Matbouly et al., 2014). On the other hand, some of these micromachined, because of miniaturization, devices suffer degradation in some critical variables like repeatability, and long and short stability.

In Section 10.4, the basis of CAT, TC, and MOX technologies are explained, so some specific aspects are only going to be discussed here. Regarding CAT sensing elements, there are available devices with dimensions of approximately $850\,\mu m^2$, while the dimensions of a micromachined TC sensing element are in the range of $10–1000\,\mu m^2$. The miniaturization of hydrogen pellistor technology results in faster response times and lower power consumption. A response time equal to 0.36 s (T_{90}) is reported by Hübert et al. (2011).

For MOX, one of the benefits of its miniaturized version is that the power requirements are lower. However, it seems that not all micromachined MOX sensing elements show improved performance metrics relative to conventional designs (El Matboly et al., 2014).

10.5.4 Cost of commercially available hydrogen sensors

The point related to the cost of these devices is included in this section, because the different prices between mature sensor technologies and emerging sensor technologies are the first fact to be studied. As a result, after consulting a few providers (Dräguer, Honeywell, H_2 Scan, etc.), it can be concluded that the first ones are cheaper than the second ones. Furthermore, Gupta (2008) provides wide ranges of cost depending on the vendor for devices which are based on the same operation principles, varying this range from 500$ to 4000$ for CAT sensors, for instance (it has to be considered that sensors include both the sensing element and the associated circuit elements).

When a sensor is needed, and in order to select the most suitable one, the requirements exposed in Section 10.3 should be taken into account, the cost being one of them. For example, in the case of a stationary application like a hydrogen detection installation in a laboratory, a CAT sensor or a MOX capacitor, which is a type of sensor based on *work function*, can be chosen. The cost of a CAT sensor is one-sixth of the cost of a MOX capacitor sensor, but the lifetime of the last one could be up to five times higher. So, as was commented previously, the ratio of today's cost/technology has to be properly assessed for each application to get the right balance.

10.6 Detection layout and maintenance of detectors

Once the different available technologies for hydrogen sensors have been studied along with the fields in which investment has focused its efforts, there are some issues that are common whatever the kind of technology/device chosen. One of the main conclusions drawn from this chapter is that means should be provided to detect the presence of hydrogen in places where leaks and/or accumulations may occur. In this way, the most important performance factors which should be considered when selecting a hydrogen sensor for a particular application can be summarized as: a suitable technology according to the detection range and response time needed, lifetime, maintenance (calibration), cross-sensitivity to other combustible or reducing gases and area coverage.

10.6.1 Stationary applications vs. portable applications

Resulting from the development of hydrogen technologies, there are two main fields of application for these devices: stationary applications and portable applications (automotive, basically).

During the last few years, the demand for hydrogen sensors has grown rapidly as the hydrogen infrastructure expands to support the production, storage, and dispensing of hydrogen in stationary applications (such as domestic combined heat and power applications and uninterrupted power supply applications) and fuel for automotive applications (hydrogen-powered fuel cell electric vehicles).

Depending on the kind of application, the performance requirements could be different. For instance, the size of the hydrogen sensor is very important when it is going to be installed on board a car. At the same time, the work conditions for these detectors (particularly in sensors with direct contact with the environment) are more difficult in comparison with a sensor placed in an electrolyzer installed for a stationary application. Other variables such as selectivity or time of response are taken into account too. Paying attention to the latter, sensors used in warehouses do not need a lower time of response in comparison with the requirement for a detector placed in a car.

10.6.2 Location of sensors

One of the most important issues which has to be considered, when a sensor is going to be used, is its location, which is related to the time of response. Regarding this, it is important to include the factor that the gas needs to accumulate and get to the detector from the leak point (gas has to travel from the leak spot to the detector). Because of the difficulty in predicting this, this parameter is not taken into consideration very often. For instance, there is no point in comparing a detector installed in a room with a detector that works in an outdoor windy installation.

So, it is clear that the required response time will depend on the location of the detectors. Nevertheless, the purpose of the system in which they are going to be installed is crucial in order to install a single detector or a whole. In this way, it can be concluded that specific rules to locate detectors cannot be provided.

Regarding considerations to keep in mind, the standard ISO/TR 15916:2004 (currently a new edition is under development), includes some suggested locations for hydrogen detectors:

- locations where hydrogen leaks or spills are possible;
- at hydrogen connections that are routinely separated (for example, hydrogen refueling ports);
- locations where hydrogen could accumulate;
- in building air intake ducts, if hydrogen could be carried into the building; and
- in building exhaust ducts, if hydrogen could be released inside the building.

Aside from these general considerations and the fact of knowing in depth the hydrogen system where a system of detection needs to be installed (forward/return pipes of ventilation systems, pipes connections, ...) it is necessary to have other tools to be able to locate detectors in a proper way.

In this scenario, computational fluid dynamics (CFD) is one of the tools available. Currently, it is useful to investigate safety issues related to the production, storage, delivery and use of hydrogen. A valuable contribution to the engineering design of safer hydrogen infrastructure and development of innovative mitigation measures and procedures can be provided.

CFD is a computer simulation tool that allows the modeling of the dynamics of fluids, that is, a numerical tool for predicting the pressure and velocity fields, and the temperature and concentration profiles on physical systems that may include chemical transformations (Gandía, 2013a: 401–435). Due to the importance, but also the difficulties and risks associated with experimentation with hydrogen in relevant situations from the safety point of view, CFD is a powerful tool for analyzing hydrogen safety scenarios that can help to design forced ventilation systems and define the number and location of the hydrogen sensors (Legg et al., 2012).

Obviously, a key issue is the previous validation against a range of relevant experiments of the mathematical models developed in order to establish confidence levels and range of validity of the simulation. Within hydrogen leaks and dispersion cases studied through CFD simulations, the regime depending on the scale of time is relevant regarding the location of the hydrogen safety sensors that should activate the corresponding alarms and the venting, decontamination or inertization systems (Gandía, 2013a: 401–435). Even though CFD is a helpful tool, there are so many variables that it is necessary to analyze each scenario individually.

As an example of CFD simulations, a work developed to check the way of working of two detectors inside an enclosure will be shown. Nieto et al. (2014) describes the applied measures for the adequacy and the implementation of an enclosure in which different equipment within the hydrogen technologies (electrolyzers and fuel cells basically) operate. The different pictures included in Figure 10.6 are courtesy of Technical and Research Department, in particular the Simulation Laboratory, of Centro Nacional del Hidrógeno—CNH2 (Spain).

Figure 10.6, as a whole, attempts to give an overview of the capabilities of the previous tool in helping to locate a detector, or some of them, inside an installation application where hydrogen is present. The most representative examples have been chosen considering extreme operation conditions for a leakage of hydrogen in order to be able to obtain useful images. In relation to the location of sensors, in images (b–f) (see Figure 10.6) two horizontal lines on two of the four walls can be seen. These lines represent the range of different available positions to optimize the best position of detectors.

10.6.3 Maintenance of sensors

Overall, as discussed in Section 10.2, the sensing element is the sensitive element responsible for converting a physical measure (e.g., gas concentration) into a useful output signal. On the other hand, a transducer (included within the concept detector/sensor) turns the output signal into meaningful information displayed by the user interface. Sensing element aging may cause drift (see Section 10.2) in time. The performance of most detectors deteriorates with time, the rate depending on the type of sensor and the operating conditions. Maintenance is therefore essential for keeping detectors at a high performance level, and is required for safe use. Regarding this, detectors should be:

- Regularly cleaned, especially the head of the detector, to allow gas to reach the sensitive element.
- Regularly inspected for possible malfunctions, visible damages or other deterioration.
- Calibrated (zero and sensitivity adjusting) with a standard gas in accordance with the procedure outlined in the instruction handbook.

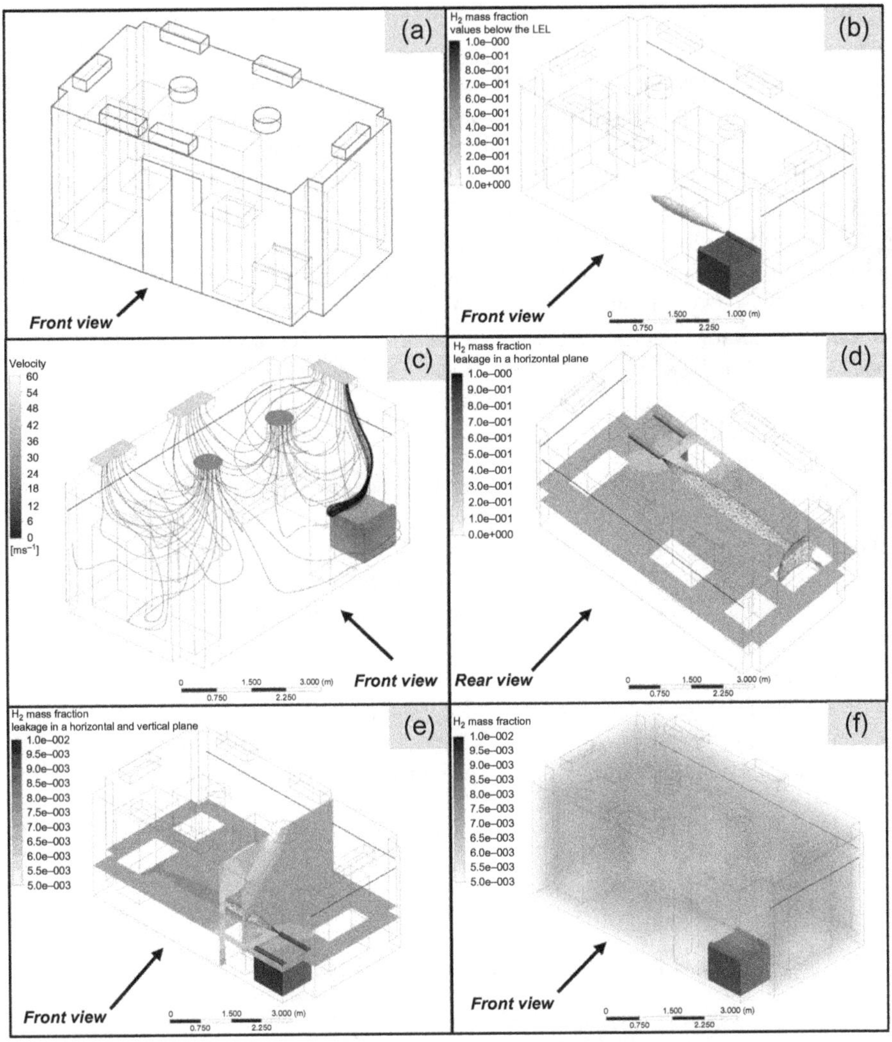

Figure 10.6 Possibilities of CFD simulations studying a real case of hydrogen leaks: (a) layout of the enclosure and H_2 equipment; (b) surface from a leak with a fixed value for LEL; (c) velocity of stream lines of air (from inlet to outlet) and H_2 from a leakage set up; (d) and (e) examples with a H_2 leak representing the H_2 mass fraction in a horizontal plane and a vertical plane; (f) H_2 mass fraction when a steady state is reached.

10.6.3.1 Calibration

Applied in this specific case, the aim of calibration (see Section 10.2) is to have a function relating the sensor signal and the concentration of hydrogen gas as well as to obtain a reference for the accuracy of the sensor.

Ideally, a linear response is expected but each technology has its own response. This fact is clear when CATs and electrochemical sensors are compared (Hübert et al., 2011).

The first one has a linear response while a nonlinear response appears for electrochemical sensors. When high hydrogen concentrations are present, a nonlinear response of a sensor causes its sensitivity to decrease. This fact has to be studied when a technology of detection is chosen. Figure 10.7 represents both situations as described here.

The most common procedure recommended by manufacturers includes carrying out the calibration with four different concentrations of the target gas (see definitions for *Span gas* and *Zero gas* in Section 10.2). Within these concentrations, the values of 20% and 40% of the LEL for hydrogen are of special interest as alarm functions and should be calibrated. However, a novel line of research which is starting to be studied involves a system with autocalibration, with end-user support in mind (see the National Renewable Energy Laboratory (NREL)-Joint Research Centre (JRC) information in Section 10.8).

10.6.3.2 Validation and certification

Validation and certification operations are beyond the scope of the end user, being compulsory and carried out by manufacturers to introduce a product into a market. So, validation and certification are discussed here from a general point of view.

Validation indicates that a product complies with a standard or meets an established set of requirements. Apart from calibration operations studied in the previous section, there are other concepts which have to be considered, such as detection range, robustness, selectivity and dynamic behavior.

Certification is the formal process under which a recognized organization verifies the soundness of the results of an evaluation and provides its official statement regarding the security functions and assurance provided by a product or system.

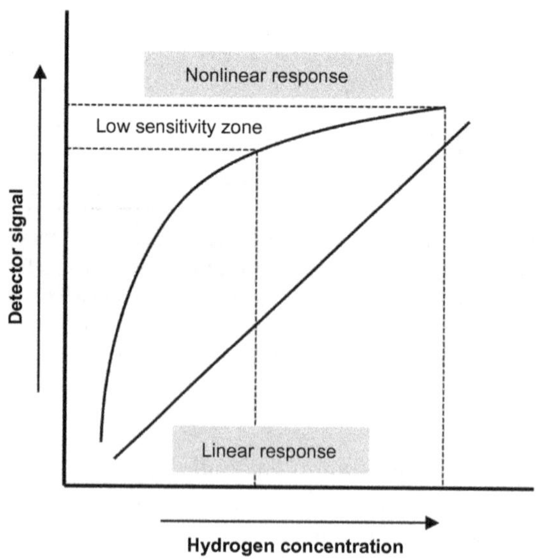

Figure 10.7 Comparison between a linear and nonlinear response for a sensor related with calibration operation.

10.7 Conclusions

A successful transition to a hydrogen economy needs public acceptance and safety is recognized as one of the main factors in achieving it. Regarding this issue, detectors play an important role as a part of a system for the early detection of hydrogen leaks before its concentration rises to hazardous levels.

Nowadays, a great number of different detector technologies are available on the market, manufactured by several providers and with a variety of capabilities and costs. In addition, new platforms are being developed to improve the existing performance requirements, such as lifetime or time of response. To select the most suitable detector among the different available technologies, attention must be paid to the technical performance requirements which define these kinds of devices. In general, they are fixed by the application in which they are going to be used. So, each application has to be studied carefully even though the common purpose is to detect a hydrogen leak.

There are some challenges related to technical requirements that this industry has to fulfill. Over this chapter, analytical, deployment, and operational requirements have been addressed but parameters such as response time, size, lifetime, and maintenance can be considered gaps needing to be covered by mature and emerging technologies.

A proper technology selection is one of the keys to having a trustworthy system. However, deployment parameters, such as the number of detectors that need to be installed and their placement inside and/or around the final application, have to be studied too. Again, each application has its own characteristics; general considerations from standards and other more specific tools like CFD are available to deal with this issue. This is another gap on which industry is currently working.

10.8 Sources of further information

Project H2Sense web page: http://www.h2sense.bam.de/en/partners/index.htm (accessed 20.08.14). The main objective of the H2Sense project (June 2013–May 2014) was to promote the effective deployment and safe use of reliable hydrogen sensors, primarily but not exclusively, for applications of hydrogen as an alternative fuel.

NREL/Hydrogen sensor testing laboratory. The Safety Sensor Testing Laboratory at NREL's Energy Systems Integration Facility aims to ensure that hydrogen technology is available to meet end-user needs and to foster the proper use of sensors. NREL is the principal research laboratory for the U.S. Department of Energy (DOE). http://www.nrel.gov/hydrogen/facilities_hsl.html (accessed 01.09.14).

JRC, the European Commission's in-house science service. With its research, the JRC aims to facilitate the commercialization of better, faster and cheaper sensors, and to contribute to the harmonization of international standards. https://ec.europa.eu/jrc/en/research-topic/hydrogen-and-fuel-cells (accessed 01.09.14).

Project Sensor INTERlaboratory COMparison (SINTERCOM). This is a collaborative project involving the Institute for Energy of the JRC and the National Renewable Energy Laboratory (NREL) of the US Department of Energy.

HySafe webpage: www.hysafe.info (accessed 29.09.14). The International Association for Hydrogen Safety (HySafe) is the focal point for all hydrogen safety related issues, focused on integrated research and information. It has been founded as an international nonprofit institution by the European Commission co-funded network of excellence.

References

Buttner, W.J., Post, M.B., Burgess, R., Rivkin, C., 2011. An overview of hydrogen safety sensors and requirements. Int. J. Hydrogen Energ. 36, 2462–2470.

El Matbouly, H., Domingue, F., Palmisano, V., Boon-Brett, L., Post, M.B., Rivkin, C., Burgess, R., Buttner, W.J., 2014. Assessment of commercial micro-machined hydrogen sensors performance metrics for safety sensing applications. Int. J. Hydrogen Energ. 39, 4664–4673.

Gandía, L., 2013a. Hydrogen technologies: production, purification, storage, applications and safety. In: Arzamedi, G., Diéguez, P. (Eds.), Computational Fluid Dynamics as a Tool for Designing Hydrogen Energy Technologies. Elsevier, Poland, pp. 401–435.

Gandía, L., 2013b. Hydrogen technologies: production, purification, storage, applications and safety'. In: Arzamedi, G., Diéguez, P. (Eds.), Hydrogen Hazards and Risks Analysis Through CFD Simulations. Elsevier, Poland, pp. 437–452.

Gupta, R.B., 2008. Hydrogen fuel: production, transport, and storage. In: Soundarrajan, P., Schweighardt, F. (Eds.), Hydrogen Sensing and Detection. CRC Press, Boca Ratón, pp. 496–532.

Hübert, T., Boon-Brett, L., Black, G., Banach, U., 2011. Hydrogen sensors—a review. Sens. Actuators B 157, 329–352.

Hübert, T., Boon-Brett, L., Palmisano, V., Bader, M.A., 2014. Developments in gas sensors technology for hydrogen safety. Int. J. Hydrogen Energ. 39, 20474–20483.

International Electrotechnical Commission (IEC), 60079-29-1:2007 Explosive atmospheres—Part 29-1: gas detectors—performance requirements of detectors for flammable gases.

International Electrotechnical Commission (IEC), 60079-29-2:2007 Explosive atmospheres—Part 29-2: gas detectors—selection, installation, use and maintenance of detectors for flammable gases and oxygen.

International Organization for Standardization (ISO), 26142:2010 Hydrogen detection apparatus—stationary applications.

International Organization for Standardization (ISO), TR 15916:2004 Basic considerations for the safety of hydrogen systems.

Joint Committee for Guides of Metrology, 2012. International Vocabulary of Metrology—Basic and General Concepts and Associated Terms (VIM), third ed. International Bureau of Weights and Measures.

Legg, S.W., Benavides-Serrano, A.J., Sirola, J.D., Watson, J.P., Davis, S.G., Bratteteig, A., Laird, C.D., 2012. A stochastic programming approach for gas detector placement using CFD-based dispersion simulations. Comput. Chem. Eng. 47, 194.

Nieto, B., Manjavacas, G., Merino, C., Quintana, R., 2014. Preliminary study for the adequacy and implementation of a hydrogen laboratory. Int. J. Hydrogen Energ. 39, 4143–4149.

Part Four

The hydrogen economy

The hydrogen economy—Vision or reality?

11

M. Ball, M. Weeda†*
*Shell Global Solutions International, HW Amsterdam, The Netherlands, †Energy Research Centre of the Netherlands (ECN), Petten, The Netherlands

Abbreviations

BEV	battery electric vehicle
CGH_2	compressed gaseous hydrogen
CNG	compressed natural gas
CCS	carbon capture and storage
CTL	coal-to-liquid
DOE	(US) Department of Energy
FCEV	fuel cell electric vehicle
GHG	greenhouse gas
GTL	gas-to-liquid
HRS	hydrogen refueling station
IEA	International Energy Agency
ICE	internal combustion engine
IGCC	integrated gasification combined cycle
ISO	International Organization for Standardization
LH_2	liquid hydrogen
LNG	liquefied natural gas
LPG	liquefied petroleum gas
NG	natural gas
PEM	polymer electrolyte/proton exchange membrane
PHEV	plug-in hybrid electric vehicle
PtG	power-to-gas
REN	renewable energies
SAE	Society of Automotive Engineers
SMR	steam methane reforming
SNG	synthetic natural gas
TWh	Terawatt hour
VAT	value added tax

11.1 Setting the context—The global energy challenge

Today's energy and transport system, which is based mainly on fossil energy carriers, can in no way be regarded as sustainable. Given the continued growth in the world's

Compendium of Hydrogen Energy. http://dx.doi.org/10.1016/B978-1-78242-364-5.00011-7

population from about 7 billion people today to over 9 billion by 2050, plus the progressive industrialization of developing nations, particularly in Asia but likewise in South America, the global demand for energy is expected to continue to grow in the coming decades as well—by up to 50% until 2040, according to the International Energy Agency (IEA) (IEA, 2014)—with fossil fuels continuing to dominate global energy use. At the same time, there is a growing global consensus that greenhouse gas (GHG) emissions, which keep rising, need to be mitigated in order to prevent dangerous GHG-induced climate change effects. Hence, security of supply and climate change represent two major concerns about the future of the energy sector which give rise to the challenge of finding the best way to rein in emissions while also providing the energy required to sustain economies.

The transport sector today accounts for nearly one-quarter of primary energy use and related global CO_2 emissions, with the vast majority of emissions coming from road transport. Transport is also responsible for about 20% of the projected increase in both global energy demand and GHG emissions until 2040. At present, there are approximately 900 million light duty vehicles globally (not counting two- and three-wheelers), and over 2 billion vehicles are projected to be on the road by 2050, according to the IEA. Oil is still the largest primary fuel and covers more than 95% of transport energy demand. Reducing the oil dependence of the transport sector would therefore improve energy security and mitigate any anxiety about the economic and geopolitical implications of possible shortages in the supply of oil as a pillar of our globalized world based on transportation.

Transport systems perform vital societal functions, but in their present state raise a number of concerns, for instance with respect to local air pollution, climate change, congestion, land use, and noise. Local air pollution (responsible for particulate matter, ozone, and acid rain), especially from road transport, is quickly becoming a major issue for urban air quality, particularly in the world's growing megacities, which calls for solutions. GHG emissions from the transport sector and from fuel production are also increasingly subject to regulation around the world, especially in Europe, Japan, and North America.

World transport energy use has doubled in the past 30 years and deep emissions cuts will be required in the transport sector, in view of the required worldwide CO_2 emissions reduction to combat irreversible and harmful climate change. However, mobility of people and transport of goods is one of the major drivers of economic growth and societal development. Reducing energy demand and CO_2 emissions from transport, especially from personal transport, therefore, poses a particular challenge.

11.2 Options for the road transport sector

A multitude of options exist to address needed reductions of emissions of CO_2 and air pollutants from road transport. The principal ones are demand-side measures aiming at reducing transport volume (e.g., by bringing about modal shift from private cars to public transport or by shifting freight transport from roads to rail), more efficient vehicles and cleaner fuels.

In the near and medium term, smaller cars, more lightweight and aerodynamic construction, efficiency improvements of internal combustion engines (ICEs), dieselization,[1] and hybridization[2] can all contribute to further improvement of the fuel economy, thus helping to reduce overall fuel consumption and transport-related CO_2 emissions. But incentives need to be given likewise to car manufacturers (via emission standards) and consumers (via taxation or subsidies) to encourage the production and purchase of more low-fuel consumption vehicles. However, there is a point beyond which further improvements in CO_2 efficiency of ICE cars are limited and also increasingly costly.

To achieve a deep decarbonization of road transport, longer-term strategies must focus on developing alternative, low-carbon fuels, and more efficient propulsion systems. This basically means the use of biomass-based fuels (biofuels) in ICEs and the use of electric drivetrains, which refers to a number of electric-drive vehicle options. The latter comprise

- pure *battery electric vehicles (BEV)*, using only electricity as "fuel," which is charged to the battery;
- *fuel cell electric vehicles (FCEV)*, using hydrogen as "fuel," which is stored on board the vehicle and converted to electricity by means of a fuel cell; and
- plug-in hybrid electric vehicles (PHEV), which combine a battery system with an ICE or fuel cell system.

Both the BEV and the FCEV, as well as the battery/fuel cell-based PHEV are truly zero-emission vehicles, as they emit no CO_2 emissions or local air pollutants during operation.

Conventional fuels like gasoline and diesel emit over 80% of their life-cycle CO_2 emissions during their combustion in the vehicle; the remainder is emitted in the fuel production process. Decarbonization of road transport is difficult, therefore, as it means having to reduce CO_2 emissions from a multitude of dispersed point sources. This is different with hydrogen and electricity: being emissions-free at the point of use in the vehicle, they centralize the CO_2 emissions and shift them entirely "upstream" to the "fuel" production stage. Provided that they are produced either from low/zero-carbon feedstock or that the CO_2 generated during their production is captured and stored, hydrogen and electricity offer an effective means to decarbonizing the entire fuel supply chain. Both hydrogen and electricity also reduce the oil dependency of the transport sector by opening it up to a much wider portfolio of primary energy sources.

The following paragraphs address the major future fuel and powertrain options in more detail. A special emphasis in this context is placed on battery electric vehicles, which have been receiving considerable attention in recent years in some countries, both from governments, car companies and media.

[1] To give a theoretical example of what an improved fuel economy of vehicles could achieve: a dieselization of the entire US light duty vehicles fleet or likewise replacing the current US gasoline vehicles fleet with more efficient European-like gasoline vehicles would result in fuel savings of as much as 2–3 million barrels of oil per day (out of a total oil consumption of about 90 million barrels per day globally).

[2] A combination of an ICE propulsion system with an electric propulsion system; there are different degrees of hybridization, depending on battery involvement in the vehicle propulsion.

11.2.1 Biofuels and unconventional liquid fuels[3]

Today, owing to policy support schemes, biodiesel, and bioethanol (so-called "first-generation" biofuels) are gaining relevant market shares in some parts of the world, such as in Europe and the United States, as a means to reduce transport-related GHG emissions and enhance supply security. Biofuels are appealing as, once produced, they require only limited changes in infrastructure, and the performance and costs of a vehicle powered by biofuel are not substantially different from those of a fossil fuel-powered vehicle. However, there are various concerns associated with the supply of biofuels, in particular "first-generation" biofuels, which challenge their overall sustainability and may constrain large-scale production: net reduction of GHG emissions, competition for water resources, use of pesticides and fertilizers, land use, impacts on biodiversity (such as loss of rainforest related to growing biofuel crops) as well as competition with food (crop) production for arable land availability, which may drive up food and fodder prices.

Biofuels, using woody biomass as feedstock, could potentially extend the feedstock base and avoid interference with the food chain, thus lessening a number of these concerns. But more R&D is needed to make these so-called "second-generation" biofuels commercially viable. On the other hand, competition with use of this biomass for heat and power generation as well as for traditional industrial applications will likely increase. Overall, biomass availability, competition for end uses as well as socio-economic and environmental implications all place limits on biofuel use.

Generally, the higher the market prices of fossil fuels, the more competitive low-carbon alternatives will become. However, it should also be mentioned that there have been significant investments in unconventional oil in recent years, such as oil sands in Canada, prompted by high oil prices, as well as in synthetic (Fischer-Tropsch) fuels on the basis of gas and coal.[4] Growth prospects for any unconventional oil source will depend to a large extent on the prices for conventional hydrocarbons and on environmental constraints and regulations. If the cost of producing unconventional oil becomes competitive with the cost of oil from conventional sources—either due to technological improvements or higher oil prices—and the environmental impacts can be kept within acceptable limits, then unconventionals will find a place in the fossil fuels market in the future.

11.2.2 Battery electric and plug-in hybrid electric vehicles

Toward the end of the 2000s, triggered among other things by the development of hybrid vehicles and the outlook of increasingly stricter vehicle emission standards, car makers took a renewed interest in electric-drive vehicles, meaning PHEVs and BEVs, as a means to reduce emissions. After previous attempts to produce battery electric cars

[3] There are other lower-carbon fuel options like LPG, CNG, and lately also the use of LNG in trucks with converted diesel engines, but while they can only achieve a limited CO_2 reduction, they are also generally considered to remain niche market fuels or restricted to specific geographies. The latter also applies for bio-methane.

[4] These fuels are referred to as gas-to-liquids (GTLs) and coal-to-liquids (CTLs), respectively.

in the 1970s and the beginning of the 1990s were not successful, as batteries fell short of achieving their development targets, the last few years have seen an ever-increasing number of electric vehicles be put on the roads across the world. In fact, today, all car manufacturers offer a portfolio of various BEV or PHEV models. At present, there are an estimated 500,000 electric vehicles (PHEVs and BEVs) globally (mainly in the United States, Europe, and China) and there is a general consensus that their share of the vehicle fleet will grow significantly in many countries over the next decades. A real boost for BEVs in the coming years is expected to come from China. The extent to which these vehicles reduce overall CO_2 emissions obviously depends on the CO_2 intensity of the electricity mix.

The main attraction of these "electric cars" and in fact the biggest advantage over hydrogen is that the "fuel distribution" infrastructure, i.e., the electricity grid, already exists. Consequently, an electric recharging network can grow organically, while the hurdle is disproportionately higher for hydrogen vehicles, since an entirely new, dedicated hydrogen refueling infrastructure with sufficient geographic coverage for customer acceptance needs to be built first. Nevertheless, an extensive network of electric recharging points would still need to be implemented, with all potential charging options having their own specific challenges, not least with respect to practicability and customer acceptance, but also economic viability. Normal (slow) charging, at home or in public, takes somewhere between 3 and 8 h; fast charging, on the other hand, allows the (partial) recharging of a battery in 15–30 min, but there is uncertainty about the impact on battery performance degradation over time and grid stability with large-scale deployment.

The capital investments for implementing an electric charging infrastructure range from 1500 to 2500 € per vehicle, which is comparable to hydrogen FCEVs (McKinsey & Company, 2010) (see also Section 11.5). Owing to their modular nature, a charging infrastructure is easier to build up initially. But at some point infrastructure costs for FCEVs drop below those for BEVs as the number of charging stations remains commensurate with the number of cars, due to the lengthy recharging time, and because the specific investment costs for hydrogen stations decrease with increasing size of these stations, as required to serve a growing fleet of fuel cell vehicles. In contrast, there are limited economies of scale for electric charging stations and in addition, once implemented, hydrogen refueling stations (HRSs) can accommodate a growing number of FCEVs without further investment (until FCEV fleet growth necessitates capacity expansions), because of the fast refueling time of only a few minutes.

BEVs are the most energy-efficient solution and superior to FCEVs as the discharge efficiency of a battery is almost double the conversion efficiency of a fuel cell. The most promising battery technology for cars today is the Lithium–Ion battery. The low energy density and, as a consequence, the size and weight of these batteries are at present a constraint on the range of purely battery-powered vehicles. For this reason, battery technology is best suited for powering smaller cars and most compatible with short-distance travel, like urban driving.

If battery performance were to improve markedly and at the same time costs could be reduced, BEVs would represent a complete solution to decarbonizing road transport, thus making the discussion about hydrogen largely obsolete. However, it has to be understood that, due to the battery chemistry, there will always be inherent trade-offs among power density, energy density, longevity, safety, and cost of batteries. For instance, owing to the low energy density and the high costs of batteries, increasing the electric range of vehicles above 150 km (under real driving conditions) will be very costly. But despite this range limitation, the size of this "market segment" can be significant, as for instance even a 60 km all-electric range—a distance sufficient for most daily commuters—would cover up to two thirds of annual mileage (as, on average, less than 20% of trips exceed 60 km in distance).

Nevertheless, market acceptance is clearly a crucial factor for the success of BEVs: the inherent limitations of full electric cars need to be accepted by customers, as they have to adjust their mobility behavior to the technical conditions of the vehicle. Customer acceptance particularly correlates with limited driving range and costs; the inconvenience of possibly long recharging times as well as restrictive charging patterns and accessibility of public recharging stations also play a role, particularly with significant car penetration. The fact that BEVs will only cover a fraction of the driving range—although sufficient for typical daily driving needs—may pose a particular challenge to their attractiveness, as consumers may still have difficulty accepting a vehicle that is range-limited and as they would likely have to afford a second car to overcome this range issue.

PHEVs provide an extended range (currently on the basis of incorporating an ICE) and overcome some of the shortcomings of full electric vehicles and can therefore act as bridging technology for passenger transport.[5] Being only partly dependent on battery power, PHEVs have lower requirements on battery performance and smaller battery capacity than BEVs, restricting their all-electric driving range to some 50–80 km. But, as indicated above, even with this range, plug-in hybrids could "fuel" most of their energy demand from the power grid, thereby drastically reducing the liquid fuel demand of the vehicle and related CO_2 emissions.

11.2.3 Hydrogen fuel cell electric vehicles

At this stage, it should suffice to say that within the portfolio of available options to address the environmental and energy security challenges, hydrogen is seen as one of the main alternative fuels for future road transport:

- FCEVs are one of only two truly zero-emission vehicle options and the efficiency of the fuel cell drivetrain is about twice that of an ICE drivetrain (50% vs. 25–30%).
- FCEVs combine the comfort and benefits of electric driving (silent and efficient) with the convenience of incumbent cars in terms of vehicle range and refueling time (typically in excess of 400–500 km on a single refueling in 3–5 min, given 700 bar onboard hydrogen storage) and thus offer an attractive proposition to customers.
- With a driving range and performance comparable to ICEs, FCEVs are the lowest carbon solution for medium/larger cars and longer trips. In Europe, for instance, these car segments account for some 50% of all cars and 75% of CO_2 emissions (McKinsey & Company, 2010).

[5] There is also the option to equip BEVs with a small ICE range extender for extended driving.

- Because of more favorable energy density characteristics compared to batteries, hydrogen, and fuels cells are better suited to electrify a wide range of road vehicles, ranging from small cars to buses and light duty trucks.
- Hydrogen can be produced from any primary energy source, and potentially CO_2-free: for instance, from water using electricity, which can be generated from a wide range of vast (locally available) renewable energy sources. Hydrogen can also be produced from fossil fuels like natural gas or coal and play a role in the decarbonization of these sources by using carbon capture and storage (CCS), to sequester the resulting concentrated CO_2 stream of the production processes.

The remainder of this chapter addresses the current status of hydrogen vehicles and the key aspects related to introducing hydrogen in the transport sector as well as the broader role hydrogen could play in the energy system for the integration of renewable energies.

11.2.4 The road ahead

Despite the worldwide efforts to develop new solutions, liquid petroleum-based fuels will retain their dominant role in the transport sector for the coming decades. Owing to their chemical and physical properties, gasoline, and diesel are excellent energy carriers for the transport sector, despite their low conversion efficiency in ICEs and associated environmental effects. Liquid fuels are simple to handle, have a high volumetric energy density, are easy to store on board a vehicle and can use the existing distribution and refueling infrastructure. The previously mentioned alternatives to today's liquid fuels all exhibit constraints and drawbacks of some kind at present, which will take some time to resolve. For instance, gaseous fuels are more difficult to handle and require a new distribution and refueling infrastructure; hydrogen and electricity additionally require new propulsion systems and technologies (fuel cells, batteries).

Particularly heavy duty and long distance road freight transport (which may account for up to 50% of total road transport fuel demand and related emissions in some countries), but also shipping and aviation will continue to rely on liquid fuels for some time, as their inherently advantageous properties make them hard to beat for these sectors. The lack of "readily" available alternatives is likely to put pressure on biofuels to be used as "low-carbon" fuel for freight transport, predominantly for trucks, and hence underpin the role of hydrogen and electricity for passenger transport. The concept of electric vehicles also addresses some of the challenges that come along with the global urbanization trend: especially BEVs can help improve urban air quality, and electric two-wheelers (scooters) for instance are becoming increasingly popular in Asian megacities.

BEVs may be perceived as a major competitor for hydrogen FCEVs, but in fact the two are rather complementary. BEVs are a partial solution only when it comes to decarbonizing personal transport. Other solutions are still needed for long-distance driving. PHEVs with ICE range extender can compensate the range limitations of BEVs, but do not offer a complete zero-emission solution like FCEVs do. As FCEVs with 700 bar hydrogen storage achieve a driving range comparable to the incumbent ICE cars, hydrogen, and fuel cells are at the same time more suitable for larger cars.

Costwise, both BEVs and FCEVs will likely have a higher purchase price than ICE cars (related to the costs of battery and fuel cell technology), but in the long run, they offer the potential for lower fuel costs (due to better efficiency) and lower maintenance costs (because of fewer rotating parts in the vehicle).

11.3 A short history of hydrogen

Neither the notion of hydrogen as an energy vector nor the vision of a hydrogen economy is new. Hydrogen has been used and produced for industrial purposes since about 1920 for ammonia synthesis, as ammonia replaced saltpetre as the basic material for manufacturing explosives and fertilizers. Until the 1960s, hydrogen was used in many countries in the form of town gas (a mixture of up to 50% hydrogen, with carbon monoxide and methane) for street lighting as well as for home energy supply (cooking, heating, and lighting).

Today, hydrogen is an important feedstock in the chemical and petroleum industry, with the production of ammonia to manufacture fertilizers accounting for around 50% of total hydrogen use, followed by the processing of crude oil in refineries with a share of close to 40%. The global hydrogen industrial gas business is significant and total production amounts to around 700 billion Nm^3 (~60,000 kt) (enough to fuel more than 600 million FCEVs), and is based almost exclusively on fossil fuels: roughly half from natural gas and close to one third from crude oil fractions in refineries. Water electrolysis accounts for some 4% only.

Emerging applications for hydrogen are mainly seen as fuel for FCEVs, for reasons outlined before, and as a potential storage medium for electricity from intermittent renewable energies such as wind power (see Section 11.7). In addition, a small market exists for portable fuel cells using hydrogen to provide electricity for electronic devices or off-grid situations. Looking further into the future, hydrogen may also be used for electricity and heat generation in stationary fuel cells, in residential or commercial applications.

Given its versatility with respect to the range of potential application areas, hydrogen is also referred to as an "energy vector." However, it must be stressed that hydrogen is not an energy source in itself but a secondary energy carrier in the same way as electricity. Similar to electricity, as far as the security of supply or GHG emissions are concerned, any advantage from using hydrogen thus depends on how the hydrogen is produced.

The idea of a hydrogen-based energy system was already formulated in the aftermath of the oil crises in the 1970s, with the term "hydrogen economy" first coined by General Motors in connection with the future fuel supply in the transport sector in 1970. While hydrogen can be utilized in different applications, the transport sector is going to play the crucial role for the possible introduction of hydrogen on a wider scale, as this is where hydrogen can offer effective solutions to both emissions control and security of supply (unlike other alternative fuels, except electricity).

But the history of hydrogen as transport fuel, like other alternative fuels, has been one of ups and downs. The first hydrogen "hype" was created in the late 1990s, when breakthroughs in fuel cell technology prompted the automotive industry to issue overly optimistic statements about market rollout of FCEVs, creating high expectations among governments, industry, and the general public. Unfortunately, those expectations were not met throughout the 2000s, and over time the debate about hydrogen was taken over first by a "push" for biofuels, and then since the end of the 2000s by the "return" of the battery electric vehicle, whose visible spread was starting to threaten the sole role of hydrogen as "the fuel of the future." Nevertheless, supported by further technical advances and cost reductions in automotive fuel cells, together with more "coordinated" public–private efforts for building hydrogen refueling infrastructures, there is currently a new wave of commercialization activities for hydrogen FCEVs taking place, with wide market introduction now expected between 2015 and 2020.

11.4 The status of hydrogen fuel cell vehicles

Over the course of the 2000s, a number of car manufacturers started to introduce prototype hydrogen cars, and today, almost every car maker possesses its own prototypes and development experience. All hydrogen cars to date exclusively use (PEM) fuel cell powertrains,[6] with hydrogen storage at 700 bar.[7] Globally, there are around 600 FECVs on the road: ~300 in the USA, ~160 in Europe, ~50 in Japan, and ~130 in South Korea (IEA, 2015). In addition, there are a little over 100 hydrogen buses used in various cities around the world, where they are favored for their low pollution as well as for reasons of raising public awareness of hydrogen vehicles.

The success of hydrogen in the transport sector will crucially depend on the development and commercialization of cost-competitive fuel cell electric cell vehicles. Today, it can be said that FCEVs, especially cars, are *technically ready* for commercialization. Through continuous R&D since the mid-1990s, automotive targets for fuel cell systems concerning cold start capability, fuel cell system size and power density, efficiency, platinum loading, and reliability have been met. With the implementation of 700 bar storage technology, hydrogen storage capacity has increased (without sacrificing volume), resulting in driving ranges that approach gasoline ICEs—a key differentiator of FCEVs vis-à-vis pure BEVs.[8] Common standards for hydrogen and FCEV equipment have also been agreed upon, further reducing their complexity and costs: for instance, standard connections, safety limits, and performance requirements for hydrogen refueling have been established by several SAE and ISO standards.

Although major cost reductions have been achieved, FCEVs are still expensive at low production volumes. The fuel cell system is the most significant cost component (next to the electric power-train and the hydrogen tank) and not yet cost-competitive.

[6] BMW was the last car manufacturer to abandon the development of cars powered by a hydrogen-fueled internal combustion engine around 2011.

[7] Fuel cells and hydrogen can also be used as range-extender for battery electric vehicles, in which case 350 bar storage may be sufficient; an example is the *Renault Kangoo ZE*.

[8] Hydrogen tanks in cars have a storage capacity of 4–5 kg of hydrogen; as a rule of thumb, hydrogen cars consume about 1 kg of hydrogen per 100 km (about 100 kg for an annual driving distance of 10,000 km).

The greatest challenge is to drastically reduce fuel cell stack costs from today's level of some 500–800 €/kW for small series by at least a factor of 10 to about 50 €/kW (which is similar to today's ICEs). Such cost reductions are expected to be largely achieved from scaling up production to exploit economies-of-scale effects, for instance by moving from batch production to continuous production. Further reductions are expected from incremental engineering improvements (technology optimization), and from competition and innovation resulting from the development of new industrial supply chains for FCEV-related technologies.[9] The cost target seems achievable with an annual production of 500,000 to 1 million FCEVs, as compared to less than 1000 units today (McKinsey & Company, 2010; US DOE, 2014).

With more than 600 passenger cars covering over 15 million kilometers and more than 90,000 refuelings (McKinsey & Company, 2010), FCEVs are now considered to have been comprehensively and satisfactorily tested in a customer environment. As a result, automotive companies are shifting their focus from demonstration to commercial deployment so that FCEVs, like all technologies, may benefit from mass production and economies of scale. Another notable and significant step has been that car manufacturers globally have started to expand their existing R&D collaborations into fuel cell vehicle production.

To achieve desired mass production levels, the cars have to pass a precommercial (market transition) phase first, which the automotive industry is currently entering. Several car producers are progressing from prototype vehicles for demonstration to producing small volumes, indicating that such developments will enable the start of commercialization in the period 2015–2020. For instance, Toyota and Hyundai have recently started the market introduction of FCEVs with their models *Toyota Mirai* and *Hyundai Tucson Fuel Cell*. Others, like Honda, Daimler, Nissan, and GM have signaled intentions to bring FCEVs on the market between 2016 and 2018. Initial sales of these models are expected to be on the order of several hundred to a few thousand annually, in the first years. But the broken promises of earlier announcements and the ever-receding dates of FCEV introduction have made everybody in the industry cautious of announcing a new imminent breakthrough and disclosing detailed market projections and rollout figures. As an indication, McKinsey & Company (2010) uses an initial market rollout scenario for Europe, assuming to reach 100,000 FCEVs within 5 years, and one million within 10 years.[10]

A successful transition from a precommercial to a sustainable market phase must be enabled by an appropriate, market-specific framework of incentives that helps overcome the "chicken-and-egg" problem of introducing hydrogen cars and developing a refueling infrastructure simultaneously, all the more as the installation of HRSs needs to precede large-scale FCEV rollout by a few years. Developing such frameworks requires a strong appreciation of the long-term societal benefits offered by hydrogen

[9] It should also be noted that BEVs, FCEVs, and PHEVs are complementary technologies as they share many similar electric drive-train components, like motor, power electronics, and battery technology. Investments and technology improvements in BEVs and PHEVs therefore also benefit FCEVs and vice versa.

[10] For example, it also took Toyota 10 years from the launch of the Prius, the world's first mass-produced hybrid vehicle, in 1997 to reach cumulative sales of 1 million vehicles.

as well as a good understanding of the market factors driving consumer choices of vehicle options. Initially, FCEV purchases will have to be "incentivized," e.g., by providing privileges for these vehicles (like use of bus lanes, free parking in cities, or exemption from congestion charges) and through subsidies or tax breaks; these measures must be specifically tailored to create a level playing field for FCEVs in each individual market.

In addition, preparation for the structural changes in industry is just as important. Qualified service technicians and skilled workers must be available to ensure that the introduction of hydrogen and fuel cell technology is managed as smoothly as possible; at the same time, new supply chains for both vehicle manufacturing and installation of HRSs need to be developed.

11.5 Building a hydrogen delivery infrastructure for the transport sector

Despite the recognition of the potential benefits of hydrogen-based mobility the pathways to achieve such a transition remain contended. A particular challenge is that the implementation of a hydrogen supply infrastructure, comprising production, distribution, and the installation of HRSs, will require considerable capital expenditures and involve a high investment risk regarding the future uptake of hydrogen demand. The following paragraphs look at the different parts of the hydrogen supply chain in more detail.

11.5.1 Hydrogen production

Hydrogen occurs naturally in the form of chemical compounds, most frequently in water and hydrocarbons, and can be produced from fossil fuels or renewable energy sources by a number of processes. Natural gas reforming, coal gasification, and water electrolysis are proven technologies for hydrogen production today and are applied on an industrial scale:

- *Steam methane reforming of natural gas* (SMR) is the most widely used and cheapest production method and has the lowest CO_2 emissions of all fossil production routes.
- *Electrolysis* is still more expensive and mainly applied where high-purity hydrogen is required, or where natural gas is not available, such as in remote locations. The CO_2 intensity of this pathway is determined by the CO_2 intensity of the used electricity: when taking electricity from "the grid," the CO_2 footprint of the produced hydrogen will decrease as the share of renewables in the electricity mix will increase. Alternatively, electrolyzers could be operated using dedicated renewable electricity (for example, when directly coupled to a wind farm) or even nuclear electricity, both of which would result in CO_2-free hydrogen.
- *Coal gasification* may be economically attractive in some countries, but comes with a significant carbon footprint, unless the CO_2 is captured and stored.
- *Biomass gasification* for hydrogen production is still at an early stage today and potentially competes with other biomass uses.
- Other hydrogen production methods such as water splitting by high-temperature heat, photo-electrolysis (photolysis), or biological processes are still at the level of basic research.

Hydrogen also occurs as a by-product of the chemical industry (for instance from chlorine-alkali electrolysis), where it is often used for heating purposes. Where available, this by-product hydrogen represents potentially a (cheap) supply source, as it can be substituted for natural gas, although investments in purification may be necessary. This option is surely relevant for supplying hydrogen during the initial start-up phase in areas where user centers are nearby and when hydrogen demand for FCEVs is still low.

Hydrogen can be produced centrally, such as from large-scale steam reforming or coal gasification, or on-site at HRSs, using small-scale reformers or electrolyzers. On-site production may be economically attractive, as it avoids distribution costs, compared to centralized production. As electrolyzers are modular and scalable and can potentially produce green hydrogen, they are the preferred technology for this application. Initially, by-product hydrogen (where available) and existing (fossil) production capacities (like SMR) will be the main sources of hydrogen supply. As demand increases, incremental capacity can be added in small quantities at reasonable cost, while there will also be a growing proportion of distributed hydrogen production at retail sites. Generally, the hydrogen production mix will be country-specific and strongly influenced by feedstock prices and CO_2 regulation; resource availability, policy support, and regulation also play a role, in particular when it comes to hydrogen from renewable energies.

It is evident that hydrogen needs to be produced in the long term from processes that avoid or minimize CO_2 emissions. Renewable or green hydrogen (e.g., produced via water electrolysis using renewable electricity) is surely the ultimate vision, but not the precondition for introducing hydrogen as a transport fuel in the first place. For instance, hydrogen from natural gas already leads to a CO_2 reduction of some 30% on a Well-to-Wheel basis compared to conventional fuels, even without CCS. This is different, however, in the case of hydrogen from coal, unless the produced CO_2 is captured and stored. Fossil hydrogen production will likely dominate initially while an infrastructure is being developed, because it is the cheapest option and because existing capacities can be used. Longer term, the competitiveness of fossil hydrogen depends on factors such as stringency of CO_2 targets, deployment of CCS, and the economics relative to renewable hydrogen.

Renewable hydrogen (from electrolysis) will initially play a role where and when it is incentivized by policies or mandated. Improvement of the efficiency to reduce electricity consumption and availability of low-cost electricity are important factors for the economic viability of electrolysis. Furthermore, the costs of electrolyzers must come down sharply. This will have to be achieved mainly through a significant scale-up of manufacturing volumes. Such volumes and hence economies of scale are likely to be achieved faster through large-scale deployment of (green) hydrogen from electrolysis in industrial applications, such as in refineries, than through incremental additions of production capacities for the transport sector. But this would require regulatory incentives that allow monetizing the potential CO_2 benefits of renewable hydrogen to become cost-competitive against incumbent hydrogen production, such as from natural gas reforming.

11.5.2 Hydrogen distribution

Different options are available for hydrogen transport and distribution, depending on hydrogen volumes, delivery distances, and local circumstances: delivery of compressed gaseous (CGH_2) and liquid hydrogen (LH_2) by trucks and of gaseous hydrogen by pipelines.

Typically, for small quantities over short distances of up to 200 km, compressed gaseous hydrogen trailers (at 200 bar) are most suitable. Liquid hydrogen trailers are most economic for smaller volumes and longer distances, and for that reason are the dominant delivery option in the United States; however, this option requires the liquefaction of hydrogen first, which is an energy-intensive process.[11] A recent achievement is the increase in pressure level for distribution of compressed gaseous hydrogen by trailer from 200 to 500 bar, thereby increasing the payload from 400 kg to more than 1000 kg of hydrogen and enabling delivery distances of up to 500 km.

Pipelines are the preferred option for large quantities and long distances. Pipelines have been used to transport hydrogen for more than 50 years, and today there are about 16,000 km of hydrogen pipelines around the world that supply hydrogen to refineries and chemical plants; dense networks exist, for example, between Belgium, France, and the Netherlands, in the Ruhr area in Germany or along the Gulf coast in the United States.

Initially, gaseous trailers are expected to be the dominant delivery method to supply HRSs, with liquid trucks bridging the gap to pipelines. Very long term, liquid hydrogen carrier ships, which are currently under development in Japan, may even open the door for ocean transport of hydrogen. The latter may enable the import of renewable hydrogen from remote locations with abundant renewable energy sources, but without large hydrogen demand centers, or of fossil hydrogen (such as from coal) from locations with good CCS potential.

11.5.3 Hydrogen refueling stations

Hydrogen-based mobility relies not only on a massive deployment of FCEVs in the market but also on the associated deployment of an adequate hydrogen refueling infrastructure to meet the expectations and needs of the FCEV owners. A satisfactory retail infrastructure is based on the implementation of a network of HRSs, the capacities of which and the density of which (i.e., reflecting the average distance separating each HRS) are compatible with the FCEV owners' requirements for refueling. As of today, there are approximately 300 public HRSs in the world, in North America, Europe, and Asia. Compared to the approximately 400,000 conventional service stations globally, this clearly demonstrates that hydrogen refueling infrastructure is still in its infancy.

Hydrogen fueling stations have been demonstrated and tested, and are basically ready for scale-up to build initial networks. The advanced developments and lessons learned from demonstration projects for refueling both hydrogen cars and buses, which have been carried out worldwide for more than a decade, have triggered standardization of functional specifications and sizes (capacities) of HRS. Major progress has been made in the field of technology for hydrogen refueling in terms of developing equipment standards and refueling protocols for high-pressure (700 bar) and fast

[11] To liquefy hydrogen, it must be cooled down to −253 °C. State of the art liquefaction technology has a power consumption of about 12 kWh/kg hydrogen, which is equivalent to 36% of the usable energy of the hydrogen.

dispensing (less than 3 min) for cars, in addition to 350 bar dispensing for buses and material handling equipment (like forklifts).

Standardizing technical components and systems is a key means to drive down the costs of hydrogen stations and to prepare a commercial market. For the roll-out of a hydrogen infrastructure for passenger cars, industry has aligned on 700 bar refueling globally. An important step in further standardization was taken by the H2Mobility initiative Germany, by developing a Functional Description of 700 bar HRSs, which serves as basis for further rollout of HRS in Germany (H2Mobility, 2010).[12] The description gives an overview of the required performance such as minimum availability, vehicle refueling process control and regulations to adopt; the scope comprises everything from the on-site hydrogen storage to (and including) the fueling nozzle. Four different station sizes are considered and three station concepts dealing with three modes of hydrogen supply described: CGH_2 trailer supply, LH_2 trailer supply, and on-site production (including pipeline supply).[13]

The rollout of a hydrogen refueling infrastructure will likely follow a tiered approach: hydrogen stations will first be installed in major metropolitan areas and along highways to connect those urban clusters, as hydrogen use will take off predominantly in densely populated areas (see Figure 11.1). During the transition phase, the network will gradually expand outwards into less urbanized areas and eventually into rural areas. As buses and fleet vehicles such as delivery vans operate locally to a large extent, run on short, regular routes and return to a central depot for refueling and maintenance, they are ideal candidates for hydrogen during the early implementation phase, as they do not need an extensive network of refueling stations.

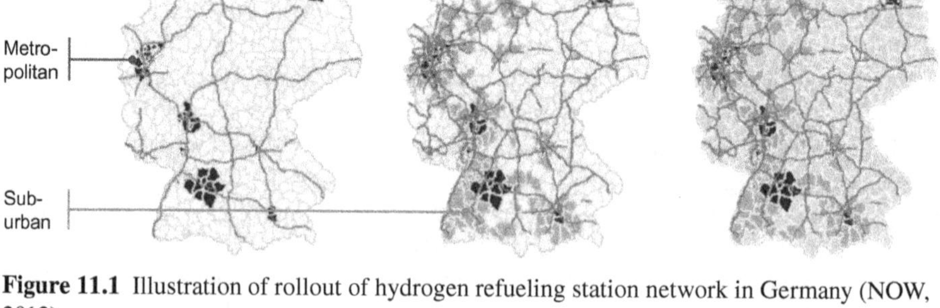

Figure 11.1 Illustration of rollout of hydrogen refueling station network in Germany (NOW, 2012).

[12] In the meantime, this Functional Description has also been adopted by the industrial gas industry for the development of HRS in other countries in Europe, like the UK.

[13] The four station sizes have a maximum daily hydrogen output of 80 ("very small"), 210 ("small"), 420 ("medium"), and 1000 ("large") kg/d, corresponding to a maximum number of 20, 38, 75, and 180 daily refuelings, respectively.

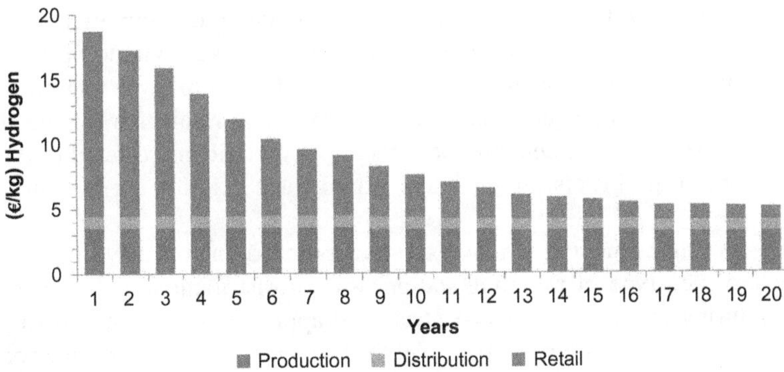

Figure 11.2 Illustration of the development of hydrogen delivery costs over time.

11.5.4 Economics of rolling out a hydrogen supply infrastructure for mobility

Figure 11.2 illustrates exemplarily for Europe how the specific costs of hydrogen delivered at the pump, i.e., the retail station, develop over time (including hydrogen production, distribution, and installation of refueling sites). During about the first decade, hydrogen costs are highest and the economics most challenging: for at least the first five years, delivery costs will exceed 10 €/kg and be dominated by retail costs.[14] This is caused by the very low utilization of HRSs (<20% in the first years), as only few FCEVs are on the road.

While there is some degree of optimization in terms of matching the HRS capacities ramp-up curve as much as possible with the expected vehicle ramp-up curve, sufficient network coverage must be available for consumers to purchase FCEVs in the first place. At the same time, the capacities of hydrogen retail stations must be chosen such that growth in the FCEV fleet can be accommodated to some extent with existing hydrogen stations. Clearly, the economic viability of hydrogen mobility from an infrastructure or retail operator's perspective can be helped with a fast roll-out of FCEVs. The introduction of hydrogen-fueled captive fleets could further help mitigate the effect of low station utilization.

Over time, hydrogen unit costs are falling, as the investments in HRS are used by a growing number of FCEVs. As a result, in the next 5 years, during the early commercial phase, when the number of stations becomes larger and utilization increases as more and more hydrogen cars come on the road, the relative share of retail costs drops, reducing delivery costs to a level of about 7–9 €/kg, at which point hydrogen, if tax-exempted, could become cost-competitive with gasoline and diesel. Thereafter, with FCEVs entering the commercial phase and station utilization approaching a mature level (around 80%), hydrogen costs are expected to start leveling off at 4–5 €/kg. In a mature market, hydrogen production dominates total delivery costs with a share of 60–70%, while retail stations and distribution make up about 20% and 10%, respectively.

[14] Costs include capital costs as well as fixed and variable operating costs.

In order to be able to offer consumers an attractive and competitive hydrogen retail price as compared to gasoline or diesel *and* to make a viable business case for station operators, hydrogen delivery costs at the pump have to be less than about 5 €/kg in the longer term, as VAT, fuel taxes and retail margins need to be added. Obviously, the cost benchmark is location-specific and related to the price of conventional liquid fuels, in particular liquid fuels taxation regimes in a given market.

There is no single blueprint or universal strategy for rolling out a hydrogen delivery infrastructure (i.e., type of refueling station, way of distribution, and production of hydrogen) during the pre-commercial phase and approaches are likely to vary from country to country. All options are possible; the best practical and economic combination of centralized or decentralized hydrogen production and way of hydrogen distribution to retail stations depends on specific national, regional and local resources, and conditions. This applies especially to the initial build-up phase that advocates making the best use of existing opportunities to be able to limit the required investments in infrastructure and the associated risk to a minimum. Factors that can have an impact on the selection of a specific delivery infrastructure include the existing hydrogen production facilities, their distance relative to the initial fueling stations network, the local energy mix and related energy prices, local codes and safety regulation, as well as specific CO_2 regulation and reduction targets.

In the long run, the investment for building a hydrogen delivery infrastructure, including production, distribution, and retail stations, is on the order of 1000–2000 €/FCEV (IEA, 2015; McKinsey & Company, 2010; Ball and Wietschel, 2009). However, during the initial 5–10 years of infrastructure rollout, the specific costs are more likely on the order of 3000 €/FCEV, mainly due to the high costs of building a new retail infrastructure, which is largely under-utilized; on that basis, a hydrogen supply infrastructure for around 1 million FCEVs would require an investment of about €3 billion.

11.6 The hydrogen infrastructure challenge and how to overcome it

Without a convenient hydrogen refueling infrastructure, no one buys a fuel cell car, and car manufacturers have no incentive to produce those vehicles in the first place. But until some choice of FCEVs is offered by car manufacturers and market conditions are created such that demand for the cars can arise, there is no point in building a network of hydrogen retail stations. This has been the classic chicken-and-egg dilemma, ever since hydrogen has been considered a potential vehicle fuel.

11.6.1 Hydrogen infrastructure development initiatives

Market development initiatives are spreading in Europe, Asia, and the USA. Frontrunners are Germany, Japan, and California; for instance, the German H2Mobility initiative has recently announced plans to establish some 400 HRSs by 2023.[15] Further market preparation and early market development initiatives are taking place in other parts of the USA, the UK, South Korea, Norway, and Denmark, while also France and the Netherlands have recently started making plans for the build-up of an initial hydrogen station network.

What has been common to these recent initiatives is that they bring together all necessary players, including infrastructure companies, like oil companies and industrial gas companies, car manufacturers and governments, and that they address the challenges of introducing FCEVs and building a hydrogen infrastructure in a collaborative effort. More specifically, what is at the heart of many of these cross-industry collaborations is to jointly develop scenarios and strategies for the implementation of a hydrogen infrastructure in accordance with expected FCEV rollout and local regulation—including hydrogen production mix and build-up of an HRS network—assess required investments and evaluate the economics. This analysis helps to create a joint understanding of the business case and its inherent risks and forms the basis for subsequent negotiations on infrastructure funding among industry partners and governments, and the operationalization of infrastructure rollout.

This collaborative approach effectively holds promise to overcome the famous chicken-and-egg problem, which so far has hampered the introduction of hydrogen. The first signs of success are now visible through the announced investment plans for building hydrogen refueling networks in Germany, Japan, and California.

11.6.2 The "valley of death"

A critical phase for hydrogen infrastructure development consists of bridging the gap between isolated demonstration sites and a pre-commercial stage where HRS network density is gradually increased. Despite the technological advances and emerging market development activities, some significant challenges related to building a hydrogen delivery infrastructure remain to be addressed. The biggest challenge to a positive business case is the long period of low utilization of stations during the transition toward a mass market, as Figure 11.3 exemplifies.

As discussed in the previous section, a growing network with relatively small numbers of FCEVs results in underutilization of (investments in) station capacity: as a consequence, hydrogen delivery costs at the pump exceed the permissible hydrogen sales price, which in combination with low sales volumes makes it impossible for retail operators to recoup their investments any time soon. Exact figures will be case-specific, but in general expectations are that it will take at least 10 years from the start of deployment for HRS operations to become cash-flow positive, i.e., with revenues from hydrogen sales exceeding operating costs for the first time. It takes another 2–5 years for the cumulative cash flow curve to become positive (see Figure 11.3), in other words a 10–15-year period for full recovery of both operating costs and capital expenditures, and to arrive at a positive business case.

[15] The six partners in the H2Mobility Germany initiative are Air Liquide, Daimler, Linde, OMV, Shell, and Total.

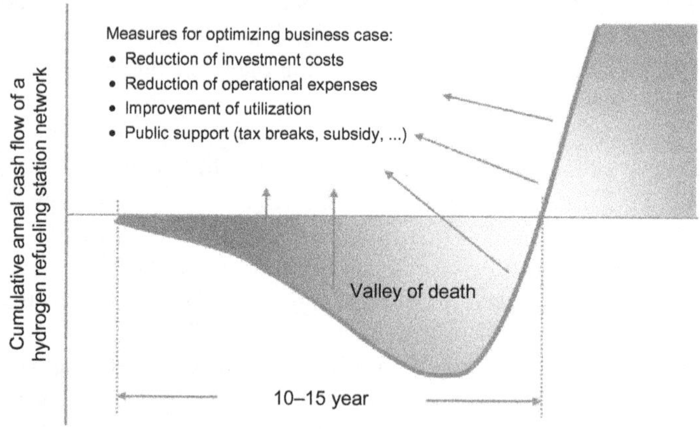

Figure 11.3 Cumulative annual cash flow illustration of hydrogen refueling station.

Nevertheless, this situation is very difficult to avoid, as sufficient initial (geographic) network coverage must be created for convenient accessibility, which is crucial for customer satisfaction and acceptance of FCEVs; in fact, for that reason, the station deployment has to precede the vehicle deployment by a number of years. As the initial stations approach a sufficient level of utilization, new larger station capacity will need to be installed to accommodate a growing demand, and these larger stations will then yet again remain underutilized for a number of years. This adverse impact of station underutilization on hydrogen costs, the so-called "valley of death," needs to be mitigated through an incentives framework to enable the selling of hydrogen at a competitive price relative to the prevailing price benchmark, i.e., gasoline or diesel fuel.

Because of the very difficult business case, especially during the early stage, the initial investment risk of building a hydrogen refueling network is too big for a single company. Infrastructure providers bear a high first-mover risk, as they have to make a significant upfront investment to build a retail station network that will not be fully utilized for some years. In fact, there would actually be a first-mover disadvantage, as not only would they develop, or rather de-risk, the market for all other infrastructure providers, who may then reap the benefits at a later stage, but in addition they face the full risk of technology delivery failure or delay, plus the impact of slower than expected market penetration of FCEVs because of a lack of consumers' interest.

The latter case could result in a potential write-off in the order of hundreds of millions. The infrastructure risk would be somewhat mitigated if several hydrogen retail infrastructure providers invest so that the risk is spread among different companies. The high investment risk, along with the lack of near-term profitability further precludes companies from attracting third-party finance, e.g., from banks and private investment funds. This is a further indication that the infrastructure rollout needs to be supported by adequate policy measures.

11.6.3 Remaining technology issues of hydrogen refueling

As for technology and standards, there remain three major issues: (1) the *reliability of hydrogen stations*, (2) *hydrogen metering*, and (3) *hydrogen quality* specifications and compliance:

1. System reliability does not yet meet the required specifications of >95%, which puts a great strain on achieving an acceptable business case. The reliability issue is related to the challenging combination of operating conditions for an HRS: 700 bar hydrogen storage on board cars to enable an operating range of 500 km and more on a single fill, and precooling of hydrogen to −40 °C to enable fast fueling in approximately 3 min. This demands high-quality components like valves, gaskets, filling hoses, and connectors, all of which increase costs; in addition, hydrogen compressors are still a relatively vulnerable piece of equipment. The availability of stations has improved over the years, but is often still low due to frequent breakdown of equipment. As a result, regular replacement of components and unplanned maintenance increase the annual operational expenditures.
2. Until now, no sufficiently accurate or standardized measuring method or device has been available for certification of the accuracy of hydrogen meters, which is required by law for commercial sales of hydrogen at retail stations: fuel retailers must guarantee that the amount of fuel that is charged to the customer corresponds to the amount of fuel put into the tank, within certain accuracy limits, typically ±1–2%. As long as certification is not possible, hydrogen cannot be commercially sold as a commodity to private customers unless exemptions are granted, or adjustments are made in the Weights and Measures Acts for early station certification, for instance such as is the case in California.
3. As fuel cells are sensitive to a number of impurities and in order to minimize risks of fuel cell degradation, practical hydrogen quality specifications, and cost efficient procedures for demonstrating hydrogen quality compliance must be developed. Implementation is needed prior to large-scale sales of hydrogen to the public in order to guarantee that the right quality is made available to the market.

11.6.4 How to overcome the challenges

As part of market development, effective initial networks of HRSs need to be established well in advance of vehicle rollout. However, potential HRS operators and, more generally, investors in hydrogen infrastructure, are faced by a difficult business case due to a long period of underutilization and high operational costs. Furthermore, they are faced with great uncertainty resulting from the current lack of certification methods for hydrogen meters, which does not allow commercial sales of hydrogen to customers, and from unknown procedures and related costs for hydrogen quality compliance, let alone the risk of consumer uptake of FCEVs. As a result, the following actions are deemed appropriate in order to overcome these challenges and kick-start a hydrogen mobility market:

- Support formation of strong consortia of industry partners (the "first movers"), comprising all relevant stakeholders, that are both able to provide the required investments and are willing to share the risks linked to these investments. There are several parties interested in the development of a market for hydrogen and FCEVs, but the investment risks are not equally divided. The biggest risks end up with the fuel retailer where the margins are smallest.

Consequently, infrastructure development is less likely to be realized if the interested parties in the hydrogen value chain do not join forces. The H2Mobility consortium in Germany is a good example of such a collaborative effort, which has been followed by a number of other EU countries.[16]
- Establish a robust framework of government interventions that contributes to mitigating the risks surrounding the business case in the early transition period. Robust, in this context, refers to level of support as well as duration of support. The latter is as important as the former, because the rollout of an HRS network and an FCEV fleet requires long-term investments that are in need of as much as possible long-term certainties. The framework preferably results from an integrated approach which looks for a balanced mix of financial and other supportive measures for the refueling stations, the fuel, and the vehicles; after all, one cannot exist without the other.
- Aim for obtaining adequate commitment of all stakeholders regarding well-coordinated roll-out for both the hydrogen refueling network and FCEVs. This includes targeted recruitment of early customers to ensure utilization of initial hydrogen stations.
- Stimulate R&D to improve critical components in HRSs, in order to increase their reliability and availability, and thus reduce operating costs.
- Stimulate the development of cost-efficient methods, procedures, and equipment for standardization of certification of hydrogen metering and quality compliance.
- Encourage the development and harmonization of standards related to designing, building, operating, and maintaining HRSs, and ensure that these standards are implemented in the appropriate regulations so that they get a legal status and can be enforced. When implemented, these standards will contribute to reducing market uncertainty and will facilitate smooth and efficient granting of permits for building and operating hydrogen retail stations.

In short, the emerging FCEV market (2015–25) requires close value chain synchronization and external stimulus in order to overcome the first-mover risk of building a hydrogen retail infrastructure. Although the initial investment is relatively low, the risk is high and therefore greatly reduced if many companies invest, coordinated by governments, and supported by dedicated legislation and funding. Once the market is successfully established, subsequent investments will present a significantly reduced risk, and with a mature market (after 2030) any potentially remaining economic gap is expected to be directly passed on to the consumer. To avoid fragmented and isolated markets which may hamper successful commercialization, timely, strategic, and coordinated planning is also needed for effective initiation, and gradual expansion and interconnection of initial markets.

11.7 The role of hydrogen for renewables' integration

From the point of view of minimizing CO_2 emissions from the energy system and reducing the dependency on fossil fuels to enhance security of supply, the ultimate vision is surely to produce hydrogen from renewable energy sources. Under such a scenario, water electrolysis using renewable electricity will become the dominant

[16] The H2Mobility Functional Description of HRS is an excellent example highlighting the outcome of the joint effort of station operators, equipment suppliers, and car manufacturers to reduce station costs.

hydrogen supply route. This inevitably leads to considerations on the interactions between hydrogen production and the electricity sector in a wider context, such as the ensuing competition for renewable energies or the identification of opportunities for synergies between the electricity sector and emerging hydrogen applications, like in the transport sector, through a closer integration between those sectors.[17] Most notably, in recent years, electrolysis technology has gained considerable attention as a means to help facilitate the large-scale integration of renewables by conversion of surplus electricity, which may otherwise be curtailed, into hydrogen; this concept is generally referred to as "Power-to-Gas" (PtG).

11.7.1 The case for electricity storage

Achieving the demanding GHG emission reduction targets of some 80% or more that prevail for 2050 will require a fundamentally different energy system, with intermittent renewables such as onshore and offshore wind and photovoltaic energy playing a crucial role in the electricity generation mix. Unlike the energy sources used in conventional power plants, these renewable sources are not dispatchable and they fluctuate over time, resulting in intermittent feed-in of electricity into the grid. As the share of volatile renewable electricity increases to very high levels, system flexibility needs, which have historically been driven by variable demand patterns, will therefore increasingly be driven by supply variability.

The greater reliance on intermittent renewable energy sources poses a number of challenges concerning their large-scale system integration. This includes having to absorb rapid and large swings in power supply, solving temporary mismatches between supply and demand, and managing growing strains on the grid resulting from a growing share of distributed electricity generation as well as electricity from wind and solar sources. This will require a mix of different solutions to maintain grid stability, like flexible power generation (e.g., from gas turbines), grid expansion, demand-side management of the structural final demand for electricity, flexible electrification of heat demand with hybrid heating systems ("Power-to-Heat"), electricity storage (including conversion to other energy carriers such as hydrogen) as well as curtailment of surplus intermittent generation peaks.

The specific characteristics of energy systems and national energy policy approaches vary significantly across countries. This will be reflected by differences regarding the challenges of integrating renewables from a capacity and regulatory perspective, and will affect the mix of applicable flexibility measures. In any case, ongoing decarbonization and increasing reliance on fluctuating renewable energy sources are expected to create an increased demand for energy storage technologies. In addition, only in combination with storage will wind and solar energy be able to become the backbone of electricity supply, and capable of providing base load power.

[17] Another link is via the co-production of "clean" electricity and "clean" hydrogen from coal in IGCC plants, under the condition that CCS is deployed at large-scale; what makes IGCC plants potentially attractive is the fact that they can deliver to two markets, the electricity market as well as the transport market, depending on price signals.

In general, the need for electricity storage is driven by the extent to which the electricity system is subject to (temporary) mismatches between supply and demand. In those cases, storage can serve as a buffer, which absorbs surplus electricity generated from renewable energy sources at times when supply exceeds demand, and can provide additional capacity in deficit situations, when the volatile generation from wind and solar energy is not sufficient to cover the electricity demand. Key parameters that determine supply–demand mismatches and thus create opportunities for electricity time shift (and price arbitrage) include, on the one hand, the amount of installed intermittent renewable generation capacity (and the voltage level at which renewable electricity is fed into the grid) and their specific generation characteristics and profiles (i.e., the wind speed and solar irradiance profiles); and on the other hand the electricity demand profile.

Initially, electricity storage will be driven by leveraging opportunities to accommodate an increasing amount of *surplus* (renewable) electricity in the energy system: if no measures were taken, annual surplus electricity in the order of several TWhs may already occur when fluctuating renewables reach a share >30–40% of national electricity demand. Longer term, however, with a very high (>60%) intermittent renewables share in the generation mix, it will increasingly become necessary to complement *deficit* situations. If conventional (fossil) back-up capacity cannot make up for shortages in the supply of renewable electricity, e.g., due to unfavorable economics and/or restrictions on CO_2 emissions, this will require TWh-scale electricity time shifts over extended periods of days to weeks, i.e., large-scale storage.

11.7.2 The role of hydrogen and electrolysis

A range of electricity storage technologies exists, such as pumped hydro storage or batteries, which essentially differ by storage capacity and discharge time. Electrolysis of water is considered the only technically viable option for TWh-scale storage of (surplus) electricity by means of conversion to hydrogen (which could further be converted into synthetic natural gas, SNG). In combination with underground hydrogen storage, where geologically feasible and acceptable to the public,[18] this hydrogen could also be used to manage deficit situations, as described above, over longer time scales in the order of days and weeks, and possibly up to months, via re-electrification in gas turbines, gas engines, or stationary fuel cells.

The unique aspect of hydrogen as an energy storage medium is its versatility with regard to a range of end-use applications. When used for re-electrification in combination with cavern storage, it can be considered a "classical" electricity storage application. If or as long as re-electrification of hydrogen from surplus electricity is not a viable option—either because underground hydrogen storage is not possible or

[18] With large-scale subsurface hydrogen storage typically only considered feasible in caverns made in salt formations, geology is a limiting factor. Salt deposits suitable for cavern construction are rare and unevenly distributed geographically and do not necessarily occur in those regions that have the highest potential for cavern storage; in this respect, the Northern parts of Germany and the Netherlands seem particularly promising in Europe (with many existing natural gas storage sites), as well as some parts of the United States, like Texas.

the supply deficit is of a scale that can be managed better by other (more economic) flexibility measures—this hydrogen can also be used in other applications outside the electricity sector. This includes the use as fuel for FCEVs, as feedstock and even as heating fuel in industry (such as the chemical and petrochemical, or the steel industry) and for the "greening" of natural gas by admixing it into the natural gas grid. Concerning the latter application an admixture of 2–5 vol% of hydrogen is generally considered to be within technically acceptable limits from an appliances perspective; this concept is attractive as it could use the existing natural gas infrastructure.[19] While the use of hydrogen from electrolysis in industrial applications would by and large mean a substitution of existing fossil hydrogen production, the other end uses would effectively create new markets for hydrogen.

11.7.3 Economics of electrolysis

The hydrogen production costs from electrolysis are influenced by the capital costs of the electrolyzer, its utilization and the (average) electricity purchase price during the time of operation.[20] High electrolyzer utilization reduces the specific share of electrolyzer capital costs in hydrogen production costs; on the other hand, a higher utilization increases electricity costs, as hours of expensive electricity will increasingly be included. Hence, in order to minimize hydrogen costs, electrolyzer utilization has to be balanced with the electricity price.

The optimum utilization of an electrolyzer resulting in lowest hydrogen production costs is in the order of 3000–6000 h, with a relatively flat cost profile in this range. At less than about 2000 operating hours, the capital costs start to dominate the production costs, making hydrogen from electrolysis increasingly expensive; therefore, the lower the utilization, the lower the capital costs of the electrolyzer need to be. But production costs are also very sensitive to electricity prices: for instance, for a utilization of 40–50% and average electricity prices of 40–50 €/MWh$_{el}$, electricity costs account for more than 50% of hydrogen costs.

In case of PtG applications, electrolyzer utilization becomes a key parameter. To illustrate the resulting utilization of electrolyzers when run on "surplus" electricity *alone* it is helpful to plot the residual load duration curves for different penetration levels of intermittent renewables.[21] Figure 11.4 depicts two typical residual load duration curves for a 30% and 80% share of generation from intermittent renewables of total electricity demand.

The part of the curves below the *x*-axis is the residual load and a measure for the amount of surplus electricity that is produced, but for which there is no immediate demand, i.e., when generation exceeds the load. (The part above the *x*-axis is a measure

[19] There is further the option of methanation, whereby hydrogen is reacted with CO_2 to produce synthetic natural gas (SNG).

[20] Where underground storage is included, electrolysis still dominates the total specific hydrogen delivery costs of an integrated hydrogen storage facility with over 80%, despite the significant investment for the cavern.

[21] Residual load is defined in this case as the difference between the load and the actual generation, which is the sum of must-run capacity and fluctuating renewable generation from wind and solar energy.

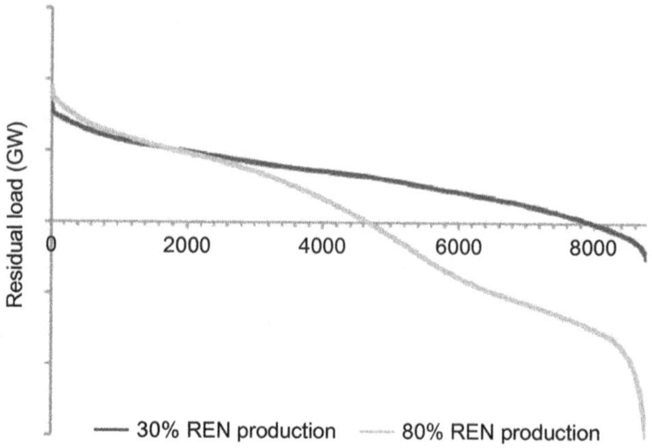

Figure 11.4 Exemplary residual load duration curves for different renewables shares.

of the deficit which has to be covered by either flexible conventional generation or storage.) Clearly, with a low (<30%) share of fluctuating renewable electricity generation, the amount of surplus is very limited, occurring in the order of 500–1000 h per year. With an intermittent renewables share of some 80% or more, substantial amounts of surplus electricity may occur during a total of 3000–4000 h during the course of a year. It is important to note, though, that these scenarios do not take into account the deployment of other storage technologies over time, such as batteries, which would reduce any surplus available for conversion to hydrogen.

The left-hand side of the following figure illustrates exemplarily how the specific hydrogen production costs of an integrated electrolysis and underground storage facility depend on the actual utilization of the electrolyzer; the bandwidth shown indicates variations in electricity prices (ranging from average 25 to 40 €/MWh) and electrolyzer capital costs (between about 1200 €/kW$_{el}$ currently and a projected future level of 500 €/kW$_{el}$). At a low utilization of less than 1000 h, hydrogen costs are prohibitively high, on the order of at least 10 €/kg. Between 3000 and 4000 h hydrogen costs start leveling off in a range of 2–6 €/kg. Comparing this cost level with what would be commercially acceptable for different end uses (see right-hand side), it becomes obvious that only hydrogen sales to the transport sector make a positive economic case (without taking into account CO_2 prices or regulatory measures).

As it may take quite some time before situations with significantly more than 2000 h of surplus are reached, storage of surplus electricity as hydrogen does not seem to represent a near-term economically viable case. Increasing the operating hours by purchasing additional (low-cost) electricity would obviously improve the economics of electrolysis, but depending on the CO_2 intensity of the grid mix, this may effectively increase the CO_2 footprint of the hydrogen produced. Under certain circumstances, electrolyzers may be able to generate additional revenue streams by providing services for grid stability, such as frequency control, which would improve the economics of hydrogen production, though marginally only.

Figure 11.5 further illustrates that electrolysis as a means to convert renewable surplus electricity to hydrogen (with or without underground storage)—while

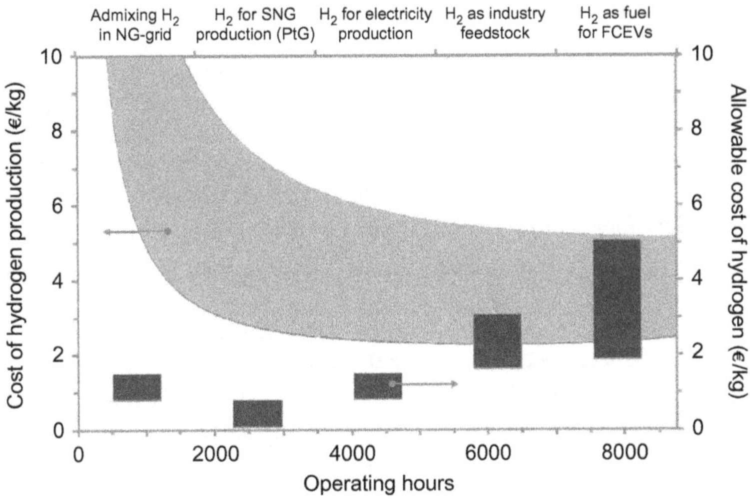

Figure 11.5 Range of actual and allowable hydrogen production costs for different end uses.

prohibitively costly at low utilization, i.e., until a reasonably high share of intermittent renewables in the generation mix—remains an economically challenging route even at higher utilization. This is mainly due to the fact that hydrogen from electrolysis struggles to be cost-competitive with other hydrogen production routes, all the more in the absence of regulation that enables the monetization of its potential CO_2 benefits (provided it is produced from "green" electricity).

In the short term, except for the use as fuel in FCEVs, no other application will result in a positive business case for hydrogen from electrolysis in the absence of favorable policy support measures or a "willingness-to-pay a premium" by the end user. Generally, what makes the mobility sector the economically most attractive market for hydrogen (from electrolysis) is the fact that the sales price benchmark for hydrogen as vehicle fuel is set by the price of conventional liquid fuels, i.e., gasoline or diesel, and hence linked to the oil price; plus the much higher efficiency of fuel cell-based drivetrains compared to ICE systems.

Where hydrogen from electrolysis has to compete on a heating value basis with natural gas, i.e., where the hydrogen sales price benchmark is set by the price of natural gas, such as in the case of admixture to the natural gas grid or when used as feedstock in industry (such as in refineries), it is not cost-competitive under the current energy and CO_2 price regimes and existing regulation. For example, for the use of green hydrogen in industrial applications to become an economically viable CO_2 mitigation option would require CO_2 prices in the order of 100–200 €/t. The same holds true for the use of hydrogen to enable large-scale electricity time shifts over weeks and months via re-electrification, for which electrolysis in combination with cavern storage seems the only technically feasible option to date; however, as mentioned previously, re-electrification may become a necessity to deal with deficit situations in case of a very high renewables share and with very stringent CO_2 emission limits, which would then require appropriate policy incentives.

11.7.4 In short...

It can be said that hydrogen may facilitate the large-scale integration of intermittent renewable electricity, offering solutions for both situations of electricity supply surplus and deficit: surplus electricity can be converted to hydrogen via water electrolysis, and re-electrification of hydrogen can be used to make up for deficit situations by enabling electricity time-shifts over extended timescales. Electrolysis may become an economically viable route for electricity storage in places with

- electricity generation from intermittent renewables and surplus in the order of tens of TWhs over some 3000–4000 h annually,
- low electricity prices during a significant part of the year, and
- a favorable and sustained policy framework that also creates a market pull for clean hydrogen from sectors such as mobility or industry.

What would additionally be required to make re-electrification of hydrogen a viable option is

- the need for large-scale electricity time shifts over weeks and months,
- suitable geology for cavern construction, and
- an appropriate electricity market design to deal with a high share of intermittent renewables.

Interestingly, the question remains how else to enable the integration of a very high share of intermittent renewables in electricity generation, if and where underground hydrogen storage is neither economically nor geologically feasible.

The unique aspect of electrolysis as compared to other electricity storage technologies like batteries is that it is not constrained to a power-to-power application only, but converts electricity to an energy carrier that can be used in a range of different applications. But in the "merit order" of flexibility solutions to integrate fluctuating renewable energies, hydrogen is certainly not the first option that will be applied. The option could, however, "enter the system" in an alternative way, if and when new markets for green hydrogen and electrolysis emerge, enabled by an appropriate incentives framework—be it in the transport sector or in industry such as refineries. These electrolyzers, when *not* run at full capacity, would then in fact represent a flexibility measure in their own right for the integration of surplus renewable electricity, through production of additional hydrogen. But in this case renewable hydrogen production, rather than the surplus issue would be the starting point. Clearly, the competiveness of hydrogen as storage option would be improved by its introduction as low-CO_2 energy carrier in other sectors.

Although residual electricity storage per se does not justify the construction of hydrogen caverns in the short term, cavern storage may enable other hydrogen applications and business cases in relation to its use in industry and transport, such as for (back-up) supply and trading, as distribution hub as well as for hydrogen import/export; given the volatile nature of the hydrogen production profile from surplus electricity, large-scale hydrogen admixture to the natural gas grid would also require buffering capacity in order to ensure a constant gas composition. Incentivizing the construction of hydrogen caverns for these applications, which may represent more economically viable use cases in the shorter term, would put the required infrastructure in place that could later on also be used for large-scale storage of electricity.

11.8 Perspectives and outlook

When looking at future energy systems, hydrogen offers a range of benefits as a clean energy carrier, and these benefits are receiving great attention as policy priorities. Hydrogen is a flexible energy carrier, which together with electricity could provide the backbone of a future, highly decarbonized energy system. While offering great promise as a low/zero-CO_2 energy vector, it is primarily as an alternative fuel in the transport sector that hydrogen will find its way into the energy system. This is because FCEVs have the potential to simultaneously respond to all major transport energy policy objectives, i.e., CO_2 emissions reduction, energy security, and reduction of local air pollution. A particular advantage of FCEVs, as compared to pure battery electric vehicles, which offer the same benefits, is that they achieve a comparable performance in terms of refueling time and driving range as incumbent ICE cars. The latter makes hydrogen and fuel cells well suited to electrify a wide range of road vehicles, ranging from small cars to buses and light duty trucks.

11.8.1 Hydrogen and fuel cells in transport

The widespread introduction of hydrogen as a vehicle fuel faces two major challenges: developing cost-competitive FCEVs and developing an infrastructure for hydrogen production, distribution, and refueling. Both the hydrogen supply side and the demand side must simultaneously undergo a fundamental transformation, as one will not work without the other. In addition, it is difficult to predict how fast consumers will accept and buy hydrogen cars, all the more in light of the current focus on BEVs and PHEVs, and the fact that there will not be a significant number of FCEVs for another couple of years.

However, considerable progress has been made toward commercialization of hydrogen as an alternative fuel over the last decade: several large hydrogen and FCEV demonstration programs have proved *technical* feasibility, while market studies and automaker projections have confirmed a favorable outlook for *economic* feasibility of FCEVs with volume production, after the initial phase of pre-commercial market development. The greatest progress has been seen in the development of fuel cell cars and buses, and systems for distribution and refueling of hydrogen, including the development of related international codes and standards. The focus is currently shifting to enable market development. In recent years, several infrastructure initiatives have emerged, and the planning and build-up of initial HRS networks have started in several countries. It can be said that market development rather than technology development is currently considered to be the main barrier for the introduction of hydrogen and the rollout of FCEVs.

But the development of a hydrogen refueling infrastructure still holds large risks and uncertainties. These mainly relate to the long period of underutilization of hydrogen stations and, as a consequence, the fact that there will not be a positive business case for retail operators for some 10–15 years. To overcome this challenge, hydrogen mobility requires a joint effort between all relevant and interested stakeholders. Hydrogen needs a collaborative approach and multilateral consortia to overcome the infrastructure challenge and share the risks, as the business case is too risky for a single company. The coalition-based approach of recent market development initiatives has proved successful so far to stimulate public–private collaboration on the commercialization of hydrogen vehicles and associated infrastructure.

Despite the progress that has been made, hydrogen (like any other alternative fuel) is unlikely to emerge in future energy markets without decisive and favorable policy support measures. Such measures need to be put in place and upheld long enough to create public awareness and stimulate consumer acceptance of hydrogen and to guarantee investment security for entrepreneurs, since significant industry investments are required for vehicle manufacturing and infrastructure build-up well before a real market emerges. International cooperation will also be crucial to establish transboundary hydrogen infrastructures because vehicles are driven, imported and exported across country borders. Last, but not least, the public will need to be trained and educated in the use of hydrogen technologies, for instance in the refueling of hydrogen cars.

What would kill the prospects for hydrogen as a fuel are the "ideal battery" offering "unlimited range" (as hydrogen is less efficient than electricity, even when used in a fuel cell) and/or "unlimited" supply of sustainable biofuels, because hydrogen is more cumbersome to distribute and use than liquid fuels. However, it is safe to assume that neither of this will come true. While the fraction of driving performed by electricity will undoubtedly grow, be it electric-drive vehicles powered by a battery or fuel cell (either directly or through a hybrid drivetrain), there is unlikely to be a "silver bullet," which satisfies all key criteria for economics, performance, and the environment. The road transport sector will rather witness a much more diversified and regionally fragmented portfolio of powertrains and fuels over the next decades, as all options are subject to constraints of some kind.

In the short to medium term, hydrogen will be additional to what biofuels and electricity can offer for energy security and CO_2 emissions reduction. In the long run, however, hydrogen holds promise to overcome some of the limitations of biofuels and electricity, allowing for further decarbonization of road transport. In particular, current battery technology suggests that full battery electric vehicles will primarily be an option for short-distance (urban) travel, due to their limited driving range; for extended range, they will need to rely on the combination with plug-in hybrid technology, and therefore be a partial solution only for CO_2 reduction in the transport sector. In this respect, hydrogen FCEVs are complementary, as they cover the entire driving spectrum. Regarding the economics, there should not be any doubt that the benefits of lower CO_2 emissions, clean air, diversification of primary energy sources and the transition to renewables all come at an initial cost to society. But these may ultimately marginalize with the reduction in battery and fuel cell costs, economies of scale, and potentially increasing costs for fossil fuels and ICE specifications.

11.8.2 Hydrogen in relation to renewable energy

Via electrolysis hydrogen also provides a potential mechanism and a source of flexibility that enables extensive integration of intermittent renewable energy sources into the energy system. In the near-term, the increasing need for flexibility options to facilitate ongoing implementation of volatile renewable electricity generation is not expected to drive the deployment of hydrogen for electricity storage on a structural basis. As electricity storage medium, hydrogen is in competition with other flexibility measures to integrate fluctuating renewables, such as expansion of the interconnection capacity between electricity markets, demand-side management, and various other storage options. To use its full potential, hydrogen needs to become an integral part of the energy system as universal energy carrier next to electricity, with its additional capability of electricity storage.

11.8.3 The hydrogen transition in perspective

The extent to which hydrogen is considered to play a role in the global energy system ranges widely across various scenarios. Hydrogen penetration is generally assumed in highly industrialized countries in scenarios with a strict climate policy, where

- the transport sector has to reduce CO_2 emissions significantly,
- renewable energies and/or CCS are deployed on a large scale,
- the oil price remains above 80–100 $/barrel in the medium and long term, and gas prices are also high, and
- there is no major technological breakthrough in vehicle batteries.

Energy systems and technologies evolve slowly—for instance, the combustion engine took more than a century to be developed and improved and reach the current level of penetration. Hydrogen and fuel cells will be no different, and it will take several decades for the build-up of a hydrogen infrastructure and for hydrogen to make a significant contribution to the fuel mix. The cumulative capital needed for the introduction of hydrogen and FCEVs should not be considered a deterrent when considered relative to the estimated investments required over the next decades in the energy sector in general.

But hydrogen should not be evaluated in isolation. Instead, it should be evaluated in conjunction with the various alternatives, as assessing its potential without taking competing options into account would result in misleading conclusions. This chapter does not attempt to give a definitive answer. Hydrogen will probably mainly replace oil-based fuels in the transport sector while other energy carriers like electricity will continue to play a role. Using the term "hydrogen economy" therefore may be misleading.

Today, there are a growing number of public–private partnerships in various parts of the world aiming at developing the market for hydrogen mobility. Will these partnerships at last be able to pave the way for the commercial introduction of hydrogen vehicles? Will hydrogen remain the fuel of the future? Perhaps in a couple of years we will know more…

Acknowledgments

This article was published in International Journal of Hydrogen Energy, Vol. 40, Issue 25, Michael Ball and Marcel Weeda, 'The Hydrogen Economy – Vision or Reality?', Page No. 7903-7919, Copyright Elsevier 2015.

References

Ball, M., Wietschel, M., 2009. The Hydrogen Economy—Opportunity and Challenges. Cambridge University Press, New York.

H2Mobility, 2010. 70 MPa hydrogen refuelling station standardization—functional description of station modules. Available from: http://www.now-gmbh.de/fileadmin/user_upload/RE_Publikationen_NEU_2013/Publikationen_NIP/H2Mobility_HRS_Functional_Description.pdf.

IEA (International Energy Agency), 2014. World Energy Outlook 2014. OECD/IEA, Paris.

IEA (International Energy Agency), 2015. Technology Roadmap—Hydrogen and Fuel Cells. OECD/IEA, Paris.

McKinsey & Company, 2010. A portfolio of power-trains for Europe: a fact-based analysis. Available from: http://ec.europa.eu/research/fch/pdf/a_portfolio_of_power_trains_for_europe_a_fact_based__analysis.pdf.

NOW (Nationale Organisation Wasserstoff- und Brennstoffzellentechnologie), 2012. Slide pack H2Mobility. Internal document.

US DOE (Department of Energy), 2014. DOE fuel cell technologies office record. Available from: http://energy.gov/sites/prod/files/14012_fuel_cell_system_cost_2013.pdf.

Further reading

E4tech and Element Energy, 2014. Development of water electrolysis in the European Union. A study for the Fuel Cells and Hydrogen Joint Undertaking. Available from: http://www.fch-ju.eu/sites/default/files/study%20electrolyser_0-Logos_0.pdf.

HyUnder, 2014. Assessing the potential, actors and business models of large-scale underground hydrogen storage in Europe. Project financed via the Fuel Cells and Hydrogen Joint Undertaking (FCH JU), under Grant Agreement No. 303417. Available from http://www.hyunder.eu.

IEA HIA (Hydrogen Implementing Agreement), 2015. Large-scale hydrogen delivery infrastructure—Final Report Expert Group Task 28 (forthcoming). Available from: http://ieahia.org/.

IPCC (Intergovernmental Panel on Climate Change), 2013. Fifth assessment report (AR5). Available from: http://www.ipcc.ch.

Roland, B., 2013. A roadmap for financing hydrogen refueling networks—Creating prerequisites for H2-based mobility. A study for the Fuel Cells and Hydrogen Joint Undertaking. Available from: http://www.fch-ju.eu/sites/default/files/HRS-finance_FinalReport.pdf.

The New Climate Economy Report, 2015. Available from: http://newclimateeconomy.report.

Building a hydrogen infrastructure in the EU

12

M. Steen
European Commission, Petten, the Netherlands

Abbreviations

AFI	alternative fuels infrastructure
BEV	battery electric vehicle
CEF	Connecting Europe Facility
CEP	Clean Energy Partnership (Germany)
CGH2	compressed gas hydrogen
CHP	combined heat and power
DoE	Department of Energy (US)
EASE	European Association for the Storage of Energy
EC	European Commission
EU	European Union
FCEV	fuel cell electric vehicle
FCH2JU	Fuel Cells and Hydrogen 2 Joint Undertaking (under Horizon 2020)
FCH-JU	Fuel Cells and Hydrogen Joint Undertaking (under the 7th Framework Programme)
FCTO	Fuel Cell Technologies Office (of US DoE)
GHG	greenhouse gases
HFC	hydrogen and fuel cells
HFP	(European) Hydrogen and Fuel Cell Technology Platform
HIT	hydrogen infrastructure for transport
HRS	hydrogen refueling station
ICE	internal combustion engine
ICT	information and communication technologies
IEA	International Energy Agency
LDV	light duty vehicle
LH2	liquid hydrogen
NG	natural gas
NOW	Nationale Organisation für Wasserstoff (Germany)
NPC	National Petroleum Council
NREL	National Renewable Energy Laboratory
ORNL	Oak Ridge National Laboratory
RES	renewable energy sources
SHHP	Scandinavian Hydrogen Highway Partnership
SNG	synthetic natural gas
TEN-T	Trans-European Networks for Transport

Compendium of Hydrogen Energy. http://dx.doi.org/10.1016/B978-1-78242-364-5.00012-9

12.1 Introduction: which hydrogen infrastructure(s) is/are required?

12.1.1 Historical overview

The realization of the need for and the subsequent development of a hydrogen infrastructure in the European Union have been triggered by the 2003 vision report (EC, 2003). This document recognized the potential that hydrogen as an energy carrier, coupled to the high conversion efficiency of fuel cell technology, offers to reach a number of goals that have been prioritized by EU climate and energy policy makers. Following the adoption of the EU integrated climate and energy policy in 2007, hydrogen and fuel cell (HFC) technologies have been consistently included in subsequent European Union policy documents (EC, 2007a, 2010, 2011a,b, 2012, 2013a,b). According to all these policy documents HFCs are expected to play an important role in achieving the EU vision of reducing greenhouse gas (GHG) emissions by 80–95% compared to 1990 levels by 2050 and in transitioning the EU to a low-carbon economy. Recently, the potential of hydrogen for energy security and energy efficiency, particularly as an alternative fuel in transport, has received more attention (Council Conclusions, October, 2014). In addition to contributing to energy, climate and transport policy objectives, fuel cells and hydrogen also contribute to industrial policy goals (FCH-JU, 2013a).

As a consequence of the policy push, the European industry and research communities have outlined a common vision for the future deployment of HFC technologies. The 2007 Implementation Plan of the European Hydrogen and Fuel Cell Technology Platform (HFP) presented a stakeholder-consented comprehensive long-term road map for Europe, which identified vehicle refueling stations and a portfolio of cost-competitive hydrogen production, storage and distribution processes as infrastructural priorities (EC, 2007b). This roadmap served as a guiding document for the industry-led public–private partnership "The Fuel Cells and Hydrogen Joint Undertaking" (FCH-JU) that was subsequently set up to substantially accelerate the development and market introduction of these technologies. Now in its second phase (labeled FCH2JU), its overall objective of FCH2JU is to implement an optimal research and innovation program at the EU level to develop a portfolio of clean and efficient solutions that exploit the properties of hydrogen as an energy carrier and fuel cells as energy converters to the point of market readiness by 2020 (Council Regulation (EU) No 559/2014).

12.1.2 Outlook

There is a wide consensus that HFC technologies can already provide benefits in all energy sectors—power, transport, heat and industry—through their use in a variety of applications, including distributed energy and combined heat and power (CHP), backup power, portable power, auxiliary power for trucks, aircraft, rail, and ships, specialty vehicles (such as forklifts), and passenger and freight vehicles, including cars, vans, and buses. An important specific advantage of hydrogen-based solutions is that they are extremely scalable in terms of capacity, which enables their use in applications from the watt to the megawatt range. In recent years, systems for storing and transmitting renewable energy have received increasing attention.

At present, hydrogen is mainly used as an industrial scale commodity product for petroleum refining, fertilizer production, metals processing, plastics manufacturing, etc. These large users generally produce the hydrogen on site (captive) or are supplied by nearby producers (merchant). Full realization of hydrogen's potential benefits, both as an energy carrier (fuel) and as an energy storage medium, relies on high-volume penetration of renewable hydrogen in an increasing number of applications in the power, heat, industry and transport sectors.

The "Snapshot 2020" of the HFP projects a hydrogen demand in Europe ranging from 0.86 to 2.5 Mt (EC, 2007b). A recent study predicts a global hydrogen fuel consumption of 1.4 Mt in 2020 and of 3.5 Mt in 2030, out of which about 0.15 Mt, respectively, 0.6 Mt in Europe (Navigant Research, 2013). This demand for hydrogen fuel is much lower than the overall 2014 demand of 69 Mt/year for hydrogen as fuel and as a commodity product (NPC, 2012, chapter 15). A recent projection of the future hydrogen market in Europe identifies mobility and industry as major growth areas as of 2020, with consumption reaching over 10 Mt/day in 2030 and above 30 Mt/day in 2050 (Shell Global Solutions International, 2013). In the longer term, there are opportunities in distributed CHP. For mobility, next to its use in fuel cell electric vehicles (FCEV), hydrogen has been earmarked as a suitable alternative propulsion fuel for other transport modes, with the exception of long-distance heavy-duty road, aviation and sea shipping (EC, 2013a). For industry, growth lies mainly in industrial combustion where hydrogen (potentially blended with natural gas (NG)) reduces emissions at a similar cost and with less complication than postcombustion retrofit carbon capture and storage.

12.1.3 Infrastructure needs

To enable high-volume penetration in future, infrastructure(s) that allow(s) for the distribution of larger amounts of increasingly renewable hydrogen to meet the needs of a higher number and a more diversified set of end users over larger and more geographically diversified areas than before is essential. Such infrastructure(s) must meet requirements of safety, energy efficiency, and cost-effectiveness.

Infrastructure for use of hydrogen as a fuel covers the equipment and facilities needed to deliver hydrogen from where it is produced to the point of end use. This includes pipelines, trucks, railcars, ships, and barges used for transporting it, as well as the facilities and equipment needed for loading and unloading. It also covers storage facilities, compressors, and dispensers for delivering the hydrogen. For energy storage purposes, infrastructure additionally covers the facilities and equipment needed for feeding and retrieving hydrogen to and from its storage location. Depending on the storage location and technique, this may include compressors, purifiers, separators, heat exchangers, etc.

The distance between production and end-use sites greatly affects the design and extent of a hydrogen infrastructure. In this respect it is convenient to differentiate between centralized and distributed production. The production of small volumes of hydrogen at the point of end use is known as distributed production and requires installations and equipment for hydrogen storage, compression and dispensing, but does not need a transport infrastructure. In the case of mobility, because demand for hydrogen will initially be low, distributed production is the most viable approach for introducing hydrogen in the near term. Distributed production methods include reforming of NG or liquid fuels and small-scale water electrolysis. In the longer term large central hydrogen production facilities (750 t/day) will be needed to meet the higher demand. Compared to distributed production, centralized production requires higher capital investment and additionally a substantial hydrogen infrastructure for transport and delivery. Intermediate-size hydrogen production facilities (5–50 t/day) located nearer (up to 200 km) to the point of use have lower transport and infrastructure costs. As is the case for centralized production, they can take advantage of economies of scale and are therefore likely to play a role when hydrogen demand increases.

This discussion shows that for hydrogen as a transport fuel an optimum system trade-off between production (feedstock, technology, location) and delivery options has to be struck. Additionally, the characteristics/components of a hydrogen infrastructure depend on the physical form in which hydrogen is transported. This is discussed further in the following paragraphs.

Enhanced concerns about EU energy security have not only stressed the importance of transitioning toward alternative transport fuels including hydrogen, but moreover have recently increased the attention that is being paid to hydrogen for energy storage purposes and to the considerable advantages that fuel cells offer in terms of efficiency, reliability, and quality, particularly for back-up power applications. A timely and successful integration of HFCs in appropriate locations of the energy, transport and industry chains, and in their contribution in facilitating the interconnection of these chains (e.g., power-to-gas) can contribute greatly to energy security. The identification and exploitation of the integration potential of HFC technologies in linking these chains require a regionally diversified systems approach and consideration and exploitation of other technologies, in particular information and communication technologies (ICT).

In the more remote future, high-temperature thermochemical reduction of water and of carbon dioxide or their coelectrolysis into hydrogen and carbon monoxide offer great promise for higher energy independence. When not used directly as a fuel, hydrogen can be combined with carbon monoxide to form synthetic natural gas (SNG), synthetic liquid fuels and chemicals that can be transported using existing infrastructures. Also, because of the versatility of hydrogen as an energy carrier, long-distance shipping from regions with a surplus of renewable electricity (hydro, wind, sun) that allows cheap hydrogen production by electrolysis to densely populated regions with high energy demand is being investigated (e.g., Canada–EU and Norway–Germany maritime corridors and Norway–Japan polar route). All these evolutions are expected to lead to additional requirements for a hydrogen infrastructure, besides that required for its production and delivery as a fuel for transport.

12.2 Current status of hydrogen infrastructure

At present, most hydrogen is produced at or near the location of its use. For industrial use at a distance from the production location, hydrogen is transported by pipeline because this is the most cost-effective method at large volumes. Europe currently has around 1600 km of hydrogen pipelines, with the largest network covering northwest France, Belgium, and part of the Netherlands. Next to pipelines, hydrogen is also transported by road, by rail, or by barge in cryogenic liquid tanks or gaseous tube trailers. However, the capacity of the existing infrastructure, similarly to the current production capacity, is by far not sufficient to support the widespread use of increasingly renewable hydrogen as a fuel or as an energy storage medium.

12.2.1 *Hydrogen infrastructure for fuel delivery*

Compared to fossil fuels, hydrogen has a lower volumetric energy density, which makes its transport, storage, and delivery less efficient. Transport, storage, and delivery hence constitute a major part of the overall cost of hydrogen fuel.

A hydrogen fuel delivery infrastructure has to ensure the correct balance between production locations and delivery options, considering the availability of feedstocks and the proximity of use centers. Hence, the type and size of market (urban, rural, cross-border) will affect the delivery infrastructure, which will likely include a number of technologies. In particular an FCEV refueling infrastructure needs to cover large geographical areas and will initially consist of relatively low-capacity fueling stations. Early infrastructure development is hence expected to be dominated by gaseous and liquid truck delivery or on-site production, rather than by pipeline delivery to the refueling stations. For stations without on-site generation, the different delivery options—trucking-in compressed gas (cylinders for 0.5–50 kg, tube trailers for 100–300 kg), trucking-in liquid hydrogen (2000–2400 kg) and pipelines—require different equipment at the station. As the demand for hydrogen grows and as the performance of transport, storage and delivery technologies improves, infrastructure options for hydrogen delivery for FCEVs are expected to evolve. A summary overview of delivery options with their potentials and limitations is given in Figure 12.1 (IEA, 2013).

Whereas hydrogen-fueled vehicles are expected to be powered by fuel cells because of their much higher efficiency, the EU nevertheless considers that in view of the likely dominance of internal combustion engines (ICE) in road vehicles in the short and medium term, adapting ICE to hydrogen or to mixtures of hydrogen and NG can contribute to a smooth transition (Commission Regulation (EU) No 630/2012).

In areas and conditions where NG is cheap and electricity is expensive, trigeneration fuel cell systems fed by NG can generate high CHP revenues, which result in low-cost hydrogen that can be used as vehicle fuel. This reduces the dependence on the number of FCEVs in the initial deployment phase of hydrogen refueling stations (HRS). However, the lack of scalability of this approach limits its potential for wider use.

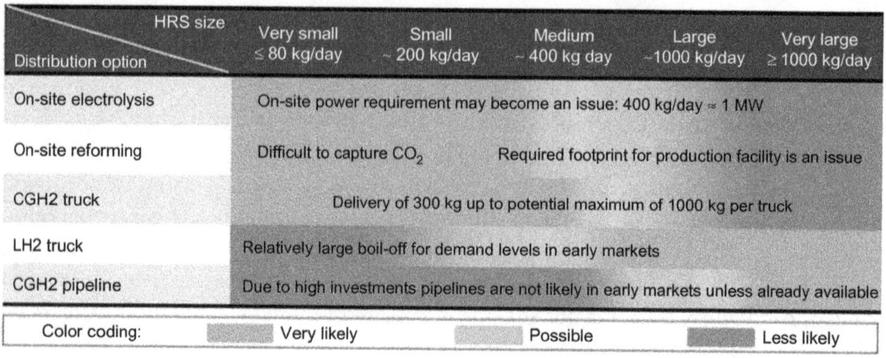

Figure 12.1 Overview of delivery options for a hydrogen infrastructure for road transport (IEA, 2013).

12.2.1.2 Compressed hydrogen gas

Transporting gaseous hydrogen via existing pipelines (40–80 bar) is the cheapest option for delivering large volumes of hydrogen. Pipeline transport exploits the high energy transmission capacity of hydrogen (four to five times that of electricity using high-voltage direct current, respectively 27 and 6 GWh/h) and of the same order of NG (38 GWh/h). This high transmission capacity is of particular advantage for hydrogen production from renewable energy sources because many of these are located far from major use centers. The high capital costs of pipeline construction (including right-of-way) and of the compressors, however, constitute barriers to expanding hydrogen pipeline delivery infrastructure. In addition, more reliable and lower-cost hydrogen compression technology is needed. Hydrogen pipeline delivery also meets with several technical barriers, including hydrogen embrittlement, the need for improved sealing and joining technology, and techniques to detect and control hydrogen permeation and leakage to ensure safe operation.

In compressed gas form, hydrogen can also be transported in seamless steel vessels at pressures up to 200–300 bar. This method is most cost-effective for delivery over distances of a few hundred kilometers from centralized production sites to end users with a small hydrogen demand (order of tens m^3/h). The use of increased pressure (up to 700 bar) and of composite materials for the containers allows the increase of trailer payload from about 350 kg to more than 1000 kg. This threefold pressure increase reduces delivery frequency by the same factor, thereby decreasing transport-related CO_2 emissions and reducing delivery costs. Such high capacity trailers are more cost-effective for large delivery distances (>100 km) and rather large HRSs (>300 kg/day). Higher pressures also reduce compression needs at the end-use site. However, this modular delivery option cannot benefit from economies of scale when demand increases.

12.2.1.3 Liquefied hydrogen

With increasing transport distance, liquid hydrogen becomes more cost-effective because for the same volume its higher density enables a much larger mass to be stored and transported. Liquefaction requires cooling gaseous hydrogen to below −253 °C.

Liquid hydrogen is transported in cryogenic tanks of around 3500 kg capacity to consumption sites where it is stored in insulated tanks. For subsequent use it is compressed and vaporized to high-pressure gaseous hydrogen. Liquid hydrogen suffers from losses through evaporation ("boil off"), especially in small tanks with large surface-to-volume ratios. Additionally, liquefaction is expensive because the energy needs are high (12 kWh/kg), which also reduces the efficiency of the full chain from hydrogen generation to end use ("well-to-pump") and increases the carbon footprint. However, recent research has been enabled nearly halving these energy needs (FCH-JU, 2013b). The energy penalty for liquefaction is also partially offset by reduced energy use for compression and for precooling before dispensing at the HRS.

12.2.1.4 Cryocompressed hydrogen

Cryocompression in insulated composite tanks exploits the combination of pressure and low temperature to increase the volumetric and gravimetric storage density and reduces the energy needs for liquefaction. Current high-pressure cryogenic tanks for on-board storage operate at 200–400 bar and can be filled with either compressed hydrogen gas (ambient to cryogenic temperatures) or liquid hydrogen (U.S. DRIVE Partnership, 2013). Filling can be much faster with cryocompressed hydrogen than with gaseous, and much of the liquefaction energy can be recovered through autogeneous pressurization of the tank (US DoE FCTO, 2014). The 2.5 times higher storage density than that of compressed hydrogen enables larger driving ranges and decreases the number of filling pressure cycles. Also the fiber loading of composite vessels can be reduced by up to a factor of four, resulting in a considerable cost reduction.

12.2.1.5 Liquid hydrogen carriers

Another delivery process consists of central production of a liquid hydrogen carrier, its distribution using existing infrastructure (pipelines, trucks, ships), and its processing at the end-use site to produce hydrogen. Some carriers may require handling precautions and the additional end-processing step consumes energy. Both factors limit the applicability of liquid carriers for onboard vehicle hydrogen storage. However, they are suitable for transporting hydrogen to filling stations or as a buffer in renewable energy systems. A specific example is the use of liquid organic hydrogen carriers, as currently investigated in Japan (Okada and Shimura, 2012).

12.2.1.6 Solid hydrogen carriers

Hydrogen can be stored in atomic form in solid materials such as metal hydrides and carbon or other nanostructures. Such storage requires cooling to adsorb hydrogen and heating to release it. Because stationary off-board storage does not have the same weight and volume restrictions as onboard vehicle storage, solid carrier systems that do not meet the goals for onboard storage might be effective for stationary storage. To increase the volumetric density, solid carriers can be used within a gas storage tank.

12.2.2 Development of hydrogen delivery infrastructure for road transport

The highest visibility of hydrogen as a fuel is its use to power fuel cells in light-duty vehicles (LDVs). The use of fuel cells to power other road vehicles, speciality vehicles, trains and boats, is less demanding for a delivery infrastructure because these vehicles are mostly operated in fleets. On the other hand, for FCEVs the development of a network of hydrogen HRSs is a prerequisite.

Key challenges for the deployment of a hydrogen infrastructure for road transport are the significant capital requirement for equipment and the large physical footprint (including safety distances) of refueling stations. The economic viability depends on the scale and utilization of the installed fueling capacity. Technology advancements in compression and storage at stations are necessary to reduce capital and operating costs, to decrease land requirements, and to increase fueling capacity. Also, the economics of renewable hydrogen production methods needs considerable improvement (NPC, 2012, chapter 4), including for small-scale electrolyzers at HRSs (FCH-JU, 2014a).

Currently efforts are ongoing in a number of regions worldwide to establish hundreds of HRSs to support tens of thousands of FCEVs, backed by hundreds of millions in public funds and billions in private investment (Ogden et al., 2014). This interest was triggered by the joint statement of seven large car manufacturers (EU, US, Japan, Korea) in 2009 that they intended to commercialize a significant number of FCEVs from 2015 onward. To enable this, they urged the oil and energy industries and government organizations to support the development of hydrogen infrastructure.

Over the past years, HRSs have been demonstrated in Europe at different sizes from stations supplying small demonstration fleets, to stations capable of serving highly frequented public locations. Where hydrogen is not locally produced, in the short term it will be trucked in. In the medium term liquid hydrogen storage and distribution are needed for a commercially viable expansion of the area supplied by central hydrogen production sites.

For LDVs 700 bar refueling technology is well established, while 350 bar is used for buses and forklifts. A standardized refueling interface ensures interoperability between the FCEV and the HRS. Precooling of the hydrogen and infrared communication between the vehicle and the station have allowed refueling times of 3–5 min to be reached. The validation of alternative protocols (e.g., mass capacitance method) is expected to improve refueling rates and reduce energy and equipment costs.

Although current hydrogen compression technologies meet most of the performance requirements, they lack significantly in reliability (Sandia, 2014). Compressor redundancy is hence required at commercial HRSs, which adds considerably to their costs. Research targets increased reliability of compression technologies, as well as new compression concepts, such as ionic liquid piston compression and electrochemical compression, easily scalable technologies that promise a step-change in terms of energy efficiency and costs. When hydrogen is produced locally by high-pressure electrolyzers, the amount of required postproduction compression can be reduced.

Hydrogen storage at the HRS represents a second major challenge. Improved understanding is needed of the effects of high-pressure charge/discharge cycles as well as of environmental effects (heat, moisture, etc.) on the integrity of gaseous storage vessels. Currently, Type I or Type II vessels are mostly used. In future, with cost reduction of tank manufacturing and of carbon or alternative fibers, Type III and Type IV high-pressure tanks will become more cost-effective for higher-pressure buffer storage at the HRS.

The stringent hydrogen purity requirements for fuel cell stacks on board FCEVs necessitate quality assurance of the dispensed hydrogen, which is hampered by the lack of simple and low-cost instrumentation. Moreover, the maximum allowable impurity levels need revision to strike the appropriate balance between the cost of purifying hydrogen produced at the HRS and the lifetime expectancy of the fuel cell stack. Modern pressure-swing absorption systems can remove most contaminants to safe levels at a reasonable cost (U.S. DRIVE Partnership, 2013). Liquefied hydrogen results in very high purity.

Another remaining issue is the metering accuracy of hydrogen dispensers because current accuracy levels ($\pm 3\%$) are not sufficient for billing purposes.

The technical hurdles identified here are exhaustively described in a recent report (Sandia, 2014) and are currently tackled through a global collaborative effort in a dedicated forum composed of industrial stakeholders and government institutions from Germany, Japan, United States and the EU. In two workshops held so far (Berlin, June 2013 and Los Angeles, May 2014), a gap analysis of the required standards has been made and prioritization of and collaboration in the needed prenormative research agreed upon.

12.2.3 Hydrogen infrastructure for bulk energy storage

Gaseous pressure vessels are the most common means to store hydrogen for buffering against the mismatch between supply and demand. Storage pressures range between 130 and 1000 bar. Although more expensive, underground cryogenic liquid hydrogen storage offers advantages of larger storage capacity per unit volume and of reduced footprint. For an HRS, underground storage is inherently safer and hence can reduce safety distances, which is particularly advantageous for urban HRSs.

A fully deployed hydrogen fuel delivery infrastructure will require large-scale bulk storage to cope with demand variations. A different need for large-scale hydrogen storage stems from the requirement to maintain grid stability with increasing amounts of intermittent renewable electricity sources in the power generation mix. This requires capabilities for energy storage throughout the power chain, next to dispatchable power and demand-side management. Hydrogen offers considerable potential in all three respects. Because of its high energy density, it is indeed one of the very few options available for meeting the high-capacity, longer-term energy storage needs (e.g., EASE, 2013). When produced by electrolysis, it can contribute to grid stability through both supply management (by providing dispatchable power when coupled with large scale fuel cells or hydrogen gas turbines) and demand management (through fast response time and good partial load performance). The latter is particularly attractive for small-scale electrolyzers sited at refueling stations and has the added advantage that it does not require a distribution infrastructure. For optimum exploitation of the increasing amounts of intermittent renewable electricity, efficient, flexible, and cost-effective electrolysis is essential. A recent study has identified that this requires major further developments in electrolyzer technology and scale (FCH-JU, 2014a).

Similarly as for storage of NG, a number of geological formations (salt caverns, aquifers, depleted oil, or gas reservoirs), as well as specially engineered rock caverns may be used for hydrogen storage. Most of these sites use storage pressures of 80–160 bar. Salt caverns are particularly suitable because hydrogen does not dissolve the salt and depleted reservoirs can benefit from available transport infrastructure. Aquifers are also attractive, due to their natural occurrence, availability, and low capital cost requirements. Mined caverns require an impermeable layer to prevent gas escape and are hence much more expensive. Additional research is needed on the chemical interaction between hydrogen and confinement materials to allow improved assessment of the integrity of the storage and identification of possible contaminants in the hydrogen upon its discharge from the storage.

For hydrogen storage at a large distance from the production location, a dedicated pipeline to the storage site is likely required. Because it has a 30% lower energy content by volume than NG, hydrogen transport by pipeline requires higher pressures, with implications for the construction materials. Since current compressor technologies do not meet the efficiency, scale and cost requirements to enable the widespread use of hydrogen as an energy carrier, performance improvements in compressor technology are critically needed. Moreover, whereas NG pipelines operate mostly at constant pressure, hydrogen pipeline transport for energy storage applications is characterized by large pressure variations. Pipeline construction materials are hence subject to fatigue loading, which may exacerbate their susceptibility to hydrogen embrittlement. An overview of the issues and of the state-of-the-art for hydrogen pipelines is given in US DoE FCTO (2014).

An alternative way of exploiting the grid-stabilization potential of hydrogen is to produce it by electrolysis from renewable electricity and inject it in the NG grid ("power-to-gas"). This allows increasing the share of renewable energy sources (RES) in the NG grid, and from there in the end-use applications of transport, heat and industry, where achieving higher RES shares is technically more difficult and more expensive than for power generation. The power-to-gas concept allows integrating electricity, heating, transport and industrial processes, thus adding flexibility in the energy system as a whole, reducing its vulnerability and increasing its overall efficiency. The concept can be extended to include grid electrolysis, thereby allowing both supply and demand management.

The injection in the NG grid exploits the huge storage potential of that grid and avoids the necessity for dedicated hydrogen storage facilities and their related infrastructure. When the electrolyzer is located close to the NG grid, the infrastructural requirements are limited to the equipment needed for controlled injection at a given pressure and flow-rate. The electrolytic hydrogen can either be injected directly up to certain amounts or after combination with carbon dioxide into SNG. In the former case, apart from safety issues at and downstream of the injection location, the major issue is the compatibility of the gas blend with the materials of the pipelines, seals and valves. Compatibility issues have been extensively studied in a number of projects (e.g., Altfeld and Pinchbeck, 2013). When pure hydrogen is needed at the point of end use, high yield, efficient, reliable and low-cost separation technologies are required. When SNG is injected in the NG grid, no modifications to existing transmission and distribution grids or to appliances are needed.

12.2.4 Safety

Infrastructure enabling the use of hydrogen as a fuel or as an energy storage medium requires means to detect hydrogen leakage over the whole chain from production to end use. This is important from both safety and economic perspectives. The use of odorants is not suitable because of the high hydrogen purity required for operation of low-temperature fuel cells. Hence, hydrogen pipelines, storage, refueling sites, and any enclosed area where hydrogen may be stored and/or used should be assessed for the need of hydrogen detection sensors.

12.3 Costs for setting up the hydrogen infrastructure

12.3.1 Capital costs for hydrogen production

The expected evolution of typical capital costs for hydrogen production compiled from different sources is shown in Figure 12.2 (EC, 2013c), together with the targets of the FCH2JU which consider production from renewable sources (FCH2JU, 2014a). The projected cost reductions result from incremental technology performance improvements and from scale effects of large-number manufacturing.

12.3.2 Hydrogen delivery costs

Delivery costs compiled from a number of sources are shown in Figure 12.3 (EC, 2013c). The FCH2JU targets refer to the overall refueling cost of hydrogen (production, delivery, compression, storage and dispensing, but exclusive of taxes).

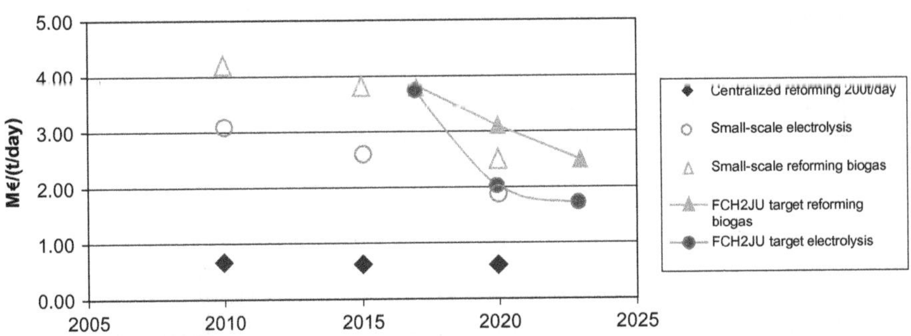

Figure 12.2 Evolution of capital costs for hydrogen production.

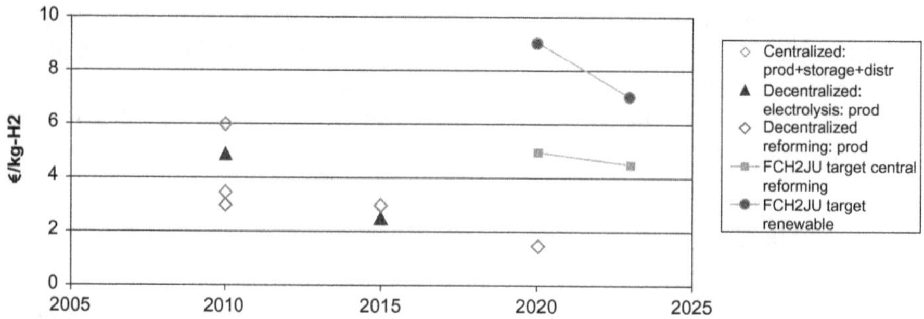

Figure 12.3 Cost evolution for hydrogen delivery.

Comparing the targets with the actual production costs reveals that compression, storage and dispensing at the HRS constitute a major part of the delivery cost. These HRS-related costs depend on the size of the HRS, decreasing with increasing station capacity. For a large HRS (700 kg/day) the projected cost for 2030 is €0.8/kg, compared to €1.5/kg for a 168 kg/day station (FCH-JU, 2013c). The relative importance of the share of the HRS in the overall delivery costs is also shown in Figure 12.4 which summarizes US analyses for a number of production and delivery pathways (US DoE Fuel Cell Technologies Office, 2012 and 2013). Breakdown of costs at the station into compression, storage and dispensing cost are given in NREL (2014).

US-DoE FCTO (2012) has set the delivery cost target at $2/kg ($0.7/kg attributable to the HRS) and the overall hydrogen cost at the pump at $4/kg to become competitive with gasoline in hybrid electric vehicles in 2020 (untaxed). Cost targets in the EU are

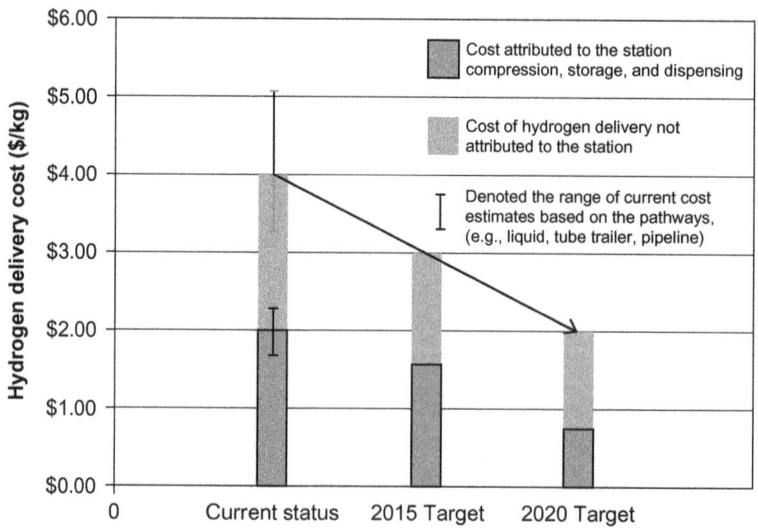

Figure 12.4 Average hydrogen delivery cost based on nine 700 bar delivery scenarios and 2015 and 2020 targets (US DoE FCTO, 2012).

less stringent than in the US because of the lower US petrol tax, which imposes lower cost targets for new technologies to become competitive.

12.3.3 Hydrogen fuel infrastructure investment costs

Scenarios of FCEV roll-out in the EU have indicated that the required capital for hydrogen infrastructure development is of the order of tens of billions of Euros during the first 5–10 years. A 2007 study indicates that the cumulative investment costs until 2027 for the deployment of a hydrogen refueling infrastructure in 10 EU member states based on a range of production and delivery options, amount to 60 B€, half of which is for hydrogen production (Hyways, 2007). According to a more recent study (FCH-JU, 2010), the deployment of 1 million FCEVs by 2020 would require 3 B€ investment for production, distribution and retail, out of which 1 B€ is for retail. The total investment for roll-out of a hydrogen supply infrastructure that meets the needs of 25% of FCEVs in the LDV fleet in 2050 (about 70 million FCEV) amounts to 100 B€, with approximately half for production (see Figure 12.5).

To put these costs in perspective, it is noted that the average annual investment needs of 2.5 B€ (100 B€ over 40 years) are far lower than those for other industries, such as oil and gas, telecommunications and roads, which each amount to 50–60 B€. Translated into the lifetime cost of ownership of a FCEV, the costs for distribution from the production site to the HRS and for operational and capital costs for HRS amount to about 2000€, i.e., approximately 5% of overall FCEV costs (FCH-JU, 2010).

Similar results are obtained for the United States where studies of a national roll-out of FCEVs indicate that the hydrogen infrastructure development (including production) during the first 5–15 years will cost tens of billions of dollars (ORNL, 2008; NRC, 2008, 2013). The range of investment for hydrogen distribution and dispensing

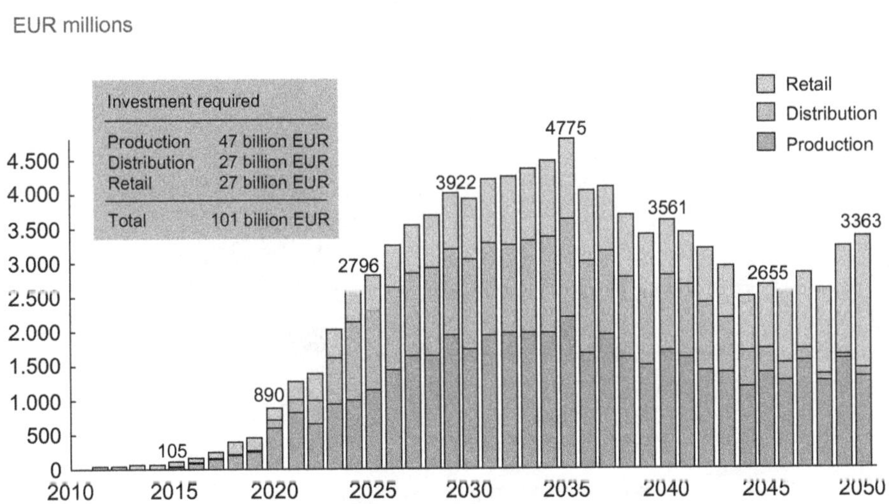

Figure 12.5 Evolution of required investments for a hydrogen infrastructure enabling the deployment of 70 million FCEVs in the EU by 2050 (FCH-JU, 2010).

infrastructure needed to displace one-third of 2012 US gasoline consumption is estimated at 275–430 B$, with the cost for hydrogen production 30–90 B$ (based on centralized steam methane reforming). Assuming full utilization of HRS capacity, the fraction attributable to dispensing infrastructure ranges between 7% and 8.5% of the total cost of driving FCEVs in 2050, increasing to 11–15% for a utilization rate of 30% (NPC, 2012, chapter 5).

12.3.4 Large-scale energy storage infrastructure costs

The role of hydrogen storage is to accommodate the intermittency of renewable electricity. The electrolyzers hence have to operate discontinuously and have to be of large capacity for grid support purposes. Low capital costs and the availability of large amounts of excess renewable electricity (in quantity and in time) are hence prerequisites for a favorable business case for hydrogen production. As indicated previously (FCH2JU, 2014a) capital costs for renewable electrolysis are targeted at 2.0 M€/(t/day) and 1.5 M€/(t/day) for 2020 and 2023, respectively, from a current cost of 8.0 M€/(t/day).

Total capital costs as well as levelized costs per kg for compressed hydrogen storage in different types of geological formation are shown in Figure 12.6 for storage of 7000 ton hydrogen at 140 bar (Sandia, 2011). The FCH-JU target costs for storage at pressures above 80 bar are € 6000/t.

The overall business case for hydrogen as a renewable energy storage medium depends on the value generated by its use after storage. Converting hydrogen back into electricity for the grid, using fuel cells, suffers from low efficiency (30–35% power-to-power) and capital cost of fuel cell systems. Because of this, competitiveness of hydrogen as a storage option depends on its successful application as a low-CO_2 energy carrier in other sectors, most importantly for hydrogen mobility. This is also apparent from an ongoing study (FCH2JU, 2014b) which indicates that conversion of

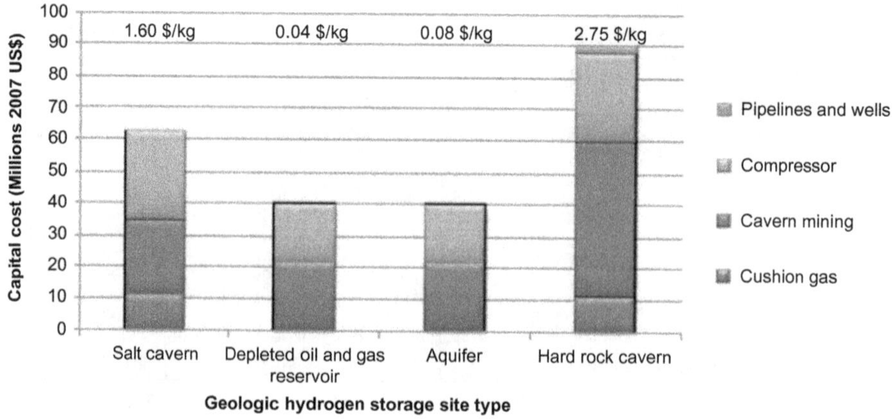

Figure 12.6 Capital cost components and overall levelized costs for hydrogen storage in geological formations (Sandia, 2011).

renewable electricity to hydrogen for use outside the power sector, i.e., in the gas grid (power to gas), for mobility or in industry has the potential to exploit nearly all excess renewable electricity that would otherwise (have to) be curtailed, thereby substantially contributing to the decarbonization of these sectors. Realizing this potential requires that there is either local demand for hydrogen at the production site or that the hydrogen can be economically transported to a demand center. Because of the volumes involved, pipeline transmission is the obvious choice.

12.4 Status and outlook of EU hydrogen infrastructure initiatives

12.4.1 Hydrogen as a fuel for road transport

Many of the HRSs in operation in the EU have been cofinanced by regional and local authorities running or financing captive fleets. A detailed overview of such early efforts is provided in "Impact Assessment Accompanying the Proposal for a Directive on the Deployment of Alternative Fuels Infrastructure" (EC, 2013d). The first industry initiatives to establish national networks of stations are the H2 Mobility initiative in Germany (NOW, 2011), with similar initiatives in the United Kingdom (FCH-JU, 2012), France and Switzerland mostly focused on refueling passenger cars. The German program is the first and most ambitious in the EU because the deployment of a hydrogen fuel infrastructure for road transport is considered critical for reaching the transport-related objectives of the German Energy Policy (BMVBS, 2013). To reach these goals, an increase of the renewable energy share in transport, electrification of the drive train and large-scale hydrogen storage are key (NOW, 2014).

To provide FCEVs with a sufficient number of HRSs in the early years of infrastructure deployment, the concept of geographically confined clusters of HRSs, with a number of additional HRSs along major traffic corridors interlinking the clusters has been adopted worldwide. In the majority of cases, the identified initial clusters are densely populated areas. This supply-driven linking-of-clusters approach results in a hydrogen availability that exceeds early demand, which causes HRSs to operate initially at a loss because of low utilization. The duration of the loss period depends, among many other factors, on the number of FCEVs, hydrogen costs, and price of conventional fuel. When at a later stage a sufficient number of vehicles is deployed and infrastructure utilization is high, further infrastructure deployment and operation is projected to be profitable and sustainable (NPC, 2012, chapter 5). This is illustrated in Figure 12.7, which shows the evolution of cash-flow versus time for the deployment of the HRS infrastructure in the United Kingdom (Hayter, 2013).

In a similar exercise for France, the magnitude of the financing need is estimated at 600 M€ over the first 10 years of HRS deployment, with cashflow break-even similarly in the late 2020s (AFHYPAC, 2014). However, next to supply-driven cluster concepts, an HRS build-up starting from demand-driven clusters is also possible, such as that followed by France. Because of the low-CO_2 electricity supply and lower electricity prices, battery electric vehicles (BEVs) have a larger market share in France

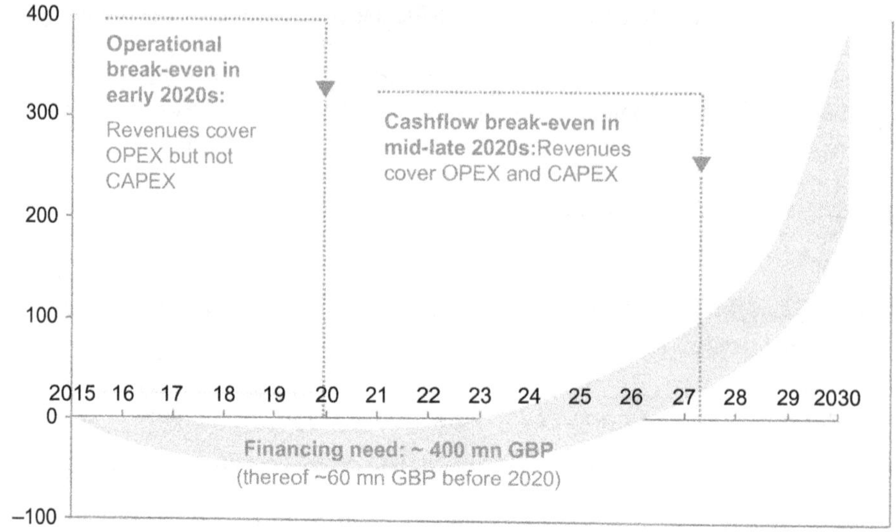

Figure 12.7 Illustrative free cash flow development from HRS investments and operations (£ millions) (Hayter, 2013).

than in other EU countries. By equipping BEVs operating in captive fleets with fuel cells, their range can be extended to serve a higher number of customers in sparsely populated areas. Such captive fleets (which may also include taxis, delivery vans, etc.) provide a predictable hydrogen demand around a few large HRSs with high utilization that can be exploited for fueling FCEVs. The high utilization greatly reduces the initial investment requirement for HRS, whereas the smaller fuel cells and lower pressure onboard storage tanks considerably reduce the vehicle costs. France also has experience with low-pressure onboard hydrogen storage using metal hydrides, which reduces the complexity of the fueling infrastructure but results in significantly longer fueling times. The linking of demand-driven clusters in France reduces the investment needs from 600 to 115 M€ over the first ten years of deployment (AFHYPAC, 2014).

An overview of the planned efforts for establishing public HRSs in a number of Member States is given in Table 12.1[1]. The cumulative number of HRSs is expected to increase from around 250 in 2020, over 1100 in 2025 to 3000 in 2030.

12.4.2 Hydrogen infrastructure for stationary applications

Hydrogen infrastructure initiatives for the energy sector are still primarily in the R&D phase in Europe. A number of power-to-gas demonstrations is ongoing, particularly in Germany (Iskov and Rasmussen, 2013), and feasibility studies for large-scale underground storage have been performed (FCH-JU, 2014b). Some of the power-to-gas projects include injection into the NG grid. In some Member States hydrogen

[1] Based in part on information provided at meetings of the Informal Government Support Group on European H2 mobility, July and November 2014.

Table 12.1 Projected number of HRSs in time period prior to full deployment and associated financing needs

	HRS in operation (2014)	Targeted number of HRSs (by year)	Initiatives	Financing need
DE	15	50 (2015), 400 (2023), 1000 (2029)	H2 Mobility, CEP, NOW	350 M€ for 400 HRSs by 2023
UK	9	65 (2020), 330 (2025), 1150 (2030)	UK H2 mobility	418 M£ (62 M£ by 2020)
Scandinavia	9	45 (2015), 150 (2020), 300 (2025)	SHHP, HIT-I, HIT-II	
F	8	15 (2019), 319 (2024), 247 (2030)	France H2 Mobility, HIT-I	115 M€ for 247 HRSs by 2030
NL	2	3 (2015), 30 (2017)	H2 Mobility NL, HIT-I, HIT-II	
AT	1	3 (2015), 15 (2020), 100 (2030)		

Note: Acronyms are explained in Abbreviation list.

infrastructure for stationary applications is included in the transition planning to a low-carbon economy. At the EU level, FCH2JU has prioritized research and innovation into hydrogen production from renewable electricity for energy storage and grid balancing, as well as hydrogen production with low carbon footprint from other resources and waste hydrogen recovery (FCH2JU, 2014a).

12.5 Moving toward full deployment

12.5.1 *Road transport*

12.5.1.1 *Overcoming barriers/challenges*

A recent coalition study (FCH-JU, 2013d) identifies four major challenges for HRS roll-out:

- High initial investment needs and underutilization, leading to an unattractive business case
- Late breakeven in Net Present Value (NPV: difference between discounted cash in and outflows accumulated over time)
- Uncertainty about FCEV ramp-up (delayed or lower numbers)
- Competition between station providers in the late roll-out phase

To overcome these challenges and jump-start HRS rollout, public support is needed, covering both financial and legislative measures. Such support is well justified, as HRS rollout contributes to higher-level transport, energy, and climate policy goals. Indeed, as shown in Figure 12.8 for the United States (NRC, 2013), the estimated net present value of a policy-induced transition to hydrogen FCEVs according to a given

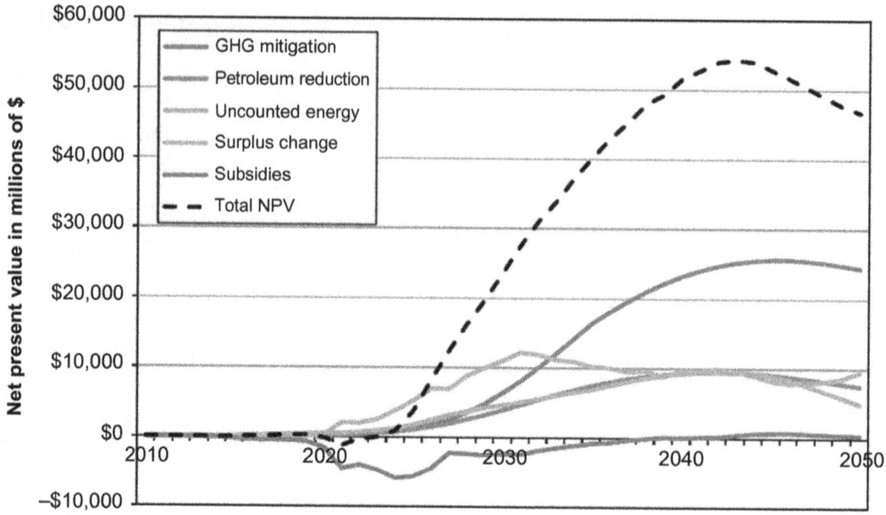

Figure 12.8 Present value cost and benefits of a transition to hydrogen fuel cell vehicle (NRC, 2013).

scenario is on the order of $1 trillion. The long-term benefits of a mature hydrogen infrastructure for transport by far exceed the transition costs for establishing it: by 2050, the total monetized benefits outweigh transition costs tenfold and are roughly equally composed of societal benefits (reduction of fossil fuel imports and of GHG emissions) and private benefits (fuel savings and consumers' surplus gains). Related figures for the United Kingdom are reduced damage to human health and to the environment by the avoidance of emissions of nitrogen oxides amounting to 100–200 M£ annually in 2050 and annual benefits of 1.3 B£ by 2030 resulting from switching from imported fossil fuels to domestic hydrogen production (UK Government report, 2013). For France, the figures are reduced societal costs of 140 M€ annually by 2030 by reduced CO_2 emissions, pollutants and noise and value creation of 700 M€ annually by 2030 from hydrogen sales by 2030 (AFHYPAC, 2014).

Public financial support should aim at providing a financial advantage to first movers over others who enter the market later when the HRS network is already profitable. To promote action by first-movers this advantage should be limited in time. Secondly, authorities should help bear costs of financing, for example by issuing guarantees for bank loans. Policy measures should address the external risk of delayed or small-volume FCEV ramp-up, which can be achieved by regulations and incentives that stimulate market uptake of FCEVs, including, for example, tax benefits and financial premiums for FCEV purchase. Also, authorities should ensure that HRS network rollout is and remains attractive to strategic investors by issuing legislation that recognizes and defines a role for hydrogen in transport.

In Europe, because of the diversity of vehicle and fueling station markets and the differences in national policies, the financing approaches for the transition period need to take into account the specific characteristics and requirements of the national markets. Therefore, governments and strategic investors have to join forces at the national and regional scale to develop the frameworks for financing the initial HRS networks. The amounts of public financial support necessary to initiate the launch of a hydrogen refueling infrastructure have been estimated in the range of 50–144 M€ for the United Kingdom and 15–342 M€ for Germany, depending on the specific financing approach chosen (FCH-JU, 2013d).

During the start-up phase of HRS roll-out, a hydrogen-specific policy support scheme is needed to make hydrogen competitive with alternative (nondisruptive) options and binding commitments from investors are a condition for public financial support. In later roll-out stages, financial incentives can be progressively reduced and hydrogen-specific policy measures replaced by support schemes promoting further market penetration of zero-emission transport and infrastructure. Table 12.2, adapted from Ogden et al., gives an overview of possible policies and measures and their adoption by the EU and European countries.

12.5.1.2 EU-level efforts

The Transport White Paper (EC, 2011a) concluded that without significant uptake of alternative fuels, including hydrogen, the targets of the Europe 2020 strategy and the EU climate goals for 2050 cannot be achieved. It therefore announced that the Commission will develop "*a sustainable alternative fuels strategy including also the appropriate infrastructure*" and ensure "*guidelines and standards for refueling infrastructures.*" This covers the following actions:

- Ensure appropriate cross-border linking between the national alternative fuel networks
- Lift the technical barrier of lack of common European standards which obstructs the creation of a single market and prevents cost reduction through economies of scale on the supply side and network effects on the demand side.

The Directive on the deployment of alternative fuels infrastructure (AFI-directive) (Directive, 2014/94/EU) addresses these issues by requiring Member States to develop national policy frameworks for the market development of alternative fuels and their infrastructure by end 2016. In those Member States that decide to include hydrogen refueling points accessible to the public in their national policy framework, an appropriate number of HRSs including cross-border links where appropriate, has to be established by the end of 2025.

The directive further includes requirements for common technical specifications for refueling points, and for user information, including a clear and sound price comparison methodology. The minimum needs for technical specifications of publically accessible hydrogen refueling points for road transport are described in Annex III.2 of the Directive. Pursuant to the new regulation on European standardization (Regulation (EU) No, 1025/2012) the Commission has requested the European standardization organizations to develop and adopt European standards, or to amend existing European standards, needed to implement the technical specifications set out in Annex III.2 of the Directive, specifying that these standards should be based on current international standards or ongoing international standardization work, where applicable, and be available by the end of 2016 (Table 12.3).

Table 12.2 Public policies and measures related to FCEVs in a number of European countries

		EU	DK	FR	DE	NL	UK	IS	NO
Consumer	Vehicle purchase subsidy		■		■		■		■
	Tax exemption/reduction (vehicle, road traffic)		■		■		■	■	■
	Vehicle perks (preferential lanes, free parking, inner city access, etc.)		▒						■
	H_2 fuel subsidy		■						■
FCEV OEM	Zero emission vehicle regulation				■		■		■
	CO_2 emission targets for vehicles	■	■	■	■	■	■	■	■
Energy/fuel supplier	H_2 Infrastructure subsidy		■		■				
	Renewable H_2 regulation	▒							
	Renewable fuels regulation	▒						■	
Policies	Targets on share of FCEVs				■		■		
	Targets on share of FCV in corporate/public fleets	▒			■	▒	■		
	Preferential public procurement	▒							
	Carbon policy	■	■	■	■	■	■	■	■
	End fossil fuel use by 2050		■						
Enabling measures	Public–private partnerships for H_2/FCVs	■	■	■	■		■		■
	Drafting of standards and regulations	■							
	■ (being) implemented								
	▒ under consideration								

Table 12.3 Requirements for European standards identified in Annex II.2 of the AFI directive

	European standards (ENs) on hydrogen supply (target date December 31, 2016)
1	A European standard containing technical specifications with a single solution for outdoor hydrogen refueling points dispensing gaseous hydrogen, complying with ISO/TS 20100
2	A European standard containing technical specifications with a single solution for hydrogen purity dispensed by hydrogen refueling points, complying with ISO 14687-2
3	A European standard containing technical specifications with a single solution for employing fueling algorithms and equipment, complying with ISO/TS 20100
4	A European standard containing technical specifications with a single solution for connectors for vehicles for the refueling of gaseous hydrogen, complying with ISO 17268

The directive must be transposed into national law of the Member States which may implement it by making use of a wide range of regulatory and nonregulatory incentives and measures, in close cooperation with private-sector actors. However, a number of related activities cannot be covered by the directive because they fall under national authority. To promote investors' confidence and interests in the internal transport market, the Commission therefore considers taking a leadership role in convening national, regional, and local authorities, the private sector, and other stakeholders to harmonize HRS siting, permitting, and regulatory processes. In this respect, particular attention should be paid to safety, as the deployment of a hydrogen fuel infrastructure will expose the general public to technologies that were dealt with before by specialized users in secluded environments. Within the German H2 Mobility project industry stakeholders and authorities have developed a guideline for HRS permitting aimed at establishing a uniform approach across the country (NOW, 2013). The guideline reflects the steps needed to comply with German legislation, safety standards and administrative responsibilities. This approach warrants follow-up at EU level, in a similar manner as EU-wide type approval for hydrogen-powered vehicles has been implemented (Regulation (EC) No 79/2009).

12.5.1.3 EU-level financing of hydrogen infrastructure for transport

EU-level financing is not foreseen for deploying the hydrogen infrastructure required under the AFI directive. However, in recent years the Trans-European Networks for Transport (TEN-T) Program (Regulation (EC) No 680/2007) has provided a grant for the Hydrogen Infrastructure for Transport project (HIT, 2013) with partners from Denmark, France, The Netherlands, and Sweden. HIT-I aims at stimulating the deployment of refueling infrastructure along a 1000-km corridor from Gothenburg to Rotterdam and at demonstrating state-of-the-art refueling technology, whereas expansion of the network around the Baltic Sea through Poland is proposed under HIT-II.

The National Implementation Plans[2] of the participants are being integrated with the roll-out scenarios of Germany and the United Kingdom within a Synchronized Implementation Plan to formulate recommendations to EU and national policymakers for efficient planning and for the use of effective support schemes for hydrogen infrastructure build-up.

As of 2014, the EU transport infrastructure policy can provide more financial support through the TEN-T Regulation (Regulation (EU) No, 1315/2013), which aims at closing the gaps between Member States' transport networks, at removing bottlenecks that still hamper the smooth functioning of the internal market and at overcoming technical barriers such as incompatible standards. Up to 2020 a budget of 26 B€ is foreseen to cofund TEN-T projects in the EU Member States with funding originating from well-established instruments as the Cohesion Fund (Council Regulation (EC) No, 1085/2006) and the European Regional Development Fund (Regulation (EC) No, 1080/2006), as well as from the new Connecting Europe Facility (CEF) Regulation (Regulation (EU) No, 1316/2013). The aim of the CEF is to accelerate investment in the field of trans-European networks (transport, energy, and ICT), to leverage funding from the public and the private sectors, and to enable synergies between the transport, telecommunications and energy sectors.

To be able to keep up with future technology developments, the TEN-T Regulation explicitly stipulates that projects that support and promote decarbonization of transport or introduce alternative propulsion systems and provide corresponding infrastructure are eligible for EU funding. The indicative funding foreseen for such market-sided innovation (not research) targeting technologies and processes ranges between 250 and 400 M€ (EC, 2013e).

12.5.2 Infrastructure for energy

The Commission has estimated the investment needs in energy infrastructure in the EU up to 2020 to amount to 1000 B€, including investment of approximately 200 B€ in electricity and gas transmission and storage infrastructures considered to be of European relevance. Following close consultations with Member States and stakeholders, it has identified 12 strategic trans-European energy infrastructure priorities, the implementation of which by 2020 is considered essential for the achievement of the EU energy and climate policy objectives (Regulation (EU) No 347/2013). These priorities cover different geographic regions as well as thematic areas in the fields of electricity transmission and storage, gas transmission, storage and liquefied or compressed NG infrastructure, smart grids, electricity highways, carbon dioxide transport, and oil infrastructure. Whereas hydrogen is not explicitly mentioned, it can nevertheless contribute to the thematic priority of building an electricity highway system across the EU. Such a system should be capable of accommodating ever-increasing wind surplus generation in and around the Northern and Baltic Seas and increasing renewable generation in the East and South of Europe and also North Africa. It should also allow connecting these new generation hubs with major storage capacities in the

[2] NIPs can be considered as elements required for the implementation of the AFI Directive.

Nordic countries, the Alps and other regions with major consumption centers, and cope with an increasingly variable and decentralized electricity supply and flexible electricity demand.

EU-level funding for the priority actions identified in the regulation on trans-European energy infrastructure is available from within the energy envelope of the Connecting Europe Facility (Regulation (EU) No, 1316/2013), amounting to 585 B€ over the 2014–2020 period.

12.6 Conclusions

The chapter has identified the policy needs and has described the status and outlook for the deployment of hydrogen infrastructures in the European Union to enable widespread use of increasing amounts of renewable hydrogen as a fuel and as an energy storage medium. Although steady technology progress has resulted in improved performance of hydrogen production, storage and delivery technologies and associated cost reductions, there are still a number of technology and nontechnology barriers that prevent industry and private investors from engaging in infrastructure deployment. Consequently, public support, covering policy as well as financial measures is needed in the early phases of infrastructure build-up. EU-level approaches and instruments for promoting and contributing to the financing of the establishment of EU hydrogen infrastructures, both for transport and energy applications, are identified.

References

AFHYPAC, 2014. Association Française pour l'Hydrogène et les Piles à Combustible. http://www.afhypac.org/images/documents/h2_mobilit_france_en_final_updated.pdf.

Altfeld, K., Pinchbeck, D., 2013. Admissible hydrogen concentrations in natural gas systems. Gas Energy (3) (ISSN 2192-158X).

BMVBS, 2013. The Mobility and Fuels Strategy of the German Government. Bundesministerium für Verkehr, Bau und Stadtentwicklung, Berlin.

Commission Regulation (EU) No 630/2012 as regards type-approval requirements for motor vehicles fuelled by hydrogen and mixtures of hydrogen and natural gas with respect to emissions, and the inclusion of specific information regarding vehicles fitted with an electric power train in the information document for the purpose of EC type-approval (retrieved from).

Council Conclusions October 2014. Retrieved from http://www.consilium.europa.eu/uedocs/cms_data/docs/pressdata/en/ec/145397.pdf.

Council Regulation (EC) No 1085/2006 establishing an Instrument for Pre-Accession Assistance (retrieved from http://eur-lex.europa.eu/legal-content/EN/TXT/PDF/?uri=CELEX:32006R1085&from=EN), additional info on http://ec.europa.eu/regional_policy/thefunds/cohesion/index_en.cfm.

Council Regulation (EU) No 559/2014 establishing the Fuel Cells and Hydrogen 2 Joint Undertaking. Retrieved from http://eur-lex.europa.eu/legal-content/EN/TXT/PDF/?uri=OJ:JOL_2014_169_R_0007&from=FR

Directive 2014/94/EU of the European Parliament and of the Council on the deployment of alternative fuels infrastructure, OJ L307/1, October 28, 2014.

US DoE Fuel Cell Technologies Office, 2012. Record 12001. Retrieved from http://www.hydrogen.energy.gov/program_records.html.

EASE, 2013. Joint EASE/EERA recommendations for a European Energy Storage Technology Development Roadmap towards 2030. European Association for the Storage of Energy. Retrieved from http://www.ease-storage.eu/Technical_Documents.html.

EC, 2003. Hydrogen energy and fuel cells—a vision of our future. European Commission, EUR 20719 EN, ISBN 92-894-5589-6. Retrieved from http://www.fch.ju.eu/sites/default/files/documents/hlg_vision_report_en.pdf.

EC, 2007a. A European Strategic Energy Technology Plan (SET-PLAN). European Commission, COM(2007)723. Retrieved from http://eur-lex.europa.eu/legal-content/EN/TXT/PDF/?uri=CELEX:52007DC0723&from=EN.

EC, 2007b. European Hydrogen and Fuel Cell Technology Platform—Implementation Plan-Status 2006. European Commission. Retrieved from http://www.fch-ju.eu/sites/default/files/documents/hfp_ip06_final_20apr2007.pdf.

EC, 2010. Energy infrastructure priorities for 2020 and beyond—a blueprint for an integrated European Energy Network. European Commission, COM(2010)677. Retrieved from http://eur-lex.europa.eu/legal-content/EN/TXT/PDF/?uri=CELEX:52010DC0677&from=EN.

EC, 2011a. Roadmap to a single European transport area—Towards a competitive and resource efficient transport system. European Commission, COM(2011)144.Retrieved from http://eur-lex.europa.eu/legal-content/EN/TXT/PDF/?uri=CELEX:52011DC0144&from=EN.

EC, 2011b. Energy Roadmap 2050. European Commission, COM(2011)885. Retrieved from http://eur-lex.europa.eu/legal-content/EN/TXT/PDF/?uri=CELEX:52011DC0885&from=EN.

EC, 2012. Research and innovation for Europe's future mobility—developing a European transport-technology strategy. European Commission, COM(2012)501. (http://eur-lex.europa.eu/legal-content/EN/TXT/PDF/?uri=CELEX:52012DC0501&from=EN).

EC, 2013a. Clean power for transport: a European alternative fuels strategy. European Commission, COM(2013)17. Retrieved from http://eur-lex.europa.eu/legal-content/EN/TXT/PDF/?uri=CELEX:52013PC0017&from=EN.

EC, 2013b. Energy technologies and innovation. European Commission, COM(2013)253. Retrieved from http://eur-lex.europa.eu/legal-content/EN/TXT/PDF/?uri=CELEX:52013DC0253&from=EN.

EC, 2013c. Technology assessment accompanying the communication on energy technologies and innovation. Commission Staff Working Document SWD(2013)158. Retrieved from http://eur-lex.europa.eu/LexUriServ/LexUriServ.do?uri=SWD:2013:0158:FIN:EN:PDF.

EC, 2013d. Impact assessment accompanying the proposal for a directive on the deployment of alternative fuels infrastructure. Commission Staff Working Document SWD(2013)005. Retrieved from http://eur-lex.europa.eu/resource.html?uri=cellar:fbc30100-319d-45d6-b4d6-684edcc98a3e.0001.03/DOC_2&format=PDF.

EC, 2013e. Building the transport core network: core network corridors and connecting Europe facility. European Commission, COM(2013)940. Retrieved from http://ec.europa.eu/transparency/regdoc/rep/1/2013/EN/1-2013-940-EN-F1-1.Pdf.

FCH2JU, 2014a. Multi-Annual Work program 2014–2020 of FCH2JU. Fuel Cell and Hydrogen 2 Joint Undertaking (not yet available from www.fch-ju.eu).

FCH2JU, 2014b. The role of energy storage to enable intermittent RES in Europe. Study ongoing for FCH2JU.

FCH-JU, 2010. A portfolio of power-trains for Europe: a fact-based analysis. Fuel Cells and Hydrogen Joint Undertaking. Retrieved from http://www.fch-ju.eu/sites/default/files/documents/Power_trains_for_Europe.pdf.

FCH-JU, 2012. http://www.fch-ju.eu/news/launch-uk-h2-mobility-new-governement-and-cross-industry-programme-make-hydrogen-powered-travel.

FCH-JU, 2013a. Trends in investments, jobs and turnover in the fuel cells and hydrogen sector. Fuel Cells and Hydrogen Joint Undertaking. Retrieved from http://www.fch-ju.eu/sites/default/files/Investment%20jobs%20%26%20turnover%20in%20FCH%20Sector.pdf.

FCH-JU, 2013b. Programme review 2013. Fuel Cells and Hydrogen Joint Undertaking, p. 112. Retrieved from http://www.fch-ju.eu/sites/default/files/14-03-30_FULL%20PRD%20Report%20with%20cover%20%28ID%20376127%29_3.pdf.

FCH-JU, 2013c. DeliverHy project deliverable 6.2 impact of high-capacity CGH_2-trailers. Fuel Cells and Hydrogen Joint Undertaking. Retrieved from http://www.deliverhy.eu/publics/publications/public-deliverables.html.

FCH-JU, 2013d. A roadmap for financing hydrogen refuelling networks—Creating prerequisites for H2-based mobility. Fuel Cells and Hydrogen Joint Undertaking. Retrieved from http://www.fch-ju.eu/sites/default/files/HRS-finance_FinalReport.pdf.

FCH-JU, 2014a. Development of water electrolysis in the European Union, a study for the FCH-JU. Fuel Cells and Hydrogen Joint Undertaking. Retrieved from http://www.fch-ju.eu/sites/default/files/study%20electrolyser_0-Logos_0.pdf.

FCH-JU, 2014b. HyUnder project deliverable 8.1 executive summary. Fuel Cells and Hydrogen Joint Undertaking. Retrieved from http://www.hyunder.eu/images/Deliverables/D8%201_HyUnder%20Executive%20Summary%202.pdf.

Hayter, D., 2013. A rollout strategy for hydrogen transport in the UK. In: Presented at International Hydrogen Infrastructure Challenges Workshop, Berlin, June 2013.

HIT, 2013. Hydrogen infrastructure for transport project. http://www.hit-tent.eu/.

Hyways, 2007. HyWays—an integrated project to develop the European hydrogen energy roadmap. Retrieved from http://www.hyways.de.

IEA, 2013. Hydrogen refuelling stations and role of utilization rates: key messages and issues. International Energy Agency. Retrieved from http://www.iea.org/media/workshops/2013/hydrogenroadmap/Session2.1WeedaECN.pdf.

Iskov, H., Rasmussen, N.-B., 2013. Global Screening of Projects and Technologies for Power-to-Gas and Bio-SNG, A Reference Report. Danish Gas Technology Centre. ISBN: 78-87-7795-373-6.

Navigant Research, 2013. Market Data: Hydrogen Infrastructure.

NOW, 2011. Nationale Organisation für Wasserstoff. http://www.now-gmbh.de/en/mobility/mobility-of-tomorrow/cars-buses-and-public-fuelling-stations/h2-mobility-development-of-a-network-of-hydrogen-refueling-stations-in-germany.html.

NOW, 2013. Approval Guidelines for Hydrogen Refuelling Stations. Nationale Organisation für Wasserstoff. Retrieved from, http://www.h2-genehmigung.de/Index/Index?lang=1.

NOW, 2014. HRS Infrastructure in Germany and Europe—Current Activities. In: Nationale Organisation für Wasserstoff, presentation at DoE Annual Merit Review, Washington, June 2014.

NPC, 2012. Advancing Technology for America's transportation Future. National Petroleum Council, Washington. Retrieved from http://www.npc.org/reports/trans.html.

NRC, 2008. Transitions to Alternative Transportation Technologies: A Focus on Hydrogen'. National Research Council. Retrieved from http://www.nap.edu/catalog/12222.html.

NRC, 2013. Transitions to Alternative Vehicles and Fuels. National Research Council. Retrieved from http://www.nap.edu/catalog/18264.html.

NREL, 2014. Hydrogen station compression, storage and dispensing technical status and costs. Report NREL/BK-6A10-58564, National Renewable Energy Laboratory.

Ogden, J., Yang, C., Nicholas, M., Fulton, L., 2014. The Hydrogen Transition. UC Davis Institute of Transportation Studies, Davis.

Okada, Y., Shimura, M., 2012. Development of large-scale H_2 storage and transportation technology with Liquid Organic Hydrogen Carrier (LOHC). Fuel Cell 11, 56–62.

ORNL, 2008. Analysis of the transition to hydrogen fuel cell vehicles & the potential hydrogen energy infrastructure requirements. Report ORNL/TM-2008/30, Oak Ridge National Laboratory.

Regulation (EC) No 1080/2006 on the European Regional Development Fund. Retrieved from http://eur-lex.europa.eu/legal-content/EN/TXT/PDF/?uri=CELEX:32006R1080&from=EN, additional info on http://ec.europa.eu/regional_policy/thefunds/regional/index_en.cfm.

Regulation (EC) No 79/2009 of the European Parliament and of the Council on type-approval of hydrogen-powered motor vehicles. Retrieved from http://eur-lex.europa.eu/LexUriServ/LexUriServ.do?uri=OJ:L:2009:035:0032:0046:en:PDF.

Regulation (EU) No 1025/2012 of the European Parliament and of the Council on European standardisation. Retrieved from http://eur-lex.europa.eu/LexUriServ/LexUriServ.do?uri=OJ:L:2012:316:0012:0033:EN:PDF.

Regulation (EU) No 347/2013a of the European Parliament and of the Council on guidelines for trans-European energy infrastructure. Retrieved from http://eur-lex.europa.eu/LexUriServ/LexUriServ.do?uri=OJ:L:2013:115:0039:0075:EN:PDF.

Regulation (EU) No 1315/2013b of the European Parliament and of the Council on Union guidelines for the development of the trans-European transport network. Retrieved from http://eur-lex.europa.eu/legal-content/EN/TXT/PDF/?uri=CELEX:32013R1315&from=EN.

Regulation (EU) No 1316/2013c of the European Parliament and of the Council establishing the Connecting Europe Facility. Retrieved from http://eur-lex.europa.eu/LexUriServ/LexUriServ.do?uri=OJ:L:2013:348:0129:0171:EN:PDF.

Regulation (EC) No 680/2007 of the European Parliament and of the Council laying down general rules for the granting of Community financial aid in the field of the trans-European transport and energy networks. Retrieved from http://eur-lex.europa.eu/LexUriServ/LexUriServ.do?uri=CELEX:32007R0680:EN:NOT.

Sandia, 2011. A life cycle cost analysis framework for geologic storage of hydrogen: a user's tool. Report SAND2011-6221, Sandia National Laboratories.

Sandia, 2014. Hydrogen fueling research and station technology. Report SAND2014-19381938, Sandia National Laboratories.

Shell Global Solutions International, 2013. Hydrogen as an energy vector. In: Presented at DG ENER Workshop on H2 in the Economy. Retrieved from http://ec.europa.eu/energy/technology/doc/h2_conference2013.zip.

U.S. DRIVE Partnership, 2013. Hydrogen Delivery Technical Team Roadmap.

UK Government report, 2013. UK H2mobility phase 1 results. Retrieved from https://www.gov.uk/government/uploads/system/uploads/attachment_data/file/192440/13-799-uk-h2-mobility-phase-1-results.pdf.

US DoE Fuel Cell Technologies Office, 2012 and 2013. Records 12021 and 12022. Retrieved from http://www.hydrogen.energy.gov/program_records.html.

US DoE Fuel Cell Technologies Office, 2014. Hydrogen transmission and distribution workshop summary report. Retrieved from http://energy.gov/sites/prod/files/2014/07/f17/fcto_2014_h2_trans_dist_wkshp_summary_report.pdf).

Building a hydrogen infrastructure in the United States

13

K. Reddi[*], *M. Mintz*[*], *A. Elgowainy*[*], *E. Sutherland*[†]
[*]Argonne National Laboratory, Argonne, IL, USA, [†]Fuel Cell Technologies Office, Washington, DC, USA

13.1 Introduction

Public hydrogen fueling stations in the United States are in their infancy. As of October 2014, 10 stations are open to the public and 41 more are funded and in development. All of the stations are located in the state of California. The United States, like Europe, must develop a plan to expand infrastructure from the leading regions across the continent. Since each state may offer different incentives and funding, the climate to catalyze infrastructure development varies widely across the country. Hydrogen supply to these stations, on the other hand, is robust; more than 9 million metric tons of hydrogen were produced annually in the United States (US) in 2011, and this is expected to exceed 11 million metric tons by 2016 (Joseck, 2012). Currently, the primary use of hydrogen occurs in the petroleum refining and ammonia production industries. Hydrogen is produced in large quantities at central locations and distributed to the point of use (merchant hydrogen). It is also produced at distributed locations for onsite use (captive hydrogen). When hydrogen is produced at central locations, it is delivered to the point of use via a dedicated infrastructure that is composed of pipelines, liquid tankers, and gaseous tube trailers. Currently, about one-third of the total hydrogen produced in the US is merchant hydrogen that is mainly delivered to customers via pipeline as a compressed gas. Today there are about 1200 miles of pipelines that transport hydrogen. About 93% of these pipelines are located in the states of Texas and Louisiana, and they mainly serve the petroleum refining industry (Lipman, 2011). Pipelines are deemed economical for hydrogen transportation only for large market demands (in excess of 100 metric tons per day). A small portion (~3%) of the merchant hydrogen in the US is transported in liquid form, primarily from the eight liquefaction plants in the US and Canada that have a capacity of more than 250 metric tons per day. Liquid hydrogen is distributed via liquid tankers with a payload capacity of about 4000 kg. A marginal quantity of merchant hydrogen in gaseous form is loaded into pressure vessels (tubes) of various sizes that are mounted on trailers, known as tube trailers, for transportation to end-use applications for which the daily demand is low (<200 kg/day).

Hydrogen as a potential replacement for petroleum-based fuels in the transportation sector is now beginning to be developed. The transportation sector accounts for 28% of the total US greenhouse gas (GHG) emissions, while petroleum constitutes 97% of total transportation sector energy use (EIA, 2014). Electricity and hydrogen are two

Compendium of Hydrogen Energy. http://dx.doi.org/10.1016/B978-1-78242-364-5.00013-0

energy sources that can satisfy zero emission vehicle (ZEV) requirements when used in battery electric vehicles (BEVs) and fuel cell electric vehicles (FCEVs), respectively. In addition to resulting in no vehicle emissions, a good feature of hydrogen is that it can be produced from a variety of domestic feedstock and energy sources by using various process technologies, including steam methane reforming (SMR) of natural gas, water electrolysis, and coal and biomass gasification. This diversity of feedstock sources provides opportunities for improved energy security and also enhances the prospect of economic and employment growth in the energy sector. Currently, hydrogen is produced in the US predominantly from natural gas via SMR. Even when the hydrogen is produced from fossil natural gas, the well-to-wheels (WTW) GHG emissions of hydrogen FCEVs are 30–40% lower than those from comparable gasoline internal combustion engine vehicles (ICEVs).

Currently, the demand for hydrogen from the transportation sector in the US is driven by California's ZEV mandate. The state of California requires that automakers ramp up the sales of ZEVs from 2% in 2018 to 16% by 2025. California also mandates that 33% of the hydrogen used as a transportation fuel should be from renewable sources, which further influences the production locations, process technologies, and feedstocks selected (Boston Consulting Group, 2012). The demand for hydrogen is expected to gradually increase along with sales of FCEVs in the next few years. The deployment of hydrogen refueling stations should precede FCEV deployment, with the main goal being to give FCEV owners a fueling experience comparable to that of owners of ICEVs fueled by gasoline and diesel. In the process of achieving this goal, various strategies are being adopted to produce and deliver hydrogen to the consumers.

13.2 Current status of hydrogen infrastructure in the United States

13.2.1 Hydrogen infrastructure status in California

The California Air Resources Board (ARB) ZEV regulation requires that major automakers produce and sell increasing numbers of ZEVs in California (CARB, 2012). Six northeastern states (Connecticut, Maryland, Massachusetts, New York, Rhode Island, and Vermont) and Oregon signed a memorandum of understanding (MoU) with California to deploy 3.3 million ZEVs by 2025 (CARB, 2013). ZEVs can be BEVs or hydrogen FCEVs. The ZEV credits in California escalate with the driving range on a single fueling, with maximum credits offered for a range of 300 miles (CARB, 2012). FCEVs have the advantages of a short refueling time (about 3–5 min) and a long vehicle range (over 300 miles on a single fill). Hyundai was the first major automaker to offer a hydrogen fuel cell vehicle in showrooms in California in 2014. Toyota has recently announced a plan to deploy its much anticipated FCEV (Mirai) in late 2015 in California (Toyota, 2014). Honda and other automakers have displayed their FCEV concept cars in recent auto shows in the US and plan to deploy them in the 2015–2016 timeframe.

Before FCEVs can be deployed, an initial network of refueling stations has to be available in various markets. In 2012, the California Fuel Cell Partnership (CaFCP)—an industry–government partnership of automakers, energy companies, fuel cell technology companies, and government agencies—identified the need for an initial network of 68 strategically placed hydrogen stations to successfully support the introduction of FCEVs throughout the state of California (CaFCP, 2012). Currently, there are only nine public hydrogen refueling stations in California. Because these stations are capital intensive and will be underutilized during the early markets for FCEVs, government incentives will be required to overcome the negative cash flow during the initial period of capital underutilization, until the market develops a sustainable business case for deploying more stations.

In 2013, California passed a bill that supports programs investing in the development and deployment of advanced technologies to achieve California's air quality, climate, and energy goals. The bill includes a provision to fund at least 100 hydrogen stations, with a commitment of up to $20 million a year through 2024 from the Alternative and Renewable Fuel and Vehicle Technology Program (CaFCP, 2013). In 2014, the California Energy Commission (CEC) announced the availability of funding for building hydrogen refueling infrastructure in the state. The goal is to support the commercial introduction of FCEVs by the major auto manufacturers starting in 2015.

In May 2014, the CEC announced funding availability for projects that develop the infrastructure needed to dispense hydrogen transportation fuel before the end of 2015 (CEC, 2014a). The CEC also offered operation and maintenance (O&M) funding to support hydrogen refueling operations prior to the large-scale rollout of FCEVs. Renewable hydrogen stations can be funded for up to $3.15 million, while other stations that are built in locations specified by the state can receive funds of up to $2.125 million, and mobile stations can receive funds of up to $1.0 million. The state of California funds O&M operations up to $100,000 per year for up to 3 years (CEC, 2014b).

13.2.2 H2USA initiative

In May 2013, the US Department of Energy (DOE) launched a public–private partnership known as H2USA to address the challenges associated with the development of the initial network for a hydrogen refueling infrastructure (H2USA, 2014). The partnership is made up of companies and organizations that include automakers, gas companies, component original equipment manufacturers (OEMs), federal and state governments, national laboratories, nongovernment organizations (NGOs), and investment companies. The mission of the partnership is "to promote the commercial introduction and widespread adoption of FCEVs across America through creation of a public–private collaboration to overcome the hurdle of establishing hydrogen infrastructure" (H2USA, 2014). The goals of the partnership are defined as follows (Markowitz, 2013):

- Establish the necessary hydrogen infrastructure and leverage multiple energy sources, including natural gas and renewables;
- Deploy FCEVs across America;

- Improve America's energy and economic security;
- Significantly reduce GHG emissions;
- Develop domestic sources of clean energy and create jobs in the US; and
- Validate new technologies and create a strong domestic supply base in the clean energy sector.

H2USA formed four working groups to address various challenges associated with developing a hydrogen refueling infrastructure: (1) Hydrogen Fueling Station Working Group, (2) Market Support and Acceleration Working Group, (3) Locations Roadmap Working Group, and (4) Financing Infrastructure Working Group. The Hydrogen Fueling Station Working Group addresses the station design specifications, costs, and utilization. It also aims at improving station operation reliability and identifies required component research and development (R&D) areas. Moreover, this working group evaluates the impacts from separation distances and conducts risk-based analyses to help determine future regulation changes. The Market Support and Acceleration Working Group develops deployment timelines, identifies first responders, and develops plans for teaching the public and reaching out to local governments and authorities. The Locations Roadmap Working Group identifies initial markets, conducts market modeling to define initial clusters of station locations, and determines refueling station rollout timing. The Financing Infrastructure Working Group evaluates private sector financing and government support, and it develops business cases for hydrogen refueling stations for reaching out to the investment community (Markowitz, 2013).

In 2014, the Fuel Cell Technology Office (FCTO) in DOE's Office of Energy Efficiency and Renewable Energy (EERE) launched the Hydrogen Fueling Infrastructure Research and Station Technology (H2FIRST) project (DOE, 2014). H2FIRST was established to directly support H2USA activities. The project leverages capabilities at the national laboratories to address the technology challenges related to hydrogen refueling stations. H2FIRST, co-led by the National Renewable Energy Laboratory and Sandia National Laboratories, is supported by a broad spectrum of public and private partners to address immediate and mid-term challenges faced by the relevant industries. The H2FIRST objective is to ensure that FCEV customers have a positive fueling experience similar to that experienced at conventional gasoline stations as vehicles are introduced (2015–2017) and the transition to advanced fueling technology continues beyond 2017. The H2FIRST goal is to enable hydrogen fueling stations to become commercially viable by coordinating and leveraging the technical resources of its public and private sector participants. The H2FIRST project activities are expected to positively affect the cost, reliability, and safety of hydrogen refueling stations and the experiences of consumers there. H2FIRST involves technical work to fill critical gaps/needs in order to achieve a better performing and less expensive hydrogen fueling infrastructure. The project scope includes the development and physical testing of components and systems; modeling of station operations and performance; validation of technologies; identification and development of low-cost, high-performance materials; and systems and station architecture design (Sandia, 2014).

13.3 Initial costs of deploying hydrogen infrastructure

13.3.1 Deployment strategies

A number of researchers have investigated the "critical mass" of fueling infrastructure needed to begin serving the early FCEV market and how to deploy it (Ogden et al., 2014; National Research Council, 2013; Greene et al., 2008). They have looked at consumer fueling behavior for guidance on how frequently vehicles are fueled, how close to "empty" the tank is at refueling, and how many minutes (or miles) drivers are willing to deviate from their routes to fuel (Nicholas, 2010; Ogden and Nicholas, 2011). While some behavioral data on drivers of conventional vehicles exist, the information is not necessarily analogous to data on drivers of FCEVs. For example, with limited fueling options, will FCEV drivers behave differently from drivers of conventional vehicles, at least in the near-term, early-adoption phase? More specifically, will "range anxiety" increase how often FCEV drivers refuel, at least until they become accustomed to how often they must do so to meet routine travel needs and where they can expect to find that fuel? Data collected by the National Renewable Energy Laboratory (NREL) on refueling frequency showed that FCEV drivers have similar fueling patterns as conventional car drivers in terms of time of fill (before and after work, or before a weekend trip); however, they did fill up more often when the tank was at ~50%, indicating drivers' anxiety with driving range in early FCEV markets (NREL, 2011).

There is some research evidence that a network roughly comparable to half of all urban gasoline stations and a tenth of all rural/intercity stations would provide adequate fuel availability for a mature market. However, how robust must the *initial* hydrogen fueling network be to meet early market needs? While it cannot be as ubiquitous as the mature network, neither should it be a barrier to adoption. Since risk impedes not only the adoption of FCEVs but also investments in supporting infrastructure, station deployment must focus on risk reduction for both FCEV purchasers and hydrogen station providers. The combination of (1) known consumer preferences for multiple fueling options near or en route to trip origins and destinations and (2) OEM preferences to limit the sale or lease of FCEVs to geographic areas where they may be more easily supported has given rise to the notion that stations should be deployed in clusters. According to such a strategy, clusters provide demand benefits by increasing the choices and convenience of nearby customers (who, by virtue of their demographics and purchase behavior, are located in discrete geographic areas) and also provide supply benefits by increasing the scale (and reducing the cost) of hydrogen production and delivery (Ogden and Nicholas, 2011).

13.3.2 Networks

The state of California has used a cluster strategy to identify a generalized network of 123 stations to support a fleet of 30,000 or more vehicles. The CaFCP has an ongoing roadmap process and has recently issued an update to its 2012 report, *A California Road Map: The Commercialization of Hydrogen Fuel Cell Electric Vehicles* (CaFCP, 2012). The report *2014 Update: Hydrogen Progress, Priorities and Opportunities*

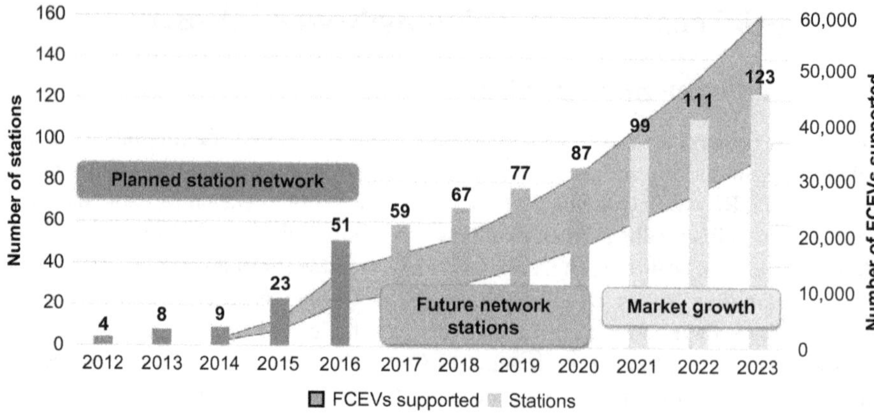

Figure 13.1 Hydrogen Station Network and FCEVs in California Rollout Scenario (CaFCP, 2014).

(CaFCP, 2014) picks up where the earlier document left off, summarizing progress achieved from 2012 to 2014 and planned for 2014 to 2016 and proposing actions to guide the rollout of FCEVs and related infrastructure for the years beyond. Figure 13.1 shows current and projected stations and vehicles in California.

13.3.3 Deployment cost

The CEC recently awarded $50 million to support the deployment of 28 hydrogen fueling stations. Scheduled to begin operation by 2016, these 28 stations will join 23 other stations to make up a 51-station network. Nineteen of the newly funded stations will be of the same design and capacity (180kg/day, expandable to 250kg/day or more) and will be operated by the same consortium of companies. This move is intended to enable the standardization of design/engineering, equipment, installation protocols, and operating and maintenance practices. In addition to the CEC's $50 million award, deployment of these stations will require additional cost sharing from the awardees, as specified in their initial proposals and contract negotiations.

Beyond 2016, the cost of the California rollout is less certain. No awards have been made for stations expected to begin operation after 2016. Thus, the characteristics of these latter stations are not known as well. However, the stations are expected to have a greater nominal capacity, either initially or as modular additions to the 2016-era stations. Table 13.1, based on an analysis of the CaFCP rollout scenario that uses data and assumptions from the 19 stations with a common design, shows the estimated investment required to deploy these and subsequent stations (Mintz et al., 2014).

In Table 13.1, the investment cost of hydrogen fueling stations per unit of capacity declines over time as learning reduces station development costs and scale reduces equipment costs. This same effect can be seen in Figure 13.2, which plots the trend in cost per unit of capacity for existing, planned, and future stations.

Table 13.1 Cost of planned hydrogen station deployment (Mintz et al., 2014)

Year	Existing hydrogen fueling stations	Hydrogen fueling stations added	Incremental investment (million $)
2016	23	28	57.4
2017	51	8	16.4
2018	59	8	16.4
2019	67	10	27.2
2020	77	10	27.2
2021	87	12	45.6
2022	99	12	45.6
2023	111	12	53.6
Total	–	100	289.4

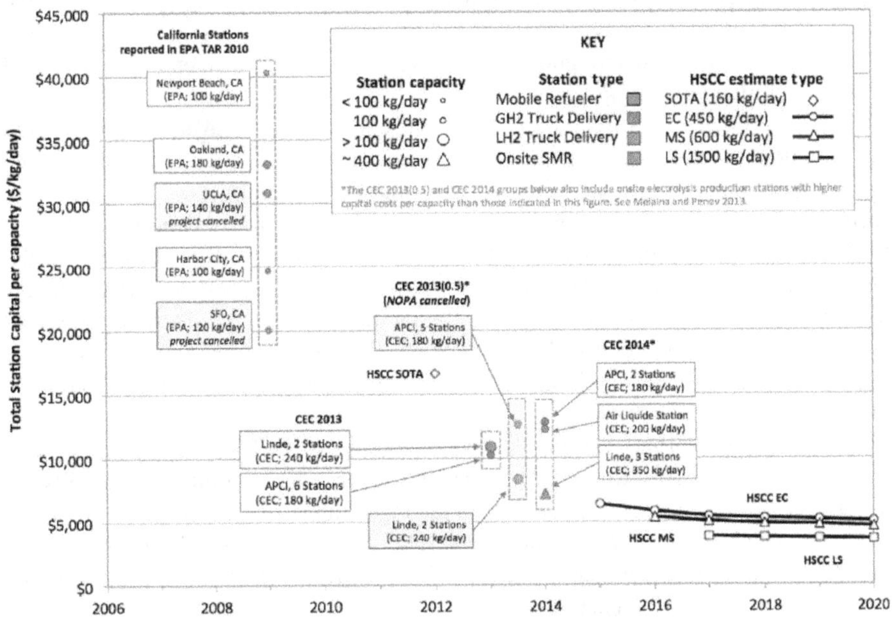

Figure 13.2 Hydrogen Station Cost Per Unit Capacity (CaFCP, 2014).
Reprinted with permission from the National Renewable Energy Laboratory, Melaina and Penev (2013).

13.4 Market trends

Given its associated costs and risks, the hydrogen infrastructure is unlikely to develop elsewhere without sustained public support. In addition to tax incentives or outright support for station development, government policies can encourage the adoption of FCEVs, creating demand not just for the vehicles themselves but also for the infrastructure to fuel them. But how and where is support needed?

13.4.1 *Vehicle commercialization*

Following years of R&D and demonstration and the commitments of various governments to eliminate tailpipe emissions, several manufacturers have begun leasing FCEVs in the US and abroad. Honda has leased the Clarity FCX to southern California customers since mid-2008 and plans to launch a second generation of the vehicle in Japan and California in 2016. Hyundai debuted the Tucson FCEV in select California markets in June 2014, and Toyota planned to introduce the Mirai to the Japanese market in December 2014 and to US and European markets in 2015. Unlike the Tucson and Clarity, the Mirai will be available for sale as well as lease.[1]

With a marketing strategy like that of many hybrids, Hyundai's FCEV is a fuel cell version of its popular Tucson crossover utility vehicle. Unlike Hyundai, Toyota and Honda are marketing their FCEVs as new nameplates, potentially competing with Tesla and other luxury models. In addition to offering free maintenance and roadside assistance; an 8-year, 100,000 mile warranty; and access to a 24-h concierge service, Toyota may offer a "power out" capability in which the electricity produced by the fuel cell can be used to power a home for 1–2 days (Voelcker, 2014). The option could enable the Mirai to qualify for additional low-emission, renewable energy credits and provides a unique value proposition.

Since FCEVs are ZEVs, a variety of incentives are already available to promote their adoption; in addition, in certain states, there are also mandates that require ZEVs to make up an increasing share of the vehicles that OEMs sell or lease.[2] In addition to California, Connecticut, Maryland, Massachusetts, New York, Oregon, Rhode Island, and Vermont have committed to developing a coordinated program to advance the commercialization of ZEVs. The governors of those states, by signing an MoU, created a multi-state ZEV Program Implementation Task Force to guide these efforts. In May 2014, the Task Force issued a Multi-State ZEV Action Plan that describes key actions that the signatory states will carry out to promote the development of the ZEV market (ZEVTF, 2014).

California's ZEV rules are particularly favorable to FCEVs having a range of 300 miles or more, permitting them to qualify for nine "ZEV compliance credits"—three times the number available to BEVs having a range of 100 miles or less. Until 2018, only the six carmakers with the largest sales in California are required to comply with ZEV rules (ZEVs must account for 0.79% of their deliveries in California for model years 2015–2017). Thereafter, ZEV rules will be extended to eight additional carmakers that account for most of the balance of California car sales, and ZEVs must account for a rapidly rising share of California deliveries (2% in 2018, rising to 16% in 2025). Covered OEMs can meet their ZEV obligations by accumulating ZEV credit balances (i.e., initially exceeding the ZEV mandate and drawing those credits down later) or by buying credits from OEMs like Tesla or Hyundai that are exceeding their ZEV obligations.

[1] With a sticker price of $57,500 minus a federal tax credit of $8,000 and a $5,000 rebate from the state of California, the Mirai will be available for roughly $44,500 or $499/month with a down payment of $3,000.
[2] Includes battery electric vehicles (BEVs) and plug-in hybrid electric vehicles (PHEVs) as well as FCEVs.

13.4.2 Infrastructure development

Some researchers have examined shifts from diesel to compressed natural gas (CNG) in transit and commercial vehicle fleets for insight on how infrastructure might evolve. Like hydrogen, CNG is a compressed gas that requires (1) special handling for which codes and standards are still under development, (2) design and engineering for which the necessary expertise is not always widely available, and (3) high-pressure components that can add significantly to capital and operating costs. While a gasoline station dispensing 4500 gallons/day (including a relatively profitable convenience store and car wash) can cost $1 million, a station able to dispense just 1000 gasoline gallons equivalent (gge) of CNG/day can require an investment that is 50% greater. Initial hydrogen stations are even more expensive; currently, they cost more than $2 million (for a station dispensing 180kg/day, roughly equivalent to 175 gge/day), which is a fifth of the size of a typical CNG station, and they support only a tenth as much potential travel as the average gasoline station.

Because supplies of low-cost natural gas are plentiful, the number of CNG stations has been growing steadily. Each month, 20–30 new stations are added, dispensing CNG for as much as $1.00–2.00/gge less than the price of gasoline (AFDC, 2014a). In addition to conventional fuel retailers, stations are now operated by gas exploration and production companies, utilities, and commercial fleets. Station operators have learned to negotiate purchase agreements with "anchor fleets," which enables them to maintain a predictable demand for their product, reduce their risks, and negotiate more favorable terms with suppliers. Thus, according to industry sources (Eichberger, 2014), CNG stations earn more profit per gge dispensed than do gasoline stations ($0.50–0.70/gge versus less than $0.05/gal), despite the previously noted higher first costs.

State regulations and incentives are spurring CNG station investment and vehicle adoption. It is no accident that states like Oklahoma, Utah, and California have the highest concentrations of CNG stations and vehicles. Oklahoma provides waivers of state tax liability to in-state suppliers of CNG station components. Utah requires that CNG be supplied by regulated utilities (effectively capping prices). California permits single-occupant CNG vehicles to use high-occupancy vehicle (HOV) lanes. Similar policies could be applied to hydrogen vehicles and fueling.

California is furthest along in applying many of these policies to encourage the adoption of FCEVs and the development of an H_2 fueling infrastructure. HOV-lane access is available to single-occupant FCEVs, as are preferential parking and tax incentives for vehicle purchase and registration (Ogden et al., 2014).

Perhaps even more important, however, are OEM efforts to reduce the effect of a limited fueling infrastructure. Hyundai already provides 3 years of fuel to lessees of the Tucson FCEV. Toyota has announced a similar plan. As noted, California is supporting the development of a 123-station network of hydrogen fueling stations, and several Northeast states are developing a coordinated program to promote ZEVs. Toyota is assisting in these efforts by providing a $7.3 million loan to First Element, a California station developer, and by collaborating with Air Liquide to develop an initial network of 12 strategically located stations to permit FCEV travel in the Boston–New York region.

13.4.3 Fuel requirements and support

In addition to requiring that a minimum percentage of vehicles delivered within the state be ZEVs, California also requires that the mix of fuel sold within the state contain an increasing share of low-carbon fuel.[3] Pursuant to California Assembly Bill AB 32 and the Governor's Executive Order S-01-07, the "low carbon fuel standard" (LCFS) requires that all transportation fuel sold within California contain 10% less carbon in 2020 than in 1990. A system of credits permits producers of petroleum-based fuels to reduce the carbon intensity of their products, beginning with a quarter of a percent in 2011 and increasing to a 10% total reduction in 2020. Petroleum importers, refiners, and wholesalers can either develop their own low-carbon fuel products or buy LCFS credits from other companies that develop and sell low-carbon alternative fuels (e.g., biofuels, electricity, natural gas, hydrogen). Special low-carbon credits in the form of energy efficiency ratios (EERs) are provided for BEVs and FCEVs because their fuel economy is so much higher than that of ICEVs.

As noted, California is also supporting the development of a fueling infrastructure that will be required for a growing fleet of FCEVs. In addition to supplying initial funds for station development, the state is supplying funds to bridge the gap between the cost to provide hydrogen at retail prices and the price that can be expected once stations are fully utilized.

13.5 Hydrogen refueling infrastructure

The primary goal of a hydrogen delivery infrastructure is the safe delivery of hydrogen from the production plant gate to the vehicle tank at the lowest possible cost. The cost of hydrogen varies with the pathway technologies and the production quantities (scale). The total cost of hydrogen at the pump includes the production cost, representing the cost incurred in producing the hydrogen from its feedstock source, and the delivery cost, representing the cost incurred in delivering the hydrogen from the production plant gate to the vehicle tank. The delivery of hydrogen from the production plant gate to the vehicle tank includes three core operations (U.S. Drive Partnership, 2013): (1) packaging, (2) transporting, and (3) fueling. These operations are carried out sequentially, thus defining the delivery pathway. Hydrogen has the advantage of having a higher gravimetric energy density (i.e., per unit mass) that is about 3 times that of diesel and 2.5 times that of natural gas (AFDC, 2014b), but it generally has a lower volumetric energy density (i.e., per unit volume). Thus, the packaging operation is aimed at improving hydrogen's volumetric energy density so that large payloads of hydrogen can be transported via truck economically. Hydrogen is usually "packaged" by being liquefied for transportation via cryogenic tankers or being compressed into tubes for transportation via tube trailers. The transportation operation then physically moves the packaged hydrogen from the loading terminals, which are next to the

[3] Since California also requires that a third of all hydrogen sold in the state be produced from renewable sources, hydrogen consumed in California is a low-carbon fuel.

production plants, to the refueling stations. The fueling operations follow standard protocols and use various processes to dispense the hydrogen into the vehicle's on-board storage (tank) at the desired rate and conditions (i.e., pressure and temperature). In Section 13.5, the refueling components and operations are discussed, while Section 13.6 discusses the packaging and transportation of hydrogen.

13.5.1 Refueling station components and processes

The refueling station system boundary starts at the hydrogen supply source and ends with the inlet to the vehicle's tank. The hydrogen can be supplied to a refueling station in either gaseous or liquid form. The components that are part of the refueling station vary and are dictated by the physical form of supplied hydrogen (i.e., gaseous or liquid) and the working pressure of the vehicle's tank. Most automakers have agreed to adopt a 700-bar vehicle storage system (Elgowainy and Wang, 2012). The primary goal of a refueling station is to refuel vehicles to a 100% state of charge (SOC) throughout the station's daily operations. Statistical data from gasoline stations indicate that the average hourly fueling demand within a day in major US markets varies, as shown in Figure 13.3. In addition, the daily demand also varies within a week (Argonne National Laboratory et al., 2010; Chen, 2008), as shown in Figure 13.4; the maximum demand occurs on Fridays. The hours when the demand is greater than the average hourly demand define the peak hours, while the hours when the demand is less than the average hourly demand define the off-peak hours. The refueling station components should be sized to satisfy the fueling demand during the peak hours of the peak day (i.e., Friday). The various refueling station components subject to sizing are shown in Figure 13.5.

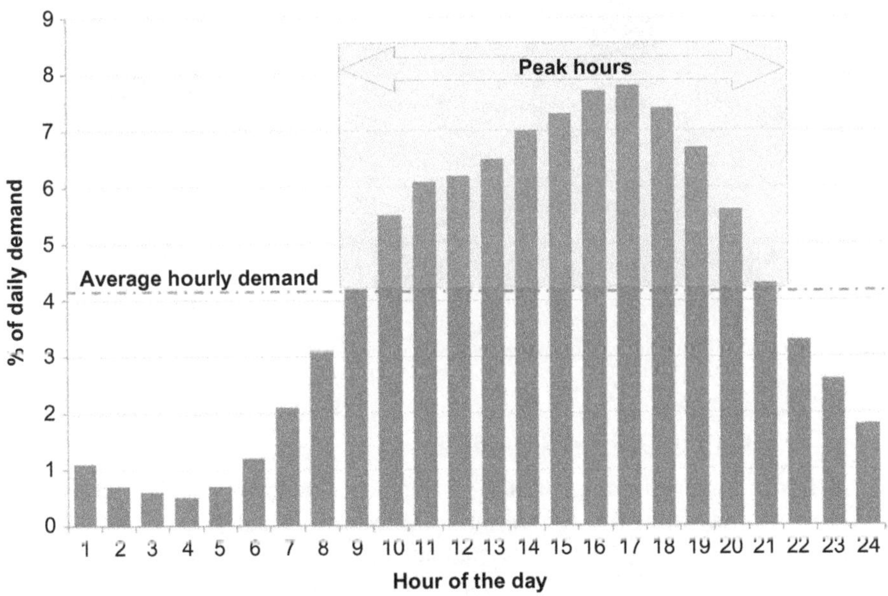

Figure 13.3 Variation of hourly demand during a day.

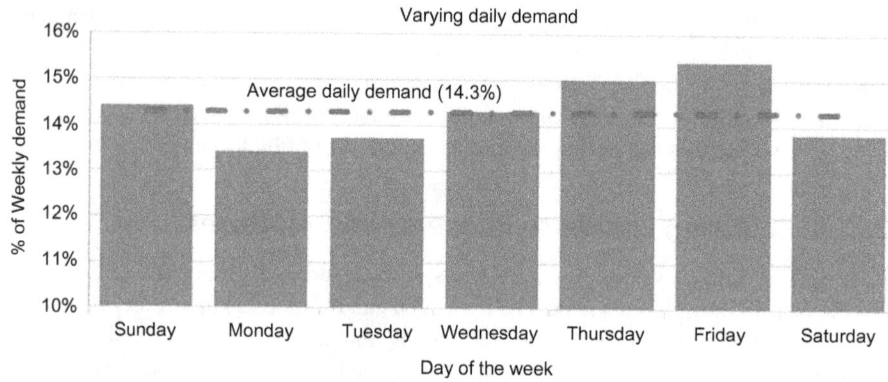

Figure 13.4 Variation of daily demand during a week.

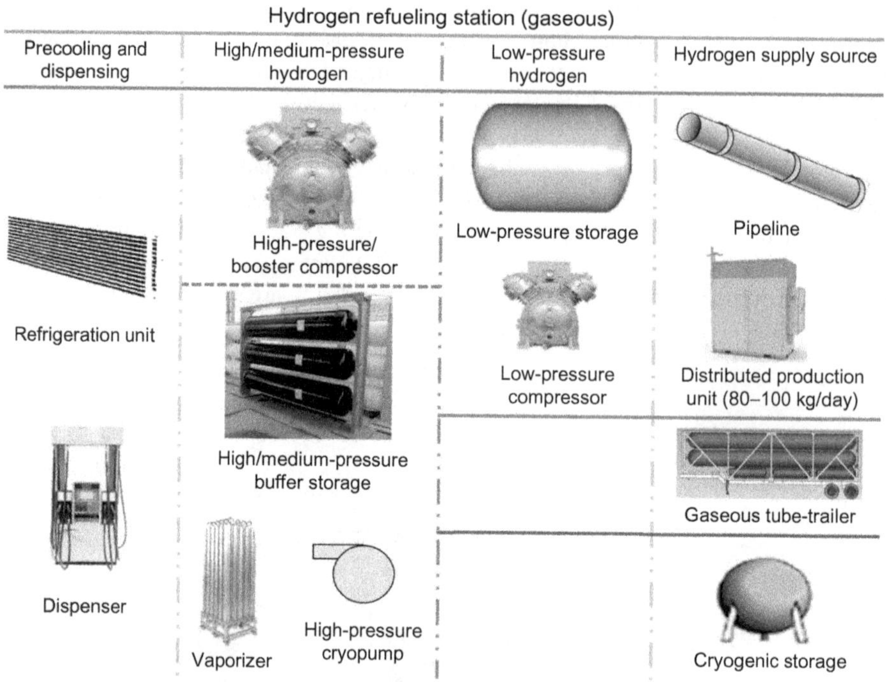

Figure 13.5 Hydrogen refueling station components.

13.5.1.1 *Refueling standardization*

The Society of Automotive Engineers (SAE) defined a fueling protocol (J2601) (SAE, 2014) for light-duty gaseous hydrogen surface vehicles by using "communication" and "noncommunication" fills. The use of communication or noncommunication fueling depends on the ability of the vehicle to transmit vehicle tank information to the dispenser. When the vehicle can transmit tank status information to the dispenser, the communication fill is allowed. In the absence of such a capability, the noncommunication protocol is followed. Protocol SAE J2601 establishes the requirements of dispensing by defining the fueling process safety and performance limits. SAE J2601 considers the tank's geometric and thermal characteristics, the station's precooling capability, and the ambient temperature to define the recommended dispensing rate and thus the fill duration. The following list briefly summarizes the SAE J2601 fueling protocol.

- At no point and place during the fill can the maximum hydrogen gas temperature within the vehicle tank exceed 85 °C.
- The maximum hydrogen gas pressure in the vehicle tank during or at the end of the fill cannot exceed 875 bar at 85 °C.
- The maximum rate of fueling at any time during the fill cannot exceed 60 g/s (equivalent to 3.6 kg/min).
- The target fueling time is 3 min for a passenger car with capacity of up to 5 kg of hydrogen at a pressure rated at 700 bar.
- The vehicle tank is considered full (i.e., 100% SOC) when the density of the hydrogen within the tank is 40.2 g/L (i.e., 700 bar at 15 °C; 875 bar at 85 °C).

13.5.1.2 Hydrogen supply source

Refueling stations with a liquid hydrogen supply include a liquid cryogenic tank (dewar), which is refilled from a liquid tanker. Stations that receive gaseous hydrogen via tube trailer usually integrate the tube trailer into the refueling station. The tube trailer is swapped with another one when it is drawn down to a specific pressure, usually between 20 and 50 bar. Alternatively, gaseous hydrogen can be supplied by pipelines or through onsite production via water electrolysis or natural gas SMR.

13.5.1.3 Storage

Storage at the refueling station is required to address the mismatch between supply and demand during daily operations. The supply and demand variations define the required amount of storage and its design pressure.

On-site storage

When hydrogen is supplied to a station by using a pipeline or an onsite production unit, it is usually provided at a uniform rate. Onsite storage is thus required to acquire hydrogen during off-peak hours and supply it back during peak-demand hours. The onsite storage should at least be sized to hold an amount equal to the excess (above average) demand over a 24-h period. The onsite storage usually consists of large steel tanks designed to store hydrogen at pressures of 200–400 bar.

Cascade buffer storage

The buffer storage consists of several banks of pressure vessels that store hydrogen at a high pressure (typically at 900–1000 bar for 700-bar dispensing) and that are used to cascade the delivery of the fuel into the vehicle's tank. The size of the high-pressure buffer storage is determined by the peak-hour demand and the throughput of the compressor that replenishes it. A medium-pressure buffer storage system (typically at 400–500 bar) is usually considered when the station is supplied with low-pressure hydrogen via a pipeline or an onsite production unit. In such a case, the vehicle's tank is initially filled from the medium-pressure buffer and then topped off by using a booster compressor that draws from the buffer storage to directly fill the vehicle's tank. High-pressure storage uses either Type II or Type IV pressure vessels, while medium-pressure storage typically uses Type I pressure vessels. Today, stations use Type I steel pressure vessels and/or Type II steel with carbon fiber reinforcement. Table 13.2 briefly describes the different types of pressure vessels.

13.5.1.4 Compressor

The compressor is considered the heart of the refueling station since it is the main component responsible for making the hydrogen flow from the supply source to the dispenser nozzle. The compressor at a refueling station usually employs a high compression ratio but has relatively low throughput. A compressor takes in low-pressure hydrogen at the inlet (~20 bar) and delivers high-pressure hydrogen at the outlet (~900–1000 bar). The high-pressure hydrogen is used either to replenish the high-pressure buffer storage system or to directly fill the vehicle's tank. The compressor is used to replenish the high-pressure buffer when the station's hydrogen is supplied by a high-pressure tube trailer. Alternatively, it can be used to replenish

Table 13.2 Types of pressure vessels (U.S. Drive Partnership, 2013)

Type of pressure vessel	Description
Type I	All metal cylinders that are capable of holding gases at any pressure and are put at a disadvantage by the extra weight resulting from high working pressures. A high working pressure increases the thickness of the wall required to contain the gas, which increases the vessel's weight. This vessel type is economical for holding gases at low to medium pressures.
Type II	Container with load-bearing metal liner with hoop-wound carbon-fiber reinforcement. Typically, the load is equally shared between the metal liner and fiber-reinforced winding. The lightweight fiber reinforcement reduces the amount of metal required to hold the gas at the same pressure, which reduces the container's weight. This vessel type is economical for holding gases at medium to high pressures.
Type III	Non-load-bearing metal (aluminum) liner and axial and hoop-wound carbon-fiber-reinforced cylinder. This vessel type has a weight advantage over Type I and Type II vessels and is economical for holding gases at low to medium pressures when weight limitation is a concern. The liner is prone to fatigue at high-pressure cycle frequencies.
Type IV	Non-load-bearing plastic liner and axial and hoop-wound carbon-fiber-reinforced cylinder. This vessel type has a weight advantage over Type I, Type II, and Type III vessels and is economical for holding gases at high pressures when weight limitation is a concern.

the low- or medium-pressure buffer storage vessels when the station's hydrogen is supplied by a pipeline or an onsite production unit. The compressor can also be used as a booster compressor to top off the vehicle's tank when the station incorporates medium-pressure buffer storage instead of a high-pressure buffer storage system. The size of the compressor is determined primarily by the size of the station and its hourly demand profile (or number of back-to-back fills); a secondary influence is the size of the high- or medium-pressure buffer storage system. There are many compressor types and technologies, including reciprocating piston and flexing diaphragm compressors. Currently, the majority of hydrogen refueling compressors in the US are diaphragm compressors, primarily because of their reliability and ability to deliver high-quality hydrogen.

13.5.1.5 Refrigeration unit and dispenser

The SAE J2601 fueling protocol requires that hydrogen be precooled to a temperature window of −33 to −40 °C before being dispensed in order to restrict the rise in temperature in the vehicle's tank to below its maximum value (85 °C) during fast fills. While the refrigeration unit cools the hydrogen to the required temperature, the dispenser controls the flow to keep the process parameters within the limits set by SAE J2601. The precooling process is important to enable fast fills (e.g., 5 kg in 3 min) while ensuring the safety and integrity of the vehicle's tank. The dispenser includes the metering equipment required to measure the amount of hydrogen dispensed into the vehicle's tank. The National Institute of Standards and Technology (NIST) specifications for fueling mass flow meters require that dispensing equipment must be accurate to within 2% (NIST, 2014).

13.5.1.6 High-pressure cryogenic pump and heat exchanger

The high-pressure cryogenic pump and heat exchanger (commercially known as a vaporizer) are used when a cryogenic storage tank is installed at the station and hydrogen is supplied to the refueling station in liquid form. The "cryopump" lifts the hydrogen pressure to the dispensing pressure; the hydrogen is then heated to the desired dispensing temperature (e.g., −33 to −40 °C) via the heat exchanger before being directly dispensed into the vehicle's tank. Alternatively, in order to enable simultaneous vehicle fills, the hydrogen may be routed from the heat exchanger into a high-pressure buffer system, which cascades the delivery of the fuel to the vehicle's tank in a manner similar to that of a gas station. In such a case, the refrigeration unit can be avoided by using the low-temperature liquid hydrogen as a heat sink to cool the gaseous hydrogen.

13.5.2 Other refueling station considerations

13.5.2.1 Sizing of refueling components

Because of the interdependency of the refueling station components (e.g., compressor and buffer storage), a careful design is needed to achieve the optimal cost and performance goals of the refueling station (Reddi et al., 2014). The sizes of the refueling station components are defined primarily by the station's daily capacity, the hourly demand profile, and the vehicle's tank capacity and operating pressure. The technologies of the refueling station components are defined by the hydrogen supply source and its physical form. Component combinations and operational strategies need to be identified to realize the least-cost fueling configuration.

13.5.2.2 Hydrogen quality

Fuel cells require high-quality hydrogen to sustain a vehicle's operational performance and efficiency over its lifetime. In order to deliver the required quality of hydrogen to the vehicle's tank, the hydrogen should be produced according to quality standards and be packaged and delivered to the vehicle tank without contamination. Contamination-free designs for the station components need to be developed in order to avoid the extra cost that would result from an additional purification step.

13.5.2.3 Compression

Compressors at refueling station reportedly suffer frequent downtime because of components that break down. Compressors must be highly reliable in order to avoid costly station downtime or the need to install additional backup compressors. Contamination-free compressor designs capable of high compression ratios, with moderate to high throughput and low maintenance, are required for refueling stations to operate economically.

13.5.2.4 Storage

A high- or medium-pressure buffer storage system primarily addresses the mismatch between the hourly demand and supply, while onsite storage (when hydrogen is supplied to the station via a pipeline or an onsite production unit) addresses the mismatch between daily demand and supply. The required setback distances for gaseous and liquid storage systems in the US are greater than those required by European standards. This results in large land area requirements and higher refueling costs. Research is needed to study and establish appropriate codes and standards that could reduce the required setback distances without compromising safety in operating the station.

13.5.2.5 Dispenser

Current metering technologies cannot meet the required measuring accuracy standards set by federal agencies for transportation fuels being sold to the public in the US. Flexible hoses that can withstand high pressures (875 bar) and low temperatures (–40 °C) need to be developed and tested for reliability and durability. Dispensers currently available in the market are costly and supplied by only a few manufacturers. New technologies and investments are needed to develop dispensers that are capable of measuring hydrogen to the necessary accuracy and that have acceptable durability.

13.5.2.6 Refrigeration

The SAE J2601 protocol requires hydrogen to be cooled to –33 to –40 °C within the first 30 s of the refueling event to avoid an increase in the refueling time. This means the refrigeration system must run more frequently in order to maintain the heat exchanger within that temperature window for immediate and fast refueling. Such semi-continuous operation can negatively affect the station's operating costs. The impact of the refueling protocol on the cost of dispensing hydrogen must be evaluated to inform further development of the refueling protocol.

13.5.3 Current hydrogen refueling station configurations

Almost all the refueling stations in operation or under development in California are supplied by a gaseous tube trailer, a liquid tanker truck, or an onsite electrolyzer. Although the target pressure for the vehicle's tank is 875 bar for all stations, the components of the refueling station vary with the supply source. The most common refueling station configurations are described in the text that follows.

13.5.3.1 Gaseous and liquid hydrogen supply

Figure 13.6 shows the refueling components and their layout when the hydrogen is supplied by a gaseous tube trailer or liquid tankers. When the station is supplied by using a gaseous tube trailer, the empty tube trailer (typically at 20–50 bar) at the station is swapped with another full tube trailer (typically at 250–500 bar) that is delivered to the station at a regular frequency. The tube trailer connected to the compressor's inlet is emptied to replenish the high-pressure buffer storage that is connected to the compressor's outlet. The compressor pressurizes the hydrogen to 950 bar and replenishes the high-pressure buffer storage to maintain it at ~900 bar. The high-pressure buffer storage is connected to the dispenser via a refrigeration unit. When a vehicle activates the dispenser, and hydrogen is channeled from high-pressure buffer storage

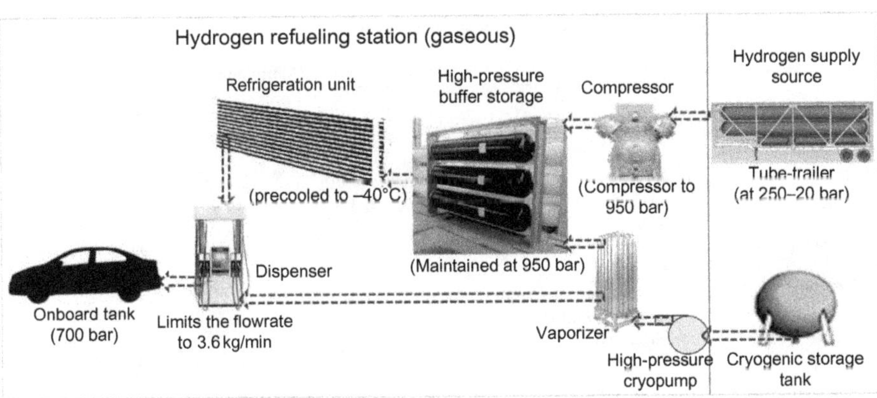

Figure 13.6 Layout of the components of a hydrogen refueling station supplied by liquid or gaseous hydrogen.

through the refrigeration unit into the vehicle's tank by the dispenser. The dispenser monitors and keeps the flow rate within protocol SAE J2601's bounds by applying backpressure to regulate the flow of hydrogen. The pressure differential between the high-pressure buffer storage and the dispenser's backpressure drives the hydrogen flow into the vehicle's onboard tank. The refrigeration system precools the hydrogen down to about –40 °C as the hydrogen passes through its system.

Alternatively, when the hydrogen is distributed to the station by using a liquid tanker, the liquid hydrogen is transferred from the tanker truck to the onsite cryogenic storage tank (dewar). The dewar is connected to the inlet of a high-pressure cryopump, the output of which is connected to the inlet of a heat exchanger (commonly known as a vaporizer). The outlet of the vaporizer is connected either to the high-pressure buffer storage (in order to enable simultaneous vehicle fills) or to the dispenser (in order to directly fill the vehicle). The vaporizer and pump combination function like a compressor to replenish and maintain the high-pressure buffer storage at 950 bar. The operation of the high-pressure buffer storage is the same as described for the tube trailer supply. For smaller stations with a single refueling position (nozzle), high-pressure storage can be avoided by routing the gaseous hydrogen at the vaporizer outlet to directly fill the vehicle's tank.

13.5.3.2 On-site hydrogen production

Figure 13.7 shows the components and layout of a hydrogen refueling station with an onsite hydrogen production unit. Currently, there are two primary technologies used for onsite hydrogen production: water electrolysis and SMR of natural gas. The onsite electrolyzer uses onsite or grid electricity to separate water molecules into hydrogen and oxygen, while the natural gas supplied by a pipeline can be reformed to produce hydrogen with high purity. Hydrogen is usually produced at a pressure of about 20 bar, but electrolyzers can produce hydrogen at higher pressures. The compressor takes the hydrogen at 20 bar from the onsite production unit and pressurizes it either to 950 bar

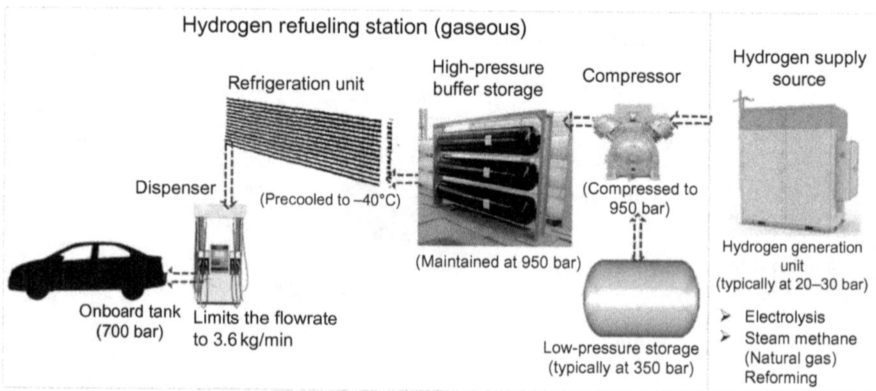

Figure 13.7 Layout of the components of a hydrogen refueling station supplied by an on-site production unit.

(in order to replenish high-pressure buffer storage) or to 200–400 bar (in order to store the produced hydrogen in low- or medium-pressure buffer storage). The low-pressure storage is used to supply hydrogen partially during the hours when the demand is above the daily average, and is replenished during the hours when the demand is below the daily average. As mentioned, hydrogen from high-pressure buffer storage is regulated into the vehicle's tank by the dispenser via a refrigeration unit that precools the hydrogen to a temperature of about –33 to –40 °C.

13.6 Hydrogen production, transmission, and distribution

When hydrogen is produced in large quantities at central or semi-central locations, a hydrogen delivery infrastructure is required to transport it to refueling stations. This delivery infrastructure needs to be reliable and safe in addition to being cost effective, in order to provide consumers with experiences similar to or better than those offered by gasoline stations. The delivery of hydrogen from central production plants to the city gate is called transmission, while the delivery of hydrogen from the city gate to the refueling stations is called distribution. The delivery of hydrogen from the hydrogen source at the refueling station to the vehicle's tank is called fueling.

13.6.1 Central/semicentral production

Hydrogen can be produced in large volumes by using various technologies (e.g., SMR of natural gas, electrolysis of water, gasification of coal and biomass). Large central hydrogen production plants produce about 100–1000 metric tons per day; this is then supplied to terminals for distribution to refueling station networks. Alternatively, a semi-central hydrogen plant might produce a smaller amount at a terminal for distribution to a local network of refueling stations. Hydrogen, due to its low volumetric energy density, needs to be packaged in order for it to be transported economically. Figure 13.8 shows the technologies used for hydrogen packaging, transmission, and distribution operations.

13.6.2 Transmission

Hydrogen is produced at a low pressure (~20 bar), which means it must be packaged and transported to terminals for distribution to refueling stations.

13.6.2.1 Packaging operations

The primary goal of the packaging operation is to improve the volumetric energy density of hydrogen so it can be transported economically. A compressor is used to "package" hydrogen for transmission. It pressurizes the hydrogen from the production plant into pipelines to enable it to be economically transported to the distribution terminals or other points of use. These compressors must be highly reliable, have high throughput, and require low maintenance but operate at relatively low compression

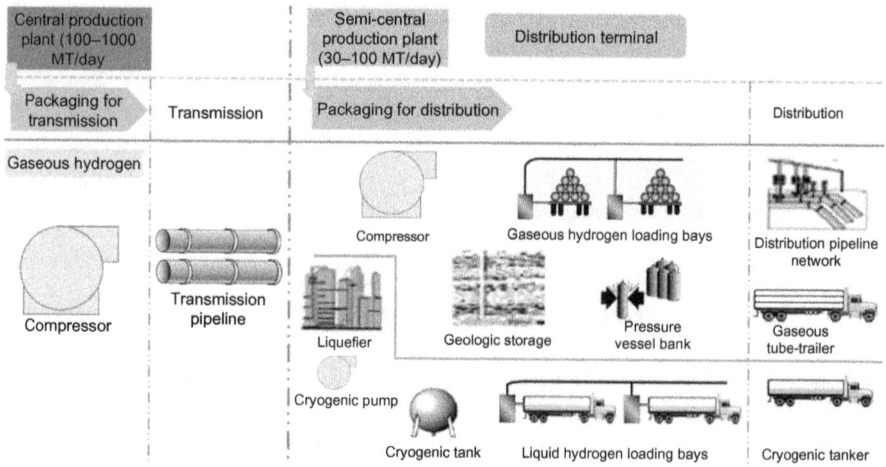

Figure 13.8 Technologies used for hydrogen transmission and distribution.

ratios. Pipeline compressors have a throughput on the order of 50,000–2 million kg/day while they compress hydrogen from 20 to about 100 bar.

13.6.2.2 Transport operations

Transport operations involve the physical movement of the packaged hydrogen from the production plant to the distribution terminals or other points of use. Hydrogen can be economically transported by pipelines when the amount that needs to be transported is large (about 100–1000 metric tons per day).

13.6.3 Distribution

A distribution terminal is expected to be able to handle about 30–100 metric tons of hydrogen each day. The hydrogen may be supplied by transmission pipelines or by a dedicated semi-central production plant. The hydrogen is packaged based on the intended form of distribution and is then transported to the refueling stations. The distribution terminal incorporates hydrogen storage to accommodate variations in the seasonal demand for hydrogen as well as scheduled and unscheduled production outages. A statistical analysis of the refueling demand at gasoline stations indicates that the demand during summer is about 10% higher than the annual average demand in the US. Storage at a distribution terminal helps address the mismatch between supply and demand during the months with demand that is above or below average (Figure 13.9).

13.6.3.1 Geologic storage

Geologic storage is an economical solution for mitigating variations in seasonal demand when it is available in the proximity of market locations. Depleted oil and natural gas wells, salt caverns, and aquifer and hard rock caverns are different forms of geologic storage. Salt caverns are the only type of geologic storage currently used for

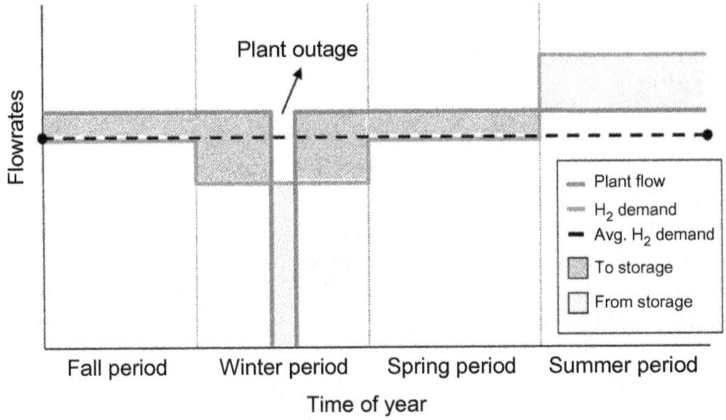

Figure 13.9 Seasonal variations in demand (Argonne National Laboratory et al., 2010).

hydrogen storage. There are three salt caverns in Texas that can be used for hydrogen storage. All types of caverns can withstand pressures between 80 and 160 bar. Salt caverns are naturally available for storage; thus, they require only a minimal capital investment and minimal effort to be processed for storing hydrogen. In the absence of salt caverns, depleted oil and gas reservoirs and aquifers may be used, and while less capital intensive, may not contain the hydrogen and maintain its purity as effectively. Alternatively, hard rock caverns could be constructed, but they are more capital intensive (Lord et al., 2014).

13.6.3.2 Terminal storage

Hydrogen can be stored at a distribution terminal in two ways to address seasonal variations in demand and supply: (1) in large banks of pressure vessels or (2) in large cryogenic liquid storage tanks after undergoing a liquefaction process.

- Pressure vessel banks: If geologic storage is not available in the vicinity of a hydrogen market, hydrogen can be stored in banks of pressure vessels. This option requires a large space for storage due to the low volumetric density of gaseous hydrogen, and so it increases the amount of land area required in the terminal. It is desirable to store hydrogen at high pressures to increase its volumetric energy density, but due to the disproportionate increase in the cost of storage with pressure, a trade-off must be made between the real estate value and the pressure vessel costs.
- Cryogenic storage tanks: When hydrogen is liquefied at the terminal, it is stored in large, vacuum-jacketed liquid tanks to address the variations in supply and demand. Due to the high volumetric energy density of liquid hydrogen, it requires a much smaller footprint than does gaseous hydrogen stored in pressure vessel banks.

13.6.3.3 Packaging operations

Packaging operations improve the volumetric energy density of hydrogen so it can be more economically transported to refueling stations. Hydrogen is "packaged" by being compressed to flow through pipelines or compressed into tube trailers to be trucked to refueling stations. Alternatively, hydrogen can be liquefied to further improve the volumetric energy density for being loaded into liquid tankers and delivered to refueling stations.

- Gaseous hydrogen technologies: Gaseous hydrogen is compressed and loaded into tube trailers or distribution pipelines to be economically transported to refueling stations. The pressure of the hydrogen in the distribution lines can be up to 100 bar. Delivering hydrogen to refueling stations at this pressure can significantly reduce the compression requirement at the refueling stations. Hydrogen can be compressed up to 500 bar to be loaded into tube trailers. Tube trailers made of Type III or Type IV tanks can carry about 1000 kg of hydrogen payload at 500 bar; they can carry 250 kg of payload when the tubes are made of steel (Type I tanks).
- Liquid hydrogen technologies: Liquefaction of hydrogen involves multiple compression and expansion cycles to cool the hydrogen to temperatures below 20 K where the hydrogen exists in liquid form. Liquefaction increases the volumetric density of hydrogen from 40 g/L at 700-bar pressure to about 70 g/L in liquid form. Improved volumetric energy density results in significant economic benefits in transporting hydrogen, especially when transportation distances are large (more than 200 miles) (U.S. Drive Partnership, 2013). The liquid hydrogen from the liquefaction plant is fed into large cryogenic storage tanks, typically sized for 5–6 days of demand. It is then transferred from the cryogenic storage tanks to the liquid tankers for delivery to refueling stations.

13.6.3.4 Transport operations

Transport operations involve the actual movement of hydrogen from the distribution terminal to the refueling stations. Hydrogen can be transported via a pipeline network, gaseous tube trailers, or liquid tankers.

- Pipelines: Hydrogen can be distributed by using a network of pipelines connecting the terminal to the refueling stations. Pipelines are commonly thought to be the cheapest means of transporting hydrogen, but due to their high construction cost, the demand for hydrogen must be high for pipelines to be economically viable. The cost of pipeline transport includes costs for materials, labor, rights of way, and other miscellaneous factors. Labor and right-of-way costs constitute the largest portions of the total pipeline cost and can vary significantly among various US regions. Currently, the cost of a natural gas pipeline in the US ranges from $2 million to $6 million per mile, depending on its diameter, the terrain, and its geographic location. The cost for labor can be significantly reduced by using fiber-reinforced polymer (FRP) pipes that can be spooled for several miles, but then pipeline diameters are restricted to 6–8 in. The material cost for FRPs is much higher than it is for steel pipes.
- Gaseous tube trailer: A gaseous tube trailer consists of pressure vessels (or tubes) packaged within an International Standards Organization (ISO) container, mounted on a trailer. The pressure vessels can be any of the types described in Table 13.2. Tube trailers are economical to use when hydrogen is being delivered to stations that are located within a radius of about 200 miles and whose capacity ranges from 50 to 500 kg/day. The capacity of a tube trailer varies with the type of pressure vessels mounted on it. Since the maximum tractor-trailer weight is restricted to 80,000 lb (36,000 kg) by the US Department of Transportation (DOT), any reduction in vessel weight will allow more room for an increased hydrogen payload. Type III and Type IV vessels weigh less than Type I and Type II vessels at any pressure rating; thus, they allow a higher payload of hydrogen to be transported in a single trip. With Type I

pressure vessels, the maximum amount of hydrogen that can be delivered is limited to about 250 kg, while with Type IV vessels, the maximum is up to 1000 kg at 500 bar.

- Liquid tankers: Liquid tankers have a vacuum-jacketed cryogenic tank installed on the trailer truck that has a capacity of up to 4000 kg. The liquid tanker transfers the liquid hydrogen to the onsite liquid storage tanks by using an onboard cryopump or through pressure transfer, in which the tank is warmed up to vaporize part of the liquid hydrogen and increase the pressure over the liquid surface in order to create a differential pressure for unloading the hydrogen into the onsite cryogenic storage (dewar). After the required amount of hydrogen is transferred to the onsite tank through pressure transfer, the hydrogen within the tanker is vented off to bring the pressure down to allowed limits for road transport. Liquid tankers can deliver hydrogen for long distances economically, primarily because they can deliver a relatively higher payload than can tube trailers that deliver compressed gas.

13.7 Hydrogen transmission and distribution barriers

13.7.1 Pipelines

Pipelines are the most economical option when market demands are large (~1000 metric tons/day), but they are also the most capital-intensive option, requiring a huge investment for their construction. In addition to their high construction cost, pipelines are constrained by the lack of (1) materials that are chemically passive to hydrogen, (2) compatible odorants for leak detection, and (3) compressors capable of delivering a high throughput at a low cost.

13.7.1.1 Construction cost

The components of the construction cost can be categorized into the material cost, labor cost, right-of-way cost, and miscellaneous cost. The labor cost is the largest component and accounts for more than 50% of the total US average pipeline construction cost. Steel pipelines, typically used to construct natural gas and hydrogen pipelines, are heavy and so have a limited span. The labor cost is driven mainly by the time it takes workers to lay and weld the pipe sections. Innovative materials, like fiber-reinforced plastic pipes, and packaging techniques, like spooling, are required to cut down the cost of labor. Other issues include the lack of clear standards to guide the officials in giving permits to construct hydrogen transmission and distribution pipelines. Codes and standards are being developed in order to address this gap; they are intended to provide clear guidelines for the officials so the permitting process can be streamlined.

13.8 Material

The interaction of hydrogen with pipeline materials, like steel, needs to be studied and better understood. The embrittlement of steel pipelines needs to be addressed by developing new coatings to avoid direct interaction between steel and hydrogen. In addition, due to the low volumetric energy density of hydrogen, it is better to transport hydrogen at higher pressures. The impact of pressure cycling (with regard to cycle depth and the frequency of cycling) on pipeline fatigue needs to be better understood.

13.8.0.1 Odorants

When odorants are added to odorless gases, they help in detecting leaks. Ideally, the odorant should diffuse faster than the gas to limit the safety risk and the amount of gas leaked. Hydrogen is a light gas and diffuses faster than currently used odorants. New odorants that are compatible with hydrogen and that can be efficiently separated (fuel cells require pure hydrogen) need to be developed.

13.8.1 Compression

Due to the low volumetric energy density of hydrogen, transporting it through pipelines or tube trailers requires high throughputs and relatively high pressures. The reciprocating compressors currently being used to compress hydrogen are not suitable for transportation applications due to contamination issues. Additional cleaning steps are needed to ensure that the high-quality hydrogen requirements by fuel cells are addressed. Lubricant-free technologies, or effective low-cost cleaning technologies, need to be developed in order to deliver high-quality hydrogen at a low cost.

13.8.2 Storage

Hydrogen must be stored at different stages of the delivery pathway in order to address the mismatch between supply and demand. The geographic location of the hydrogen and amount of it that needs to be stored define the technology that needs to be employed. When hydrogen is stored at low pressure in large quantities in geologic storage facilities, there is a significant risk of contamination from foreign matter and/or leakage, depending on the type of soil or rock formation. Unlike natural gas, hydrogen must be of high quality and must also be stored at a higher pressure due to its low volumetric energy density. The effect of pressure cycling (due to frequent filling and emptying patterns) on the storage structure is unknown and needs to be studied. Hydrogen's reactivity with surrounding rock formations is also unknown. The geologic storage option is limited by the location of the hydrogen, and engineering a geologic storage facility requires a huge investment.

13.8.3 Gaseous tube trailers

DOT restricts the weight of commercial trucks using national highways to 80,000 lb (36,000 kg). In addition, the payload should be packed within an ISO container measuring $8 \times 8 \times 40$ ft ($2.4 \times 2.4 \times 12$ m). Due to the weight restriction, it is desirable to minimize the weight of the pressure vessels while increasing the hydrogen payload. This can be achieved by selecting materials with high strengths and low densities. For example, carbon fiber composite pressure vessels can increase the hydrogen payload to up to 1000 kg (at 500 bar), while steel pressure vessels have a maximum payload of 250 kg (at 180 bar). The cost of the tubes is relatively high, at about \$900–\$1200 kg^{-1} of payload. Innovative materials and technologies need to be developed to reduce the tube costs. The effect of the pressure cycling of composite pressure vessels during loading and unloading operations is unknown and needs to be studied in order to estimate vessel lifetimes and the consequent implications on the levelized delivery cost.

13.8.4 Liquefaction

Although liquefaction increases the volumetric energy density and also the quality of hydrogen, it is an energy-intensive process. The liquefaction of hydrogen requires about 35–40% of the energy content of hydrogen and accounts for about 12–15 kWh of electricity/kg of liquid hydrogen. Novel or advanced technologies, like acoustic and magnetic liquefaction, need to be developed to reduce the energy intensity of liquefaction. Novel strategies are also needed to minimize the exothermic reaction of the conversion of *ortho*-hydrogen to *para*-hydrogen.

13.8.5 Liquid tanker trucks

Liquid tankers provide an economical means for transporting large amounts of hydrogen over long distances but result in about 6% losses due to boil-off during loading and unloading. The hydrogen boil-off is not entirely unavoidable, so strategies that can minimize it by streamlining the transfer processes or by recovering the boil-off could reduce the overall delivery cost.

References

AFDC, 2014a. Alternative Fuel Price Report (accessed November 2014). http://www1.eere.energy.gov/cleancities/toolbox/price_report.html.

AFDC, 2014b. Fuel Properties Comparison. Department of Energy (accessed December 2014). http://www.afdc.energy.gov/fuels/fuel_comparison_chart.pdf.

Argonne National Laboratory, National Renewable Energy Laboratory, Pacific Northwest National Laboratory, 2010. Hydrogen Delivery Scenario Analysis Model, 2.3 ed. Fuel Cell Technology Office, US Department of Energy (accessed December 2014). https://apps1.hydrogen.energy.gov/cfm/h2a_active_folder/h2a_delivery/05D_H2A_Current_(2010)_Delivery_Scenario_Analysis_Model_Version_2.3.xls.

Boston Consulting Group, 2012. Understanding the Impact of AB 32 (accessed December 2014). http://www.deq.state.or.us/aq/cleanFuel/docs/IndividualCommenters7.pdf.

CaFCP, 2013. State Commits Funding for First 100 Hydrogen Stations (accessed November 2014). http://cafcp.org/getinvolved/stayconnected/blog/california_announces_299M_hydrogen_fueling.

CaFCP, 2014. 2014 Update: Hydrogen Progress, Priorities and Opportunities Report, July (accessed November 2014). http://cafcp.org/sites/files/Roadmap-Progress-Report2014-FINAL.pdf.

CaFCP (California Fuel Cell Partnership), 2012. A California Road Map: The Commercialization of Hydrogen Fuel Cell Electric Vehicles (accessed November 2014). http://cafcp.org/sites/default/modules/pubdlcnt/pubdlcnt.php?file=/sites/files/A%20California%20Road%20Map%20June%202012%20(CaFCP%20technical%20version)_1.pdf&nid=1807.

CARB (California Air Resources Board), 2013. State Zero-Emission Vehicle Programs—Memorandum of Understanding (accessed November 2014). http://www.arb.ca.gov/newsrel/2013/8s_zev_mou.pdf.

CARB (California Air Resources Board), 2012. The Zero Emission Vehicle (ZEV) Regulation (accessed November 2014). http://www.arb.ca.gov/msprog/zevprog/factsheets/general_zev_2_2012.pdf.

CEC, 2014a. Notice of Proposed Award—Alternative and Renewable Fuel and Vehicle Technology Program (accessed November 2014). http://www.energy.ca.gov/contracts/PON-13-607_NOPA.pdf.

CEC, 2014b. Questions, Answers, and Clarifications—Addendum 2, Hydrogen Refueling Infrastructure Solicitation (accessed November 2014). http://www.energy.ca.gov/contracts/PON-13-607/Q_and_A_PON-13-607_Addendum_02.pdf.

Chen, T., 2008. Hydrogen Delivery Infrastructure Options analysis. Nexant Inc. http://www1.eere.energy.gov/hydrogenandfuelcells/pdfs/delivery_infrastructure_analysis.pdf.

DOE (US Department of Energy), 2014. H2FIRST: Hydrogen Fueling Infrastructure Research and Station Technology (accessed December 2014). http://energy.gov/eere/fuelcells/h2first.

EIA (Energy Information Administration), 2014. Annual Energy Outlook 2014 with Projections to 2040. US Department of Energy, Washington, D.C.

Eichberger, J., 2014. Retail considerations in the installation of natural gas fueling infrastructure. In: Annual Meeting of the Transportation Research Board, Washington, D.C, January.

Elgowainy, A., Wang, M., 2012. Sustainable vehicle fuels, well-to-wheel analysis. In: Meyers, R.A. (Ed.), Encyclopedia of Sustainability Science and Technology, Springer, New York, pp. 10502–10529.

Greene, D., et al., 2008. Analysis of the transition to hydrogen fuel cell vehicles and the potential hydrogen energy infrastructure requirements, Oak Ridge National Laboratory Report ORNL/TM-2008/30, March. http://www.fuelcells.org/uploads/ornlreport.pdf (accessed November 2014).

H2USA, 2014. To Promote Fuel Cell Electric Vehicles across America (accessed December 2014). http://www.h2usa.org/.

Joseck, F., 2012. Current U.S. hydrogen production, DOE Hydrogen and Fuel Cells Program Record Number 12014. http://www.hydrogen.energy.gov/pdfs/12014_current_us_hydrogen_production.pdf (accessed December 2014).

Lipman, T., 2011. An Overview of Hydrogen Production and Storage Systems with Renewable Hydrogen Case Studies. Clean Energy States Alliance (accessed December 2014). http://www.cesa.org/assets/2011-Files/Hydrogen-and-Fuel-Cells/CESA-Lipman-H2-prod-storage-050311.pdf.

Lord, A., Kobos, P., Borns, D., 2014. Geologic storage of hydrogen: scaling up to meet city transportation demands. Int. J. Hydrog. Energy 39 (28), 15570–15582. http://dx.doi.org/10.1016/j.ijhydene.2014.07.121.

Markowitz, M., 2013. Hydrogen and Fuel Cell Technical Advisory Committee (accessed November 2014). http://www.hydrogen.energy.gov/pdfs/htac_oct13_6_markowitz.pdf.

Melaina, M., Penev, M., 2013. Hydrogen Station Cost Estimates: Comparing Hydrogen Station Cost Calculator Results with Other Recent Estimates. National Renewable Energy Laboratory. Technical Report: NREL/TP-5400-56412, Golden, CO. (accessed December 23, 2014). http://www.nrel.gov/docs/fy13osti/56412.pdf.

Mintz, M., Gillette, J., Mertes, C. 2014. Economic impacts. In: Burr Stewart, S. (Ed.), Economic Impacts Associated with Commercializing Fuel Cell Electric Vehicles in California, Draft Report, September.

National Research Council, 2013. Transitions to Alternative Vehicles and Fuels. Washington, DC: The National Academies Press.

Nicholas, M., 2010. Driving demand: what can gasoline refueling patterns tell us about planning an alternative fuel network. J. Transp. Geogr. 18 (6), 738–749. http://dx.doi.org/10.1016/j.jtrangeo.2010.06.011.

NIST (National Institute of Standards and Technology), 2014. Specifications, tolerances, and other technical requirements for weighing and measuring devices, Section 3.37: Mass Flow Meters, Handbook 44-14. Available at: http://www.nist.gov/pml/wmd/pubs/hb44-14.cfm.

NREL (National Renewable Energy Laboratory), 2011. Composite Data Products, Fuel Cell Vehicle Range and Driving Behavior (accessed December 2014). http://www.nrel.gov/hydrogen/cdp_topic.html#driving.

Ogden, J., Nicholas, M., 2011. Analysis of a "cluster" strategy for introducing hydrogen vehicles in Southern California. Energy Policy 39 (4), 1923–1938. http://dx.doi.org/10.1016/j.enpol.2011.01.005.

Ogden, J., Yang, C., Nicholas, M., Fulton, L., 2014. The Hydrogen Transition: A NextSTEPS White Paper. UC Davis Institute of Transportation Studies, July. http://steps.ucdavis.edu/?energy-category=hydrogen (accessed November 2014).

Reddi, K., Elgowainy, A., Sutherland, E., 2014. Hydrogen refueling station compression and storage optimization with tube-trailer deliveries. Int. J. Hydrog. Energy 39 (33), 19169–19181.

SAE (Society of Automotive Engineers), 2014. Fueling Protocols for Light Duty Gaseous Hydrogen Surface Vehicles. SAE International, SAE J2601.

Sandia (Sandia National Laboratories), 2014. Widespread Hydrogen Fueling Infrastructure Goal of H2FIRST Project (accessed November 2014): https://share.sandia.gov/news/resources/news_releases/htwofirst/#.VI3IzdKjN8F.

Toyota, 2014. Announcing the Toyota Mirai Fuel Cell Vehicle, A Turning Point in Automotive History (accessed November 2014). http://www.toyota.com/fuelcell/.

U.S. Drive Partnership, 2013. Hydrogen Delivery Technical Team Roadmap (accessed March 2015). http://energy.gov/sites/prod/files/2014/02/f8/hdtt_roadmap_june2013.pdf.

Voelcker, J., 2014. 2016 Toyota Mirai power-out jack could run your home in emergencies, Green Car Reports, November 18. Accessed at http://www.greencarreports.com/news/1095522_2016-toyota-mirai-power-out-jack-could-run-your-home-in-emergencies.

ZEVTF (ZEV Program Implementation Task Force), 2014. Multi-State ZEV Action Plan, May. Accessed Nov. 2014 at http://governor.maryland.gov/documents/MultiStateZEVActionPlan.pdf.

Building a hydrogen infrastructure in Japan

14

M. Harada, T. Ichikawa†, H. Takagi‡, H. Uchida§*

*INPEX Corporation, Minato-ku, Tokyo, Japan, †Hiroshima University, Higashi-Hiroshima, Japan, ‡National Institute of Advanced Industrial Science and Technology, Tsukuba, Ibaraki, Japan, §Tokai University, Hiratsuka-City, Kanagawa, Japan

Abbreviations

CCS	carbon capture and storage
FC	fuel cell
FCEV	fuel cell electric vehicle
FY	Financial year
GOJ	Government of Japan
HTTR	high-temperature gas reactor; High-Temperature engineering Test Reactor (a type of nuclear power plant)
JAEA	Japan Atomic Energy Agency
JHFC 2	The Japan Hydrogen & Fuel Cell Demonstration Project phase 2
JPY	Japanese Yen
LOHC	liquid organic hydrogen carriers
MCH	methylcyclohexane
METI	Ministry of Economy, Trade and Industry
NEDO	New Energy and Industrial Technology Development Organization
PV	photovoltaic
SIP	Cross-ministerial Strategic Innovation Promotion Program

14.1 Introduction

The Strategic Energy Plan was formulated by the Government of Japan (GOJ) in order to show to the public the basic direction of Japan's energy policy under the Basic Act on Energy Policy. On April 11, 2014, the Cabinet made the decision to approve the new Strategic Energy Plan as the basis for the orientation of Japan's new energy policy, taking into account the dramatic changes in energy environments inside and outside Japan, including those caused by the Great East Japan Earthquake and the subsequent accidents at Tokyo Electric Power Co., Inc. Fukushima Daiichi Nuclear Power Station.

The Ministry of Economy, Trade and Industry (METI) established a Council for a Strategy for Hydrogen and Fuel Cells in December 2013, and since then the council has been studying ideal approaches to the future utilization of hydrogen energy,

Compendium of Hydrogen Energy. http://dx.doi.org/10.1016/B978-1-78242-364-5.00014-2

through collaboration between industry, academia, and government. On June 23, 2014, the Council compiled a Strategic Road Map for Hydrogen and Fuel Cells containing the measures to be taken by the people involved in realizing a hydrogen society, and METI hereby publicizes the Road Map.

In this chapter, the new Strategic Energy Plan and the Strategic Road Map for Hydrogen and Fuel Cells are first introduced to indicate the GOJ's measures for the spread and expansion of the utilization of hydrogen energy. Next, the Japanese situation regarding building a hydrogen infrastructure is discussed.

14.2 The new strategic energy plan (Strategic Energy Plan, 2014)

Positioned as the future energy vector, hydrogen plays a central role in the new Strategic Energy Plan. Hydrogen is an energy carrier that can be produced from inexhaustibly available water and various primary energy sources through a number of processes. In addition, hydrogen offers convenience of usage since it can be stored and transported in any form—whether gas, liquid, or solid, it can achieve high energy efficiency in combination with fuel cells (FCs) and low environmental burden. To realize a "hydrogen society," which will make full-fledged use of hydrogen, it is important to promote cost reduction as well as technology development activities with sufficient depth and diversity to make it possible to select technologies superior in safety, convenience, economic efficiency, and environmental friendliness from among the various technical options, under a strategy that encompasses the entire hydrogen supply chain from production, storage, and transport to end use. The Strategic Energy Plan indicates the following five measures for the acceleration of steps toward the realization of a hydrogen society:

(1) Spread and expansion of the introduction of stationary FCs (Ene-Farm, etc.). The goal of introducing 1.4 million units by 2020 and 5.3 million units by 2030 has been set. In order to achieve this goal, an environment favorable for autonomous introduction through cost reduction will be developed, and assistance will be provided with an eye to creating a self-sustaining market. In addition, research and development (R&D) of catalysts and other technologies for reducing cost and improving standardization will be continued. Also, regarding sectors in business and industry where stationary FCs have not spread so widely, R&D and demonstration activities to achieve durability and cost reduction at the level required by industrial activities are being facilitated, which in turn promotes market creation.

(2) Creating an environment for the acceleration of the introduction of FC vehicles. To promote the introduction of FC vehicles, which will go on sale commercially in 2015, hydrogen refueling stations will be constructed in about 100 places, mainly in the four major metropolitan areas, through regulatory reform and support measures, including support for introduction. If FC vehicles play an active role as a means of transportation in the 2020 Tokyo Olympic and Paralympic Games, there will be a chance to convince the world of the possibilities of hydrogen as a new energy source.[1] With that in mind, preparations for using hydrogen should surely be made from this point on.

(3) Realizing new technologies such as hydrogen power generation for full-scale usage of hydrogen. To realize the full potential of hydrogen as a universal energy vector, not only

stationary FCs and fuel cell cars but also hydrogen power generation is expected to expand. Regarding hydrogen power generation, mixed combustion where hydrogen makes up part of the fuel is possible in conventional gas turbines as long as the amount of hydrogen remains under a certain level. R&D for the practical future realization of unmixed combustion (i.e., combustion in which only hydrogen is used as the fuel) is also promoted. As for hydrogen usage technology, the GOJ is steadily promoting such strategic efforts including R&D from this point on.

(4) Promoting the development of production and storage/transportation technology for a stable supply of hydrogen. Hydrogen will be supplied through the use of by-product hydrogen from industrial processes or by the reforming of natural gas or naphtha for some time to come, but in order to utilize hydrogen on a full scale, other processes need to be developed on a large scale and commercialized as well: for example, from unharnessed coal or associated gas from oil production abroad for transport into Japan (although this method of hydrogen production will have a high CO_2 footprint unless the CO_2 can be captured and stored), and in the future, also to produce hydrogen by utilizing renewable energy such as solar power, wind power and biomass at home and abroad, e.g., via electrolysis. The GOJ is steadily promoting R&D related to the production and storage/transportation of hydrogen, such as large-scale storage or long-distance transportation by advanced technologies, including liquid hydrogen shipping vessels, liquid organic hydrogen carriers (LOHC) or conversion into chemical materials such as ammonia.

(5) Formulating a Road Map toward realization of a hydrogen society. To create a hydrogen society, it is essential to formulate a Road Map that provides a full picture of the endeavor, including the various elements related to production, storage, transportation, and use of hydrogen, such as large-scale storage and long-distance transportation of hydrogen using advanced technologies, FCs and hydrogen power generation. To implement a long-term, comprehensive Road Map, it is important for the various relevant entities to participate in the project, overcoming the barriers of entrenched interests. Therefore, a Road Pap toward realization of a hydrogen society has been formulated, and a council comprising representatives of industry, academia and government, responsible for implementation of the Road Map, has been established.

14.3 Strategic Road Map For Hydrogen and FCs (Strategic Road Map for Hydrogen et al., 2014)

As mentioned, a council for a Strategy for Hydrogen and FCs was established, meeting from December 2013 to June 2014. It compiled views on how Japan would be able to make use of hydrogen, the goals to be achieved in each step of the manufacture, transportation and storage of hydrogen, and collaborative efforts among industry, academia and government for achieving these goals, arriving at a Road Map with clear time frames attached, taking into account the significance of an initiative for disseminating hydrogen energy (Figures 14.1 and 14.2).

[1] In fact, the Tokyo metropolitan government has decided that the Athletes' Village for the 2020 Olympic and Paralympic Games will be made into a "hydrogen town" where electricity and hot water are supplied through hydrogen energy.

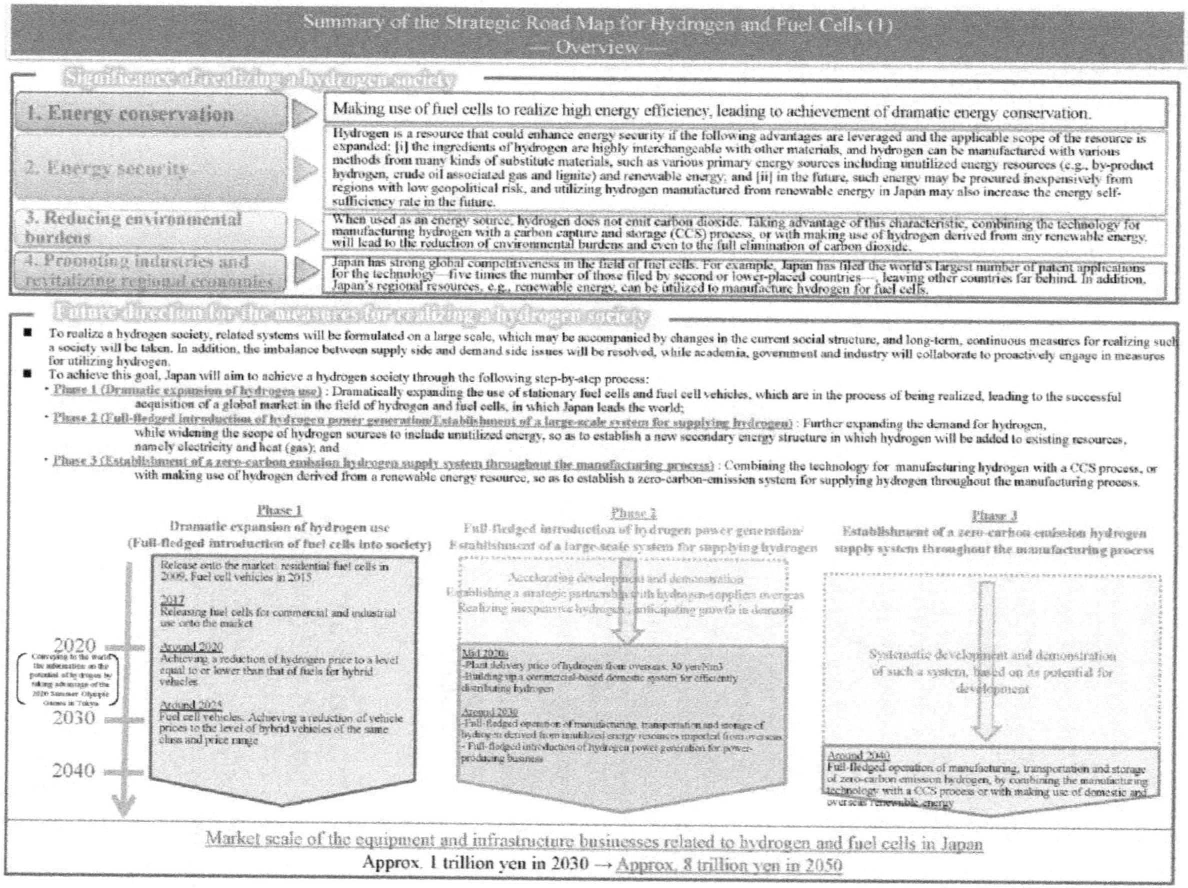

Figure 14.1 Summary of the Strategic Road Map for Hydrogen and Fuel Cells (overview).

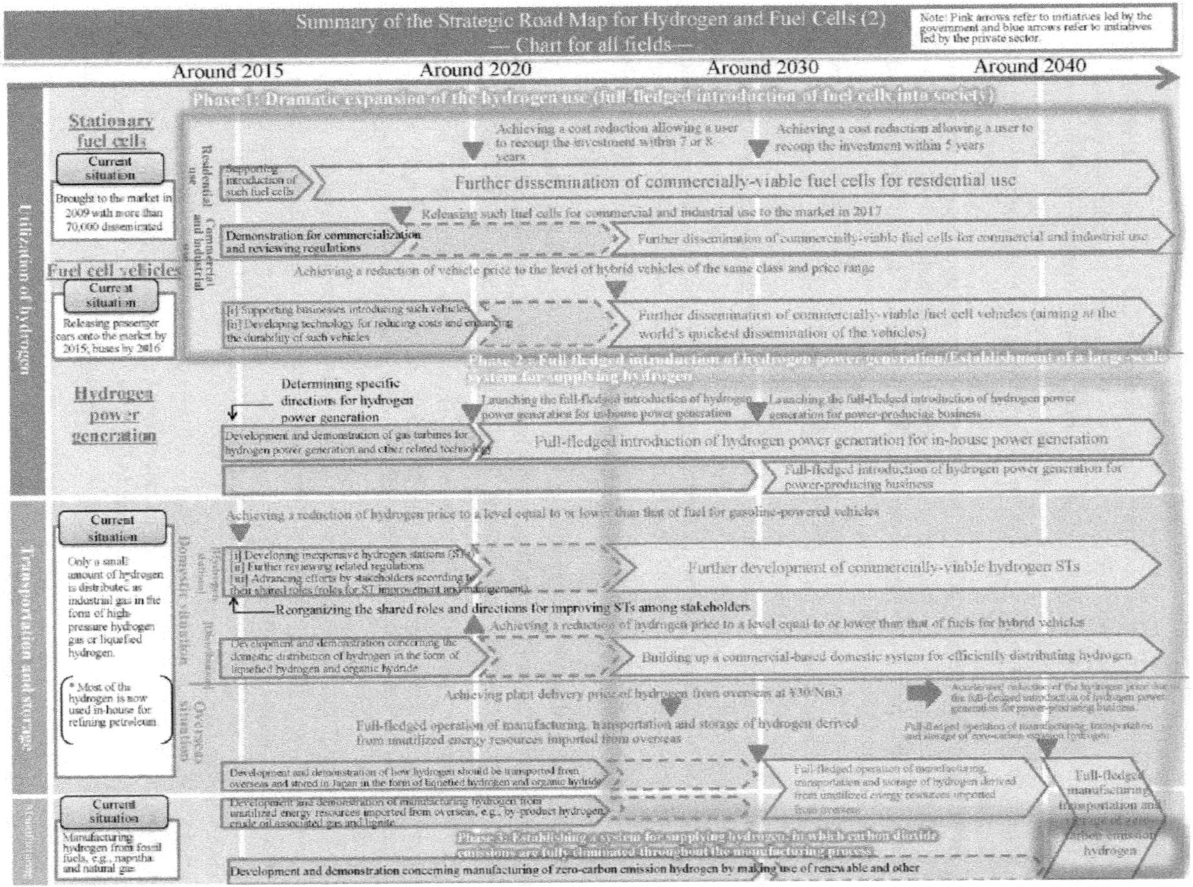

Figure 14.2 Summary of the Strategic Road Map for Hydrogen and Fuel Cells (chart for all fields).

The Road Map indicates the significance of realizing a hydrogen society as follows:

(1) Energy conservation. Making use of FCs to realize high energy conversion efficiencies and contribute to energy conservation.
(2) Energy security. Hydrogen is a resource that could enhance energy security if (i) hydrogen is manufactured from various primary energy sources including those so far not utilized in Japan (e.g., by-product hydrogen, associated gas from crude oil production or coal/lignite) and renewable energy sources; and (ii) in the future, such energy may be procured inexpensively from regions with low geopolitical risk. Utilizing hydrogen manufactured from renewable energy in Japan may also increase the country's energy self-sufficiency rate in the future.
(3) Reducing environmental burdens. During its use, hydrogen does not emit carbon dioxide. Taking advantage of this characteristic, combining the technology for manufacturing hydrogen with a carbon capture and storage (CCS) process, or by making use of hydrogen derived from any renewable energy, will lead to a reduction in the environmental burden or even to the full elimination of carbon dioxide emissions across the supply chain. However, if hydrogen is produced from fossil fuels, such as coal, without CCS, it significantly augments overall CO_2 emissions.
(4) Promoting industries and revitalizing regional economies. Japan has strong global competitiveness in the field of FCs. For example, Japan has filed the world's largest number of patent applications for the technology—five times the number of those filed by the second or lower-placed countries—leaving other countries far behind. In addition, Japan's regional resources, e.g., renewable energy, can be utilized to manufacture hydrogen.

To realize a hydrogen society, METI took into account different time periods required for solving technical challenges and for securing economic efficiency, and decided to advance efforts by categorizing them into three phases, as follows:

Phase 1 (Dramatic expansion of hydrogen use): Dramatically expanding the use of stationary FCs and FC vehicles, which are in the process of being realized, leading to the successful establishment of a global market in the field of hydrogen and FCs, in which Japan leads the world;
Phase 2 (Full-fledged introduction of hydrogen power generation/Establishment of a large-scale system for supplying hydrogen): Further expanding the demand for hydrogen, while widening the scope of hydrogen sources to include unutilized energy, so as to establish a new secondary energy structure in which hydrogen will be added to existing resources, namely electricity and heat (gas); and
Phase 3 (Establishment of a zero-carbon emission hydrogen supply system throughout the manufacturing process): Combining the technology for manufacturing hydrogen with a CCS process, or with making use of hydrogen derived from a renewable energy resource, so as to establish a zero-carbon-emission system for supplying hydrogen throughout the manufacturing process.

The Road Map has been compiled by the Council on the basis of collaborative efforts among significant relative players from industry, academia, and government. The GOJ will use it to promote the measures for realization of a hydrogen society.

14.4 Off-site (centralized) versus on-site (distributed) hydrogen production

There are two major approaches to produce hydrogen: one is H_2 production on-site at the hydrogen filling station and the other is central (off-site) H_2 production with subsequent distribution to the filling station. Currently, the major technologies of hydrogen production are steam reforming of natural gas, a by-product of soda electrolysis, and water electrolysis using (excess) power from renewable energy or conventional power plants. In addition to these sources, thermal energies obtained from solar thermal power located in the Sun Belt and from high-temperature gas nuclear reactors are expected to play a role as well. As shown in Figure 14.3, all of the fossil fuels used in Japan are imported from overseas, and then brought to thermal power plants. Their excess power could be converted into hydrogen by water electrolysis. Similarly, the excess power from hydropower, photovoltaic (PV), and light water nuclear reactors should be converted into hydrogen as well. However, the cost of electrical power transmission is currently much lower than that of the corresponding direct H_2 transportation. Hydrogen from water electrolysis could also be produced, for instance, on-site at the hydrogen filling station. Where hydrogen filling stations are connected to a city, a gas (natural gas) pipeline or Liquefied petroleum gas delivery network for on-site reforming for hydrogen production is also possible. On the other hand, depending on the hydrogen demand in the future, these fossil fuels could be used at a central production site to be directly converted into hydrogen, which would then need to be transported to the hydrogen filling station. Moreover, thermochemical hydrogen produced in solar thermal and high-temperature gas reactors could be brought to the off-site type hydrogen filling station as well.

As shown in Figure 14.4a and b, the hydrogen costs at on-site and off-site filling stations have been evaluated in "The Japan Hydrogen & Fuel Cell Demonstration Project phase 2 (JHFC 2)" (The Japan Hydrogen & Fuel Cell Demonstration Project phase 2 (JHFC 2), n.d.). The current status of the cost of hydrogen is about 105 JPY/Nm^3

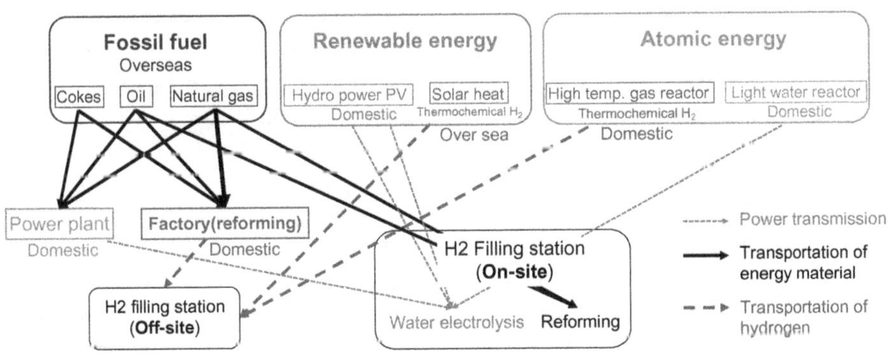

Figure 14.3 Hydrogen sources and their corresponding routes for hydrogen.

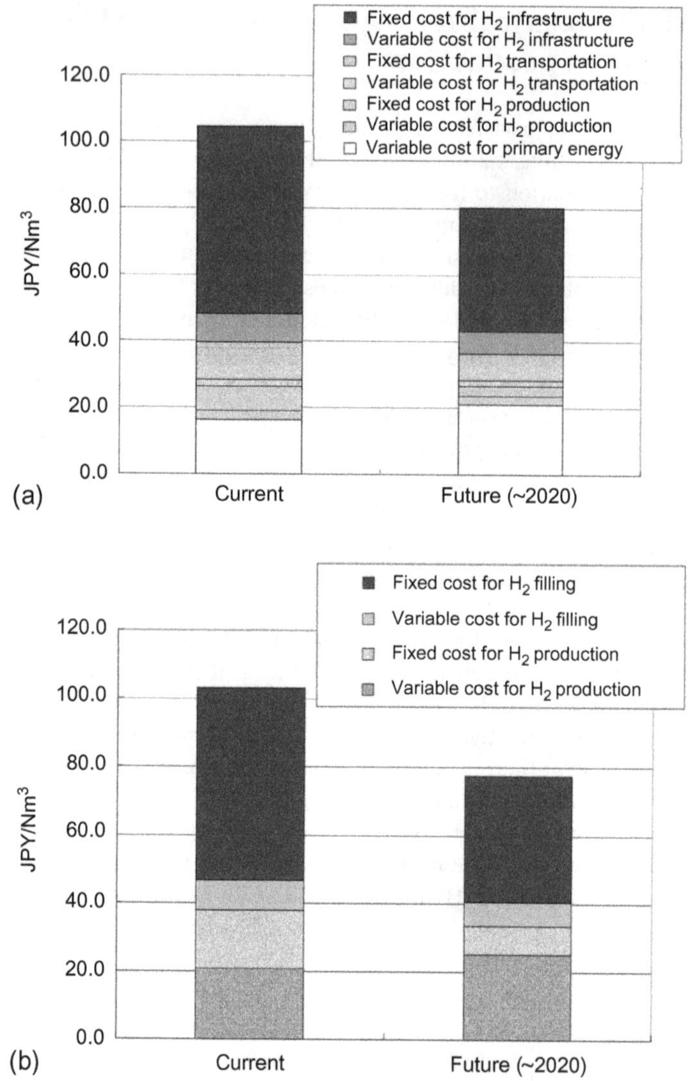

Figure 14.4 Costs of hydrogen at (a) off-site hydrogen filling station and (b) on-site hydrogen filling station (http://www.chiyoda-corp.com/technology/en/spera-hydrogen/).

(≈10.0$/kg @118JPY/$), which should be equivalent to the price of 1000–1100 JPY/kg announced by industry in press releases (http://www.iwatani.co.jp/eng/). The GOJ aims to have 100 hydrogen stations by the end of FY2015. Table 14.1[2] shows the number of hydrogen stations and their H_2 supplying capacity. Actually, in 2013, almost all the stations established were the off-site type (Next-Generation Vehicle Promotion Centre, n.d.), and half of them were mobile stations.

[2] The assumed current conversion rate as of March 2015 is 1 Yen = 0.009 US$.

Table 14.1 The number of hydrogen stations and their H_2 supplying capacity that received a subsidy (The Japan Hydrogen & Fuel Cell Demonstration Project phase 2 (JHFC 2), n.d.)

		Off-site		On-site		Mobile	
	Supplying capacity (Nm^3/h)	Number	Subsidy (million JPY)	Number	Subsidy (million JPY)	Number	Subsidy (million JPY)
FY2013	300~	17	190	1	250	0	–
	100–300	0	130	0	160	0	–
FY2014	300~	8	220	3	280	0	250
	100-300	0	150	1	180	12	180

14.5 Novel hydrogen production methods

Thermochemical water splitting is attractive as a technique for the conversion of renewable energy to hydrogen. So far, various kinds of thermochemical water splitting cycles have been proposed and investigated in Japan, as shown in Table 14.2. Two-step "ferrite and ceria" (Kodama et al., 2015) cycles require more than 1400 °C to generate hydrogen. Niigata University is currently developing two types of solar reactors, the foam device reactor type and the fluidized bed reactor type. In this case, the hydrogen production cycle should take place in a 1-day period because suitable materials for heat storage above 600 °C do not exist. In other words, the production cannot be done continuously by using these two-step processes. From the viewpoint of continuous hydrogen production, the high-temperature gas reactor (HTTR; high-temperature engineering test reactor) should be quite useful. This reactor provides more than 800 °C, so the three-step "IS cycle" can be applied

Table 14.2 Thermochemical water splitting cycles being investigated in Japan

	Name of water splitting process	Temperature (°C)	Reactions
Two-step	Ferrite cycle	~1400	$AFe_2O_4 \rightarrow 3(A_{1/3}Fe_{2/3})O + ½O_2$ $3(A_{1/3}Fe_{2/3})O + H_2O \rightarrow AFe_2O_4 + H_2$
	Ceria cycle	~1500	$CeO_2 \rightarrow CeO_{2-\delta} + \delta/2O_2$ $CeO_{2-\delta} + \delta H_2O \rightarrow CeO_2 + \delta H_2$
Three-step	IS cycle	~900	$SO_2 + I_2 + 2H_2O \rightarrow 2HI + H_2SO_4$ $2HI \rightarrow I_2 + H_2$ $H_2SO_4 \rightarrow SO_2 + ½O_2 + H_2O$
	Na redox cycle	~500	$2NaOH + 2Na \rightarrow 2Na_2O + H_2$ $2Na_2O \rightarrow Na_2O_2 + 2Na$ $Na_2O_2 + H_2O \rightarrow 2NaOH + ½O_2$

(Japan Atomic Energy Agency, n.d.). The demonstration tests are being carried out in JAEA (Japan Atomic Energy Agency), Oarai. If the operating temperature for water splitting is reduced below 600 °C, various kinds of heat storage energy can be utilized, even if solar heat is the focus. To realize the low-temperature splitting, Hiroshima University is focusing on a three-step alkali metal redox cycle, which is able to operate below 500 °C (Miyaoka, 2012). Therefore, this cycle is recognized as a thermochemical hydrogen production technique of the low-temperature type and is being investigated under the SIP (Cross-ministerial Strategic Innovation Promotion Program) in Japan (http://www.jst.go.jp/sip/k04.html).

14.6 Hydrogen distribution and storage

For large-scale transportation of hydrogen, liquid hydrogen has been considered the preferred method because of its higher energy density as compared to compressed gaseous hydrogen. In 2006, Iwatani Corporation started operating "Hydro-Edge," a hydrogen liquefaction plant in Japan, which produces about 200 kg of hydrogen per hour. In the Harima factory of Kawasaki Heavy Industries, another hydrogen liquefaction plant was established in 2014. The liquefaction capacity of this factory is 5 tons per day, which corresponds to 1000 cars assuming the filling of fuel cell electric vehicles (FCEVs). By using liquid hydrogen tank lorries or liquid hydrogen trailers 2–3 tons of hydrogen can be transported per delivery, but about 30% of the energy of the hydrogen is consumed to liquefy it and, moreover, the extremely low temperature of liquid hydrogen means that liquid hydrogen is not suitable for long-term storage, because of boil-off. Therefore, compressed gaseous hydrogen up to 40 MPa is also recognized as an important method of hydrogen transportation, even though only 200–300 kg of hydrogen can be transported at a time.

As an alternative to liquefied or compressed gaseous hydrogen, the Chiyoda SPERA hydrogen storage concept proposed by Chiyoda Corporation, in which methylcyclohexane in the hydrogenated state can desorb hydrogen by using a suitable catalyst (http://www.chiyoda-corp.com/technology/en/spera-hydrogen/), could be commercialized. This is a novel hydrogen supply chain concept developed by Chiyoda, in the form of an LOHC, using a toluene/methylcyclohexane system. In this method, hydrogen is fixed to toluene with the hydrogenation reaction and converted into methylcyclohexane (MCH), which can be offloaded onto and transported with a chemical tanker like toluene. At the demand side, hydrogen is generated from MCH by a dehydrogenation reaction, and toluene is recovered, with the hydrogen supplied to existing infrastructures. LOHC is an alternative for large-scale storage and long-distance transportation of hydrogen at ambient temperature and pressure, and a potentially competing concept to LH_2 shipping, as it does not require large capital investments for hydrogen liquefaction and vessels dedicated to hydrogen transport.

The Chiyoda R&D center demonstrated the desorption of 50 Nm^3 hydrogen within 1 h. This system should be promising for long-distance transportation and long-term storage of hydrogen. In principle, about 60 kJ should be necessary to generate 1 mol hydrogen from this methylcyclohexane, which corresponds to 25% energy of generated hydrogen.

In addition, ammonia is another possible transport carrier for hydrogen. The energy carrier project by "Cross-ministerial Strategic Innovation Promotion Program (SIP)" in Japan was launched in 2014, and focuses in particular on the development of synthesis technology for ammonia and organic hydrides with high energy efficiency. About 170 million tons of ammonia is being industrially produced all over the world by the Haber–Bosch process from hydrogen and nitrogen. Because only about 12% energy of generated hydrogen could be consumed for the cracking of the ammonia molecule, ammonia is recognized as a promising hydrogen carrier. Of course, the removal of residual ammonia would be quite an important issue because even 1 ppm ammonia seriously damages a proton exchange membrane FC.

14.7 Initial current cost of hydrogen stations

The current plan is to build about 100 hydrogen stations by the end of 2015 in four major population areas, centering on Tokyo, Aichi, Osaka, and Fukuoka. The construction cost of a hydrogen station is about 4–500,000,000 JPY (M$3.4–4.2) in the case of a fixed station of supply capacity 300 Nm^3/h. This can be compared to a typical gasoline station, which costs less than 100,000,000 JPY (M$0.85). This shows that hydrogen stations are still very expensive. Table 14.3[2] compares the cost of equipment for Japanese and European hydrogen stations (NEDO hydrogen report, n.d.). Regulations and the strict configuration of Japanese stations are the main reasons for their high cost, as an extremely high safety level is required in Japan.

A hydrogen station is an important piece of social infrastructure to support the hydrogen society. Iwatani International Corporation, one of the Japanese industrial gas giants, started to operate the first publically available hydrogen station in Hyogo Prefecture in July 2014. The company announced on August 28 that it planned to build a station near Tokyo Tower, which will be the first full-fledged station available in the Tokyo metropolitan area. JX Nippon Oil & Energy, the largest oil-dealing company in Japan, plans to build 40 stations across the country by the end of FY2015 in October.

Table 14.3 Comparison of the costs of equipment in Japanese and European hydrogen stations (NEDO hydrogen report, n.d.)

	Japan	European
Compressor	1.3	0.8
Accumulator	0.6	0.1
Pre-cooler	0.4	0.2
Tank	0.5	0.2
Total	2.8	1.3

Unit: One hundred million yen.

14.8 Residential use FC system (The Japan Gas Association, n.d.)

FC technology is at the stage of actual application and commercialization in Japan. Recent advances can be seen in the residential and mobility use of FC systems. The number of Ene-Farm sets, a residential co-generation system that produces electricity through a chemical reaction between oxygen in the atmosphere and hydrogen extracted from city gas, reached over 100,000 by 2014, as shown in Figure 14.5. This Ene-Farm system is operated with city gas (natural gas), because this gas is easily reformed into hydrogen inside each FC system. These are not pure hydrogen FC systems.

A typical Ene-Farm electric-generating capacity is 0.7 kW per system. The cost is about 1.5 million yen per system ($15,000). The efficiency of an Ene-Farm is more than 80%; the electricity and heat efficiency are about 40% and 40%, respectively. Compared to the conventional method of using electricity from a thermal power plant and hot water supply and heating using city gas, the FC system allows primary energy consumption to be reduced by approximately 35% and CO_2 emissions by approximately 48%. Ene-Farm users can save around 50,000–60,000 yen ($424–508) from annual utility bills, and reduce CO_2 emissions by approximately 1.5 tons a year.

The combination of PV and Ene-Farms is called *double generation*. In this system, electricity generation from PV power is used during the daytime, and electricity from the Ene-Farm system is used at night.

Figure 14.5 Household Fuel Cell system Ene-Farm (http://www.fca-enefarm.org/about.html).

14.9 FC vehicle

Construction of hydrogen stations for hydrogen FCEVs is in full swing at last. The trend was triggered by the announcement of Toyota Motors, which in 2014 released the FCEV "Mirai" as shown in Figure 14.6. However, there are obstacles to its acceptance and it takes time for development of the infrastructure. Following Toyota Motors, Honda will release FCEVs in 2016, and Nissan Motors will start to sell them by 2017. Initially, Toyota stated that they would sell their FCEV at a price of about 7 million yen (about $58,600US). The price is lower than that of a domestic luxury car and it is expected that FCEVs will diffuse more rapidly if the government were to allocate a subsidy for them. Currently, FCEVs are scheduled to be deployed at a rate of several hundred units per year, initially.

The government will boost its budget to increase subsidies not only for FCEVs but also for hydrogen stations. The METI requested in the budget for 2015 to increase the subsidies for the construction of the stations by a factor of 1.5 compared with those in 2014. If the safety regulations were eased, the stations could be constructed at much lower cost. Construction costs of hydrogen stations are expected to be reduced through large-scale deployment and standardization. In addition, a centralized control center for the hydrogen station is envisioned, thereby reducing the operating costs.

However, there are many obstacles to the realization of the hydrogen society in Japan. The GOJ will aim to construct 100 stations by the end of 2015, and 1000 stations by 2025 (compared to more than 30,000 existing gasoline stations). The executive of a major automaker said that the number of stations will still not be sufficient even if all the stations currently planned are constructed (NEDO, 2010).

At this point there are subsidies for construction of the stations, but station equipment and station maintenance are still expensive. It is estimated that after 2020 the business case may be positive, depending on the uptake of FCEVs. One major advantage of FCEVs compared to electric vehicles is that it takes a short time to fuel a FCEV (a few minutes) compared to a few hours for charging a battery.

Figure 14.6 Fuel Cell Vehicle "MIRAI" (http://toyota.jp/sp/FCEV/).

14.10 Current situation in Japan as regards hydrogen infrastructure

It is necessary to reduce the costs of future development through technical development. For example, it is necessary to use less expensive designs and materials. A reduction in the hydrogen station costs will also lead directly to a reduction in the hydrogen supply cost. In NEDO's 2010 hydrogen technology roadmap, although a target hydrogen supply cost of about 60 JPY (51¢)/Nm^3 in 2020 was presented, with the on-site station cost targets that can be realized, the hydrogen supply costs about 200 million yen (500 Nm^3/h). Further, in the case that it was possible to achieve a 500 Nm^3/h station at about 200 million yen at the same time, a 300 Nm^3/h station is estimated also be realized in about 150 million yen ($1,270,000).

14.11 Conclusions

Currently, there has been significant progress in Japan toward the aim of realization of the hydrogen energy society, with support from the GOJ. In the future, when the commercialization stage is reached, there is a further plan to expand business involvement. Looking toward the Tokyo Olympic and Paralympic Games scheduled to be held in 2020, a great deal of business is planned with the aim of furthering the hydrogen energy society. Japan's current hydrogen energy technology leads the world.

Japanese performance to date is as follows:

1. In 2009, Japan was the first in the world to commercialize a stationary FC system, Ene-Farm. More than 10 million units have been sold up to the present.
2. In December 2014, the FC vehicle Mirai was the first "precommercial" model sold in the world. In the future, it is assumed that the number of FC vehicles will increase.
3. Hydrogen stations have been commercialized. However, there are still only a few. It is planned that the number of hydrogen stations will increase toward 2020.

Japan will continue to maintain its world-leading hydrogen energy technology in the future.

References

http://toyota.jp/sp/FCEV/.
http://www.chiyoda-corp.com/technology/en/spera-hydrogen/.
http://www.fca-enefarm.org/about.html.
For example, http://www.iwatani.co.jp/eng/ and http://www.noe.jx-group.co.jp/english/.
http://www.jst.go.jp/sip/k04.html.
Japan Atomic Energy Agency, http://httr.jaea.go.jp/.
Kodama, T., Gokon, N., Matsubara, K., 2015. Solar thermochemical water splitting cycles utilizing concentrating solar radiation. J. Jpn. Inst. Energy 94, 35–41.
Miyaoka, H., Ichikawa, T., Nakamura, N., Kojima, Y., 2012. Low-temperature water-splitting by sodium redox reaction. Int. J. Hydrogen Energy 37, 17709–17714.

NEDO, 2010. Hydrogen and Fuel Cell Strategy Roadmap. http://www.nedo.go.jp/activities/FF_00518.html.
NEDO hydrogen report. http://www.nedo.go.jp/content/100567362.pdf.
Next-Generation Vehicle Promotion Centre. http://www.cev-pc.or.jp/index.html.
Strategic Energy Plan, April 2014. http://www.enecho.meti.go.jp/en/category/others/basic_plan/pdf/4th_strategic_energy_plan.pdf.
Strategic Road Map for Hydrogen and Fuel Cells, June 2014. http://www.meti.go.jp/english/press/2014/pdf/0624_04a.pdf.
The Japan Gas Association. http://www.gas.or.jp/newsrelease/notice20140929.pdf.
Based on the report of The Japan Hydrogen & Fuel Cell Demonstration Project phase 2 (JHFC 2). http://www.jari.or.jp/portals/0/jhfc/e/index.html.

Environmental impacts of hydrogen use in vehicles

15

C. Wulf, M. Kaltschmitt
Hamburg University of Technology (TUHH), Hamburg, Germany

15.1 Introduction

There is a global understanding within society and politics that anthropogenic climate change can only be fought effectively if the increase of global mean temperature does not exceed 2 °C (United Nations, 2009). Therefore, anthropogenic greenhouse gas (GHG) emissions released within the energy system have to be reduced severely. Thus, next to industry, the transportation sector has to cut back GHG emissions, especially on the background of high growth rates expected for future years. For this reason, alternatives to transportation fuels from fossil sources are under discussion in various countries, such as hydrogen, electricity and biofuels. Currently many demonstration projects for the use of hydrogen within the mobility sector are ongoing. Within such mobility schemes no local emissions – except of water – occur if a fuel cell electric vehicle (FCEV) is used. Nevertheless, depending on the primary energy carrier hydrogen is produced from, emissions might occur in the pre-chains. The same is true for battery electric vehicles (BEVs). Also, when driving on biofuels most of the emissions occur in the pre-chains. For these reasons, advantages and disadvantages due to GHG emissions and other emissions of different mobility options are not obvious at first glance.

Against this background, this chapter concentrates on a life cycle assessment (LCA) of mobility options regarding potential GHG emissions as well as acidifying emissions. Five different hydrogen and four other (alternative) mobility options are analysed:

1. Hydrogen from alkaline water electrolysis with electricity from an EU-mix for FCEV.
2. Hydrogen from steam methane reforming (SMR) of natural gas for FCEV.
3. Hydrogen from SMR of biomethane for FCEV.
4. Hydrogen from gasification of coal for FCEV.
5. Hydrogen from gasification of biomass from forest residues for FCEV.
6. Electricity from an EU mix for BEV.
7. Compressed natural gas for natural gas vehicle.
8. Diesel from crude oil for diesel vehicle,
9. Biodiesel from rape seeds for diesel vehicle.

Before the actual analysis is done, some methods for environmental assessment are introduced and compared to international limit values for GHG emissions.

Compendium of Hydrogen Energy. http://dx.doi.org/10.1016/B978-1-78242-364-5.00015-4

15.2 Environmental assessment

For the environmental assessment of a product or service, several methods are available that differ from others in the definition of the boundaries of the analysed systems, the level of detail, and the coverage of environmental effects (e.g., analysed types of emissions). However, most methods are based on an LCA with specific boundaries or impact categories. At the moment mainly GHG emissions are discussed. Here three different methods are introduced that all cover this field. As the LCA is the method applied for the following analysis, it is explained below in more detail, including how it is implemented for the goal outlined here.

15.2.1 Life cycle assessment

By means of an LCA, two or more alternatives of a product or service (such as mobility) can be assessed concerning their potential impact on the environment, on human health and/or on resource depletion. The LCA methodology is based on the fact that the environmental impact is not limited to the production process itself (e.g. electrolysis) because environmental effects may also occur within the pre-chains. This might be true for the provision and transportation of material needed for the production of the analysed product or service (e.g. steel for the fuel production plants). According to the given standards (i.e. ISO, 14040, 2006; ISO, 14044, 2006) an LCA is carried out in four interacting steps (see Figure 15.1), which are explained in more detail in the following sections.

15.2.1.1 Goal and scope definition

General

A clear definition of the goal of the LCA is the basic requirement for the execution of a transparent assessment. This goal pre-defines the choice of the system parameters and the complexity with respect to the level of detail of the assessment

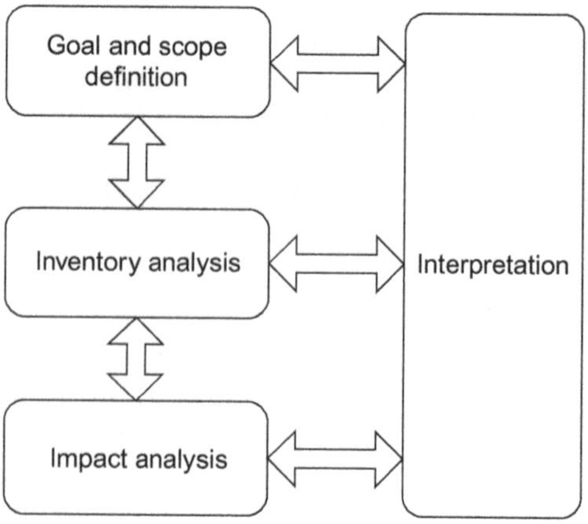

Figure 15.1 Stages of an LCA according to ISO 14040 (2006).

as well as the system boundaries. Without an explicit and transparent goal definition, practically no comparable and reliable results are achievable. Important aspects in this respect are, for example, the selection of the alternatives to be assessed leading to an identical product/service, the definition of the functional unit, the determination of the depth of the analysis, the content of the inventory and impact analysis as well as the spatial and temporal conditions. With the goal and scope definition, the content of the three following LCA steps are already basically defaults.

Specific

The goal of this LCA is to assess selected potential environmental effects within different life cycle stages as well as throughout the overall life cycle of different transportation alternatives in private mobility conducted by a passenger vehicle. The environmental effects calculated throughout the overall life cycle are related to one vehicle kilometre with a defined passenger car. The time reference is the year 2011 and the local reference is Europe. The potential environmental effects to be analysed are the anthropogenic share of global warming (GHG emissions) as well as the acidification of soil and aquatic ecosystems (acidifying emissions). The path of life includes the provision of the feedstock for the different fuel types (i.e., electricity, methane in the forms of natural gas and biomethane, biomass, crude oil and coal as well as all the transportation of the feedstock to the production plant). For systems providing sulphur and/or heat as by-products, a credit according to the system expansion is given. Used auxiliaries are treated like feedstock (e.g., electricity).

15.2.1.2 Inventory analysis

General

In this part of the analysis, which is called *life cycle inventory analysis* (LCIA), the input and output parameters (i.e. energy inputs, raw material inputs, other inputs; products, co-products, waste; emission to air and other environmental aspects) are collected and calculated considering all sectors of the product life. For that purpose, the real situation has to be translated into a model so that the assessed parameters can be quantified in the most realistic way, taking the defined frame conditions into consideration. To reduce the complexity of this assessment and to allow the inventory analysis to be handled within the given time constraints, the process chain is traced back only to the point that the neglecting of additional chain components would lead to a mistake within the overall result of a maximum of a defined percentage.

Carbon dioxide from the oxidation of sustainable provided organic matter does not count in the calculation of the overall GHG emissions. During the conversion of biomass it is only released back into the atmosphere because it has been removed from the ambient air during the growth of the plants. Thus, assuming a sustainable biomass production, the carbon cycle is closed. However, all direct emissions occurring during the oxidation of fossil fuels are included in the calculation.

Specific

Based on a literature search as well as on primary data obtained from industry, the LCIA is conducted with the help of commercially available software (Umberto NXT LCA) using existing databases (ecoinvent v3 by SWISS CENTRE FOR LIFE CYCLE INVENTORIES, 2013).

15.2.1.3 Impact analysis

General

Within this LCA step the results acquired during the inventory analysis (see Section 15.2.1.2) are associated according to certain environmental impact categories and category indicators predefined within the goal and scope definition (see Section 15.2.1.1). Additionally to this selection of appropriate impact categories, category indicators and characterisation models, this step must include the classification of the life cycle inventory (LCI) results to one or more different impact categories. For example, nitrous oxides are potentially acidifying. However, they also contribute to the potential eutrophication (accumulation of nutrients in an ecosystem). Afterwards, these results are characterised by converting them to common units and aggregating them to indicator results (e.g. all GHG emissions are converted to carbon dioxide equivalents (CO_2-eq)).

Specific

Following the explanations in Section 15.2.1.1 this is realised here for the GHG emissions with the characterisation model of the IPCC (2007) as well as the acidifying emissions according to CML 2001 (Guinée et al., 2001). The characterisation factors for these two impact categories are listed in Table 15.1. Because so many different pollutants cause GHG emissions, in Table 15.1 only the most important ones are presented.

Table 15.1 Characterisation factors for used impact categories per kilogram pollutant (IPCC, 2007; Guinée et al., 2001)

Climate change	CO_2-eq	Acidification	SO_2-eq
Carbon dioxide	1	Ammonia	1.88
Methane	25	Nitrogen oxides	0.70
Dinitrogen oxides	298	Sulphur dioxide	1.00
Sulphur hexafluoride …	22,800	Sulphur trioxide	0.80
		Sulphuric acid	0.65
		Phosphoric acid	0.98
		Hydrogen chloride	0.88
		Hydrogen fluoride	1.60
		Hydrogen sulphide	1.88
		Nitric acid	0.51

15.2.1.4 Interpretation

General

The interpretation of the results from the inventory and impact analysis is realised qualitatively by discussing the various impact categories separately. Finally, some suggestions for improvement can be given, if possible. With a background of very different impact categories (e.g. climate change or noise), this might be challenging.

Specific

Due to the assessment problem discussed previously, a discussion of the various results as well as a sensitivity analysis referring to the electricity mix used are carried out here.

15.2.2 Well-to-wheel analysis

The well-to-wheel analysis is a nonstandardised method to quantify the impact of transportation fuels and vehicles regarding energy and climate change. As more and more alternative drive trains and fuels are used whose impact on the environment does not occur during driving of the vehicle, fuel production emissions and energy consumption for fuel production also have to be assessed. Usually carbon dioxide or GHG emissions (according to IPCC, 2007), as well as other emissions, energy demand and efficiency, are investigated within such an analysis.

A well-to-wheel analysis can be subdivided into two parts: the well-to-tank (energy provision) and the tank-to-wheel (vehicle efficiency) analysis. Compared to a life cycle analysis the production, maintenance and disposal of the vehicle are not assessed. In addition, fewer environmental impact categories are taken into account (Edwards et al., 2014; Brinkmann et al., 2005).

15.2.3 Carbon footprint

The calculation of a carbon footprint is also not a fully standardised process. The carbon footprint is an environmental key parameter to describe the climate friendliness of products, services, institutions, countries and persons. Sometimes only carbon dioxide emissions are taken into account (Wiedmann and Minx, 2008). Others include only methane and sometimes nitrous oxide, while in many cases all GHGs are included according to the IPCC method (IPCC, 2007).

Due to this variety, the unit of a carbon footprint is also different within various studies. The value can be indicated as kg CO_2 or kg CO_2-equivalents (CO_2-eq). It is also possible to express this value as an area; this is where the name "footprint" arises. In this case the area describes the amount of land that is needed to neutralise the emitted carbon dioxide (GFN, 2012).

Over the last few years different countries have attempted to establish standards for the calculation of the carbon footprint. In 2008 the UK published the first standard on carbon footprint (PAS, 2050, 2011). However, it was not accepted by other countries (BMU and BDI, 2010). Therefore, an international standard ISO/TS 14067 (2013) is in development. It focuses on the carbon footprint of products, including services. Explicitly, a carbon footprint is defined by all GHG emissions and the execution of such an assessment is closely related to the standards of an LCA, e.g., definition of four stages; see Figure 15.1.

15.2.4 Limit values for GHG emissions

All over the world governments are attempting to reduce GHG emissions within the mobility sector. Thus, the limit values for these emissions for passenger vehicles in the EU, USA and Japan are explained in the following paragraphs, with regard to battery and FCEVs.

In 2014, the **EU** decided on a new regulation concerning targets for 2020 to reduce carbon dioxide emissions in new passenger cars. The new car fleet of each passenger car manufacturer that produces more than 1000 cars per year should not exceed 95 gCO_2/km in the year 2020. In the first year only 95% of the fleet have to reach this value; in the following years this threshold value has to be met by the overall fleet. There are several exceptions and facilitations of this rule. For example, passenger cars with specific emissions of CO_2 of less than 50 gCO_2/km are counted as two cars in the calculations for the year 2020 (super-credit). Not until 2023 will the super-credits be omitted. As only tailpipe emissions are counted (tank-to-wheel) battery and FCEVs have no CO_2 emissions and receive these super-credits. However, in this regulation it is mentioned that in the future GHG emissions from car manufacturing and generation of electricity or other alternative fuels to supply vehicles should also be taken into account. If a manufacturer uses innovative technologies, e.g., fuel cells, he can apply for an additional reduction of up to 7 gCO_2/km for his fleet, in addition to the super-credits (EU, 2014).

In the **USA** standards from the Environmental Protection Agency are defined from the years 2017 to 2025 which is a continuation from the 2012 to 2016 period. For each car model a certain car standards curve (changing over the years) has to be applied to get the carbon dioxide emission per mile (CO_2/mile). The cars are categorised by their footprint.[1] This results in higher CO_2 limit values for bigger cars. These standards lead to a projected fleet-wide emission target for 2020 of 113 gCO_2/km (182 gCO_2/mile). Not only carbon dioxide emissions are included in this value but also methane (0.01 g/mile) and nitrous oxide (0.03 g/mile), with the agreed characterisation factors of carbon dioxide (IPCC, 2007). For battery-electric, plug-in hybrid electric (PHEV), fuel cell electric and compressed natural gas vehicles sold in the model years 2017 through 2021, incentive multipliers are applied. This means that each "innovative" vehicle counts more than one "classical" vehicle in the manufacturer's calculation. Battery-electric and FCEVs start with a value of 2 in 2017 and reach a value of 1.5 in 2021, while CNG and PHEV start with 1.6 and reach 1.3. In the last years of this program no incentives are given. As only tailpipe emissions are counted, again EVs and FCEVs have a large advantage (US, 2012).

[1] Track width (inch) times wheelbase (inch) divided by 144, result in square feet (US, 2014).

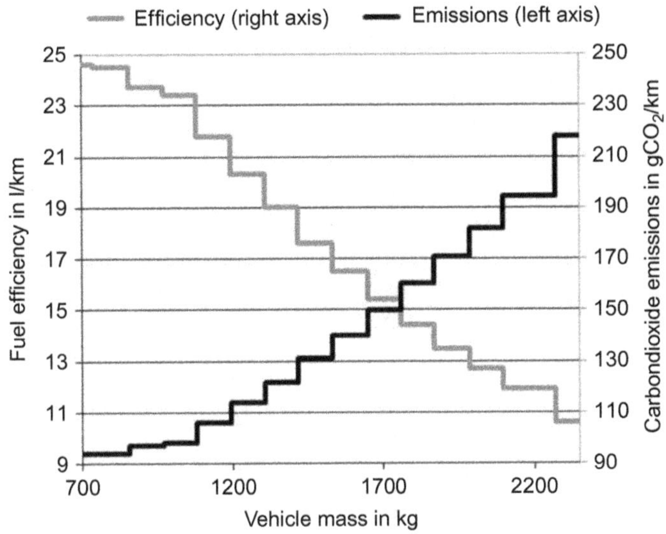

Figure 15.2 Fuel efficiency targets for passenger cars in 2020 from the Japanese Top Runner program.

Japan has used a so-called Top Runner Program since 1999 to increase energy efficiency and reduce GHG emissions. For different technology categories, the most energy efficient available product is set as a standard for the other products for the next five years. Sometimes also an extrapolation for the five years after that is done to accelerate the gain of efficiency. This is applicable to computers, air conditioning, etc., as well as heavy- and light-duty vehicles including passenger cars. On the one hand, the vehicles are grouped in weight categories, and on the other hand according to their drive trains. At the moment the categories *gasoline*, *diesel* and *liquefied petroleum gas (LPG)* are applicable. For hybrid, battery electric and FCEVs, no limits are defined yet. Each category has its own efficiency value expressed in kilometres that can be driven with one litre of fuel (km/l). The defined values for a gasoline vehicle for the target year 2020 are illustrated in Figure 15.2. To compare the efficiency to GHG emissions, they are also part of the diagram with a conversion factor of 2310 gCO_2/l gasoline.

As a weighted-harmonic average of the gasoline fleet in Japan, a value of 20.3 km/l should be reached by 2020 that corresponds to 113.8 gCO_2/km. Due to the different heating values of diesel and LPG the efficiency value of diesel has to be multiplied by 1.1 and for LPG with 0.75 (Dubbers et al., 2012; Daisho et al., 2011).

15.3 Reference systems

In this section a detailed description of the analysed pathways is presented. Figure 15.3 summarises them graphically to give a better understanding of this chapter. It is divided into four sections characterising mobility with a passenger vehicle. These are fuel production, fuel distribution, vehicle production and recycling.

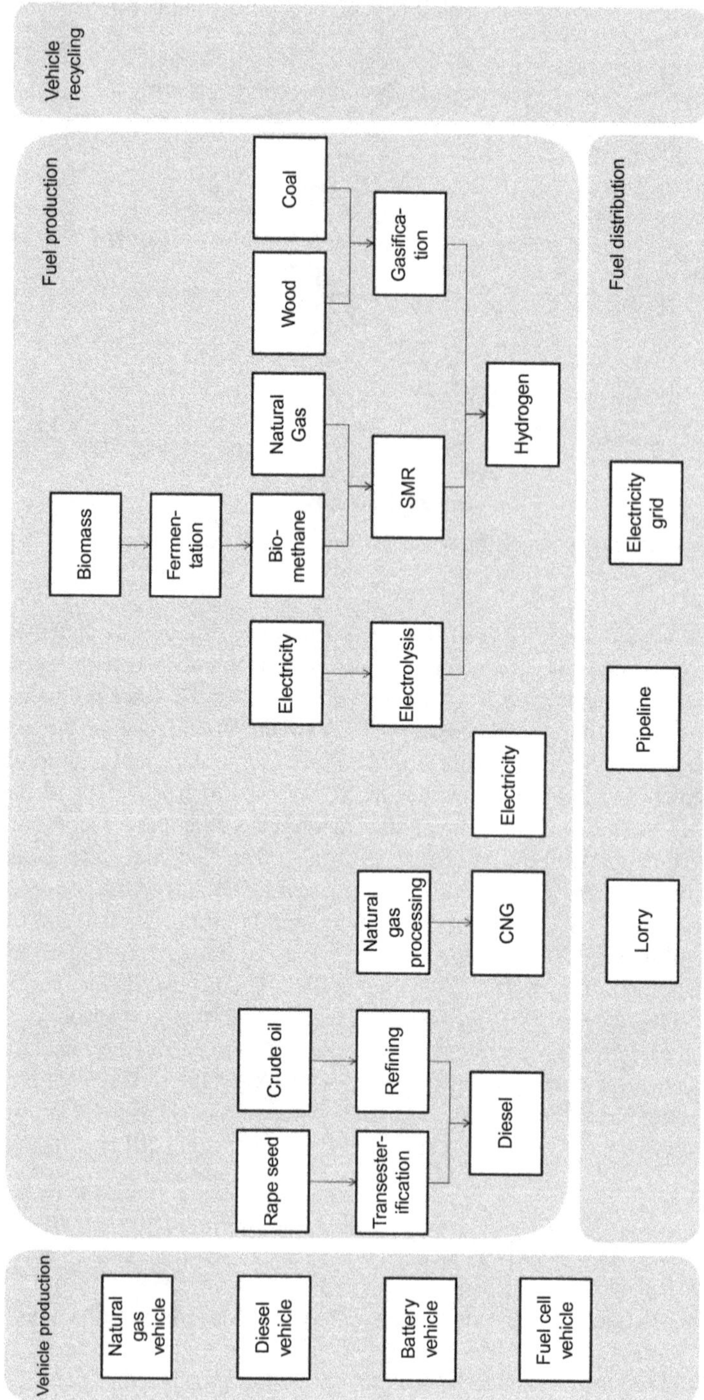

Figure 15.3 Analysed hydrogen production pathways showing all processes that were modelled here.

Against this background, first the mobility options in general are introduced in the paragraphs below. Second, as this book deals mainly with hydrogen, special attention is given to the hydrogen production pathways.

15.3.1 Mobility

For all analysed mobility options a standard vehicle (compact class) is assumed. According to the fuel used, the drive trains are selected, i.e., fuel cell, electric motor and battery for an FCEV; electric engine and battery for a BEV; diesel engine as well as a petrol engine for the natural gas vehicle. The analysis of the vehicles is based on Weinberg et al. (2013) with updated data from DAT (2014).

Electricity is the fuel for operating the BEV. Furthermore, it is basically needed in each of the hydrogen provision chains shown in Figure 15.3. Thus, the electricity mix used is a very important parameter for this LCA and is presented here in detail for a better understanding of the results. As this analysis is conducted for Europe, the gross electricity production mix for the EU-27 for the year 2011 (European Commission, 2013) is utilised. This electricity mix, composed according to Figure 15.4, causes 460 gCO_2-eq/kWh on average.

Additionally, data for the production and distribution of the other nonhydrogen fuels, i.e., biodiesel (rape seed methyl ester), diesel and CNG, is retrieved from a commercially available database (SWISS CENTRE FOR LIFE CYCLE INVENTORIES, 2013). From this data source the tailpipe emissions – carbon dioxide from diesel and CNG; methane from biodiesel, diesel and CNG; nitrous oxide and ammonia from biodiesel and diesel; nitrogen oxide from biodiesel, diesel and CNG; and sulphur dioxide from diesel – are also identified and verified partly based on the fuel composition.

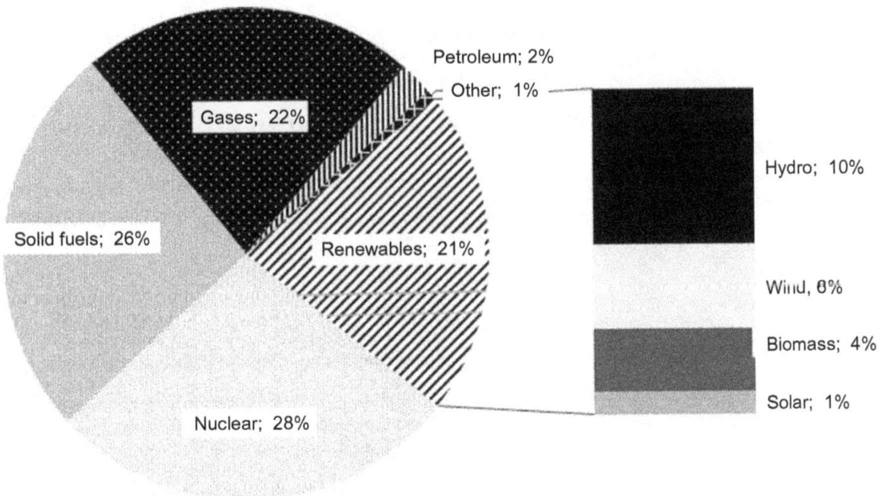

Figure 15.4 Composition of the European (EU-27) electricity mix in 2011 (European Commission, 2013).

Table 15.2 **Fuel demand of a compact car with different drive trains (DAT, 2014)**

Diesel vehicle	3.8 l/100 km
Biodiesel vehicle	4.1 l/100 km
CNG vehicle	3.5 kg/100 km
FCEV	0.9 kg/100 km
Battery-electric vehicle	18.6 kWh/100 km

For the fuel consumption of the different drive trains, values according to the New European Driving Cycle are used. That may not reflect the actual consumption but this driving cycle is very common as a standard value. For the different mobility options the fuel demand is summarised in Table 15.2.

15.3.2 Hydrogen

15.3.2.1 Alkaline water electrolysis

Hydrogen can be produced onsite at a hydrogen refueling station (HRS) based on electrolysis of water. As this process is basically emission free the environmental performance is more or less defined by the emissions due to the electricity needed to operate the electrolyzer as well as the necessary supporting processes for providing the hydrogen. These data are taken from the HRS Hamburg HafenCity[2] and listed with other necessary information in Table 15.3. Additional data regarding the materials of an electrolyzer are collected from literature (Pehnt, 2002; Hydrogenics, 2011). Also, the necessary auxiliary equipment of an electrolyzer has to be taken into consideration. This includes the preparation of the water for electrolysis by reverse osmosis and addition of potassium hydroxide (KOH), as well as the drying and cleaning of the hydrogen afterwards.

Table 15.3 **Data for the electrolysis based on own evaluations and Pehnt (2002)**

Technical data	
Lifetime	20 a
Lifetime stack	6.6 a
Full load hours	7500 h
Production capacity	120 Nm^3/h
Operating resources	
Electricity for electrolysis	4.7 kWh/Nm^3
Electricity for auxiliaries	0.1 kWh/Nm^3
Water	1.7 kg/Nm^3
Potassium hydroxide	76.5 mg/Nm^3

[2] http://corporate.vattenfall.com/news-and-media/press-releases/press-releases-import/hydrogen-station-in-hamburg-vattenfalls-latest-endeavour-in-sustainable-mobility/.

15.3.2.2 Steam methane reforming

SMR from natural gas or biomethane is another way to produce hydrogen. Within such a plant, heat is produced as a side product. Since this thermal energy can be used in other applications to increase the overall efficiency of the process, a credit is given within this assessment using heat production from natural gas in an industrial furnace.

For natural gas as a feedstock the amount of this heat is stated in Table 15.4 along with the direct emissions occurring during the reforming process. The data are gathered from different LCA studies discussing SMR (Pehnt, 2002; Boyano et al., 2011; Spath and Mann, 2001)

Basically the same process can be realised based on biomethane. With biomethane, the organic substrate (e.g. animal manure or crops) is fermented first within an anaerobic digestion plant. Here microorganisms convert biomass into biogas. Then hydrogen sulphide is removed from this biogas by adding iron(II)chloride that binds the sulphur and precipitates as iron sulphide. As this biogas also contains a high amount of carbon dioxide, it has to be removed to achieve natural gas quality. Here it is assumed that, with a pressurised water scrubbing process, most of the carbon dioxide can be separated from the biogas till basically pure methane is provided. In order to fulfil the natural gas standards, the gas has to be dried afterwards with the help of triethylene glycol that adsorbs water (Roddy, 2012). Furthermore, a small amount of propane has to be added to the biomethane to reach a similar Wobbe Index as natural gas (Scheftelowitz et al., 2013) to be allowed to feed this substitution gas into the existing natural gas grid. If the biomethane is available in the grid, it can be removed at any location and converted with the steam methane reformation technology mentioned previously.

The LCI of the analysed biomethane production is adapted from published process parameters (Kaltschmitt et al., 2012; Belau, 2012). The assumptions regarding the substrate compositions used for biogas production, as well as other important parameters

Table 15.4 Data for steam methane reforming (Pehnt, 2002; Boyano et al., 2011; Spath and Mann, 2001)

Technical data	
Duration of life	20 a
Full load hours	7500 h/a
Production capacity	8000 Nm^3/h
Operating resources	
Electrical energy	0.03 kWh/Nm^3
Natural gas	4.2 kWh/Nm^3
Water	1.3 kg/Nm^3
Steam, export credit	0.37 kWh/Nm^3
Direct emissions	
CO_2	0.81 kg/Nm^3
CH_4	1.7×10^{-4} kg/Nm^3
NOx	2.2×10^{-4} kg/Nm^3
SO_2	1.1×10^{-5} kg/Nm^3

Table 15.5 **Data for SMR of biomethane (Kaltschmitt et al., 2012; Belau, 2012; Scheftelowitz et al., 2013)**

Resource	**Substrate mix**[a]
Biomethane production	
Capacity	7 MW
Duration of life	20 a
Full load hours	7500 h
Biomass demand	90,000 t_{FM}/a
Electricity demand	4200 MWh/a
Demand of operating materials	
Water	220 kg/a
$FeCl_3$	13,000 kg/a
Triethylene glycol	2500 kg/a
Propane	170,000 kg/a
Direct emissions	
CH_4	8200 kg/a
SO_2	6500 kg/a
Direct emissions (SMR)	
NO_x	2.2×10^{-4} kg/Nm^3
SO_2	1.1×10^{-5} kg/Nm^3
CH_4	1.7×10^{-4} kg/Nm^3
CO_2	0.042 kg/Nm^3

[a] *German mix*: maize silage 38.5 m%, grass silage 8.0 m%, grain silage 5.1 m%, grain 0.7 m%, liquid manure 43.1 m%, bio waste 4.5 m%.

for the hydrogen production process, are listed in Table 15.5. The substrate mix represents the current situation in Germany (Scheftelowitz et al., 2013).

Gasification

Solid fuels containing carbon (like hard coal, lignite, and solid biofuels) can be gasified with water (among others) as a gasification agent to produce hydrogen. Here, two fuels are assessed: hard coal and woody biomass.

Within this analysis it is assumed that a hard coal mix is gasified with water to produce a gas containing, among others, hydrogen. Due to the ingredients in hard coal, a pressure swing adsorption to remove CO_2 alone is not sufficient to clean the hydrogen because also hydrogen sulphide (H_2S) produced from the sulphur content of the coal has to be removed. Thus, it is assumed that H_2S is separated from the gas and the sulphur is regained through the Shell Claus Off-gas Treating (SCOT) process (Doctor et al., 2001). Hence, a credit is given not only for the produced steam provided as a by-product but also for the regained sulphur to be sold to the chemical. Here it is assumed that heat production from a hard coal industrial furnace is substituted. The demands of coal and the other parameters are listed in Table 15.6.

In addition to coal, solid biofuels can also be gasified for hydrogen production. Within such a process usually woody biomass is used as a feedstock, which is assumed to be

Table 15.6 **Data for coal gasification (Doctor et al., 2001)**

Technical data	
Duration of life	20 a
Production capacity	15,000 kg/h
Full load hours	7500 h/a
Operating resources	
Electrical energy	0.47 kWh/Nm^3
Coal	0.75 kg/Nm^3
Water, deionised	2.4 kg/Nm^3
Steam, export credit	1.1 kWh/Nm^3
Sulphur, export credit	0.018 kg/Nm^3
Direct emissions	
CO_2	1.7 kg/Nm^3
SO_2	7.2×10^{-7} kg/Nm^3
H_2S	4.0×10^{-7} kg/Nm^3

forest residues, from spruce in this case. From forests, mainly log wood for high-value products (e.g., furniture) is harvested. Less valuable forest wood is used by other industry branches for the provision of mass products (e.g., for pulp and paper production). The remaining residues can be used for energy purposes. Allocation between these three product groups is realised by volume (allocation factors: residue wood, 1; industry wood, 2.04; log wood, 5.65 (SWISS CENTRE FOR LIFE CYCLE INVENTORIES, 2013).

Before feeding this material into the gasifier the wood has already been cut into chips within the forest. Then, based on natural air drying, a water content of 30% – originally starting from 50% of the fresh wood – is achieved (SWISS CENTRE FOR LIFE CYCLE INVENTORIES, 2013). Additionally, losses occurring during storage and transport (Seifert, 2010) are taken into consideration.

The gasification is conducted in a Fast Internal Circulating Fluidised Bed gasifier (Hofbauer et al., 2005). The gasification process parameters are taken from Gellert, (2013) (see Table 15.7).

Distribution

Additionally, the distribution (compression, transportation and dispensation) of the hydrogen is taken into consideration calculated according to Wulf and Kaltschmitt (2013) and our own evaluations. For the non-onsite hydrogen production technologies 3.9 kg CO_2-eq have to be added to the emissions of the production of 1 kg H_2 (onsite electrolysis 2.8 kg CO_2-eq/kg H_2).

15.4 Results and discussion

The life cycle calculations for the hydrogen production pathways are carried out based upon the data outlined previously (see Section 15.3). According to the discussed methodology (see Section 15.2.1), the two environmental impact indicators are related to

Table 15.7 **Data for gasification of wood (Gellert, 2013; SWISS CENTRE FOR LIFE CYCLE INVENTORIES, 2013)**

Allocation (spruce)	**Spruce (forest residue)**
Resource	***Residues – Industry – Log, volume; 1:2.04:5.65***
Technical data	
Production capacity	280 kg H_2/h
Duration of life	20 a
Full load hours	7500 h
Operating materials	
Wood (50% water)	2.8 kg/Nm^3
Electricity demand	0.37 kWh/Nm^3
Olivin[a]	0.026 kg/Nm^3
N_2	0.035 kg/Nm^3
Water	0.48 kg/Nm^3
Direct emissions	See text

[a] Bed material for fluidised bed.

one kilometre driven with a passenger vehicle. If heat, in particular sulphur, is produced in addition to hydrogen, it is credited as a by-product.

15.4.1 Results

Figure 15.5 shows the results for the **GHG emissions** for the mobility options discussed. The results are divided into emissions resulting from the vehicle production and its recycling as well as the fuel production, including distribution and tailpipe emissions of the fuel combustion (fuel usage).

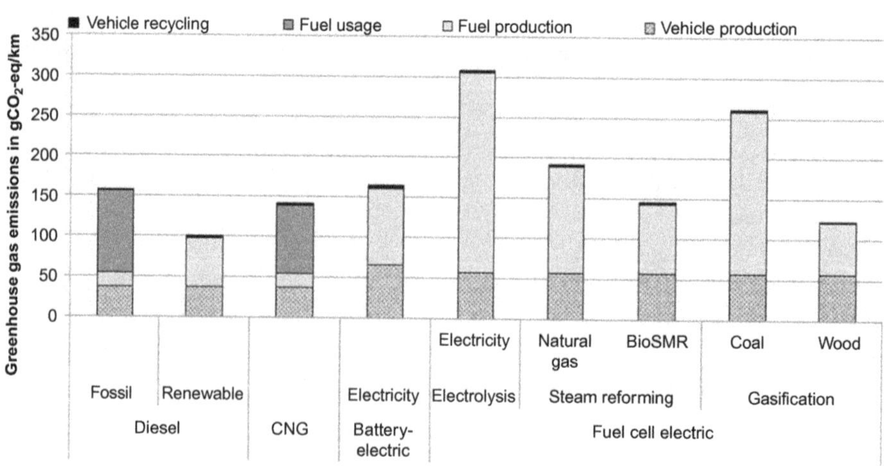

Figure 15.5 GHG emission results of the discussed mobility options.

This figure makes it obvious that the GHG emissions for the production of hydrogen differ significantly among the various mobility options. The lowest GHG emissions can be achieved by using biodiesel provided from rapeseed as a fuel. In the overall results, mobility by biodiesel achieves the lowest values. However, this is only true because the production of an FCEV produces more emissions than the production of a classic diesel vehicle. Looking only at the fuel hydrogen from biomass gasification would be more favourable than biodiesel. A FCEV, and also a BEV, requires components that are very energy intensive to produce (e.g. hydrogen tank from carbon-fibre-reinforced polymer, and the battery). Vehicles operated by fossil diesel and CNG, as well as the BEV and the FCEV driven by hydrogen from biomethane reforming, all show similar results. Due to the energy intensive biomethane production and distribution and the complex construction of the car, the fossil option CNC is comparable to this renewable hydrogen option. The differences between CNG and hydrogen from biomethane, respectively diesel and the BEV, are so small that no ranking can be made here due to the insecurities within the assumptions (e.g. fuel consumption) as well as the data. The highest GHG emissions are induced by alkaline water electrolysis. For this hydrogen production process, the emissions are released during the generation of electricity necessary to operate the electrolyzer. Furthermore, electrolysis uses a secondary energy carrier (i.e. electrical energy) and thus includes one more lossy energy conversion step as compared to other hydrogen provision routes. Altogether, this results in considerable higher efficiency losses. Also coal gasification produces a large amount of GHG emissions due to the high C-content of coal which is released during the gasification process. The recycling of the vehicles has no effect on the overall results.

With respect to a well-to-wheel analysis and EU limit values of 95 gCO_2/km, all three fuels based on biomass (i.e. biodiesel, hydrogen from biomethane reforming and from biomass gasification) could reach this. The natural gas vehicle would reach this value if only tailpipe emissions (tank-to-wheel) are counted, whereas in a well-to-wheel analysis, the emissions would be too high to meet this limit.

The emissions of the hydrogen pathways are very much influenced by the emissions of the hydrogen distribution. Figure 15.6 shows the origin of these emissions for the hydrogen from SMR. The term "Consumables" subsumes all goods that are consumed during the hydrogen production process (for SMR, i.e. water, electricity and natural gas). Thus, the distribution of hydrogen causes a considerable share of the overall GHG emissions. This includes the compression to 500 bar at the production facility (Compression I), the transport with a lorry in special, very heavy pressure tanks over 200 km back and forth, the compression at the HRS from 500 to 800 bar, as well as all processing at the HRS itself (e.g. cooling of the hydrogen to −40 °C).

This amount of released emissions stays the same for all hydrogen production technologies. That is why the "distribution" step counts for over half of the emissions for the woody biomass gasification; see Figure 15.7. Typical for gasification related to reforming is that a significant amount of electricity is additionally needed, whereas the actual wood provision accounts for very little of the overall emissions. Gasification is quite a complex procedure. For example, for the fluidised bed special bed material is needed, which here is just called "sand."

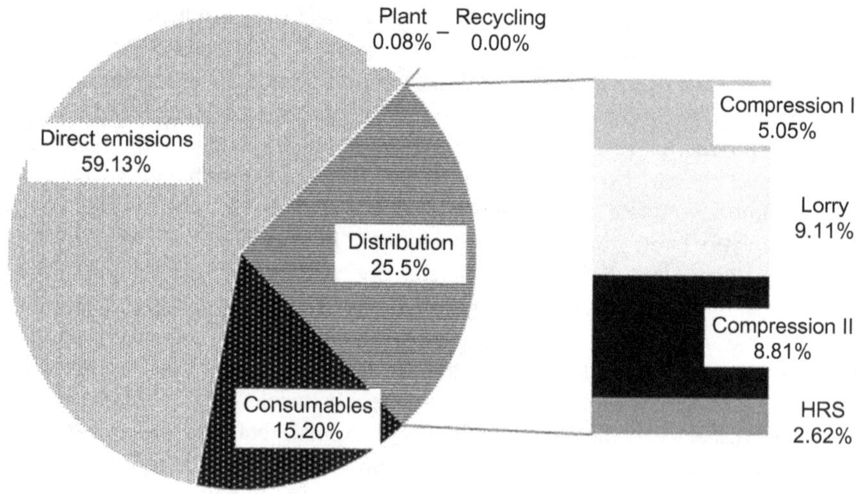

Figure 15.6 Detailed look at the GHG emissions of the hydrogen provision by steam methane reforming.

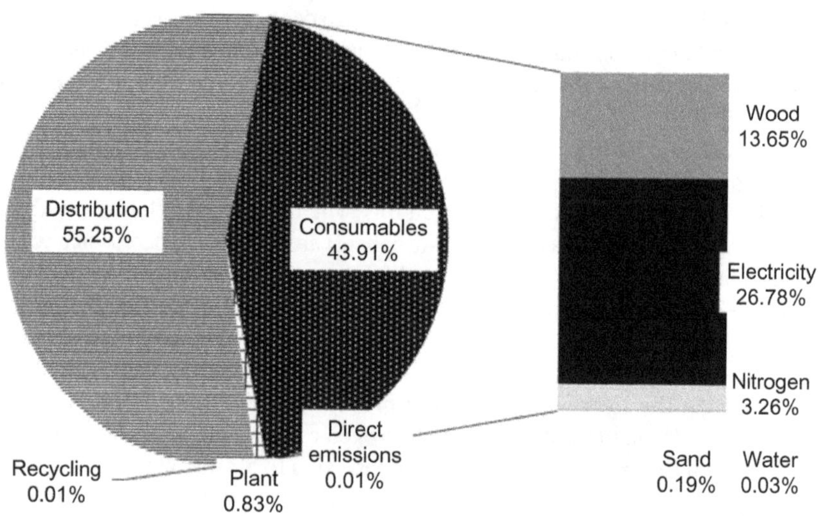

Figure 15.7 Detailed look at the GHG emissions of the hydrogen provision by biomass gasification.

The results for the **acidifying emissions** differ a great deal from the GHG emissions; see Figure 15.8. The options based on agricultural biomass have the most emissions of this type. During the cultivation of the biomass the mineral fertilisers used and digestate release a large amount of ammonia to the air. This means, for example, for the production of hydrogen from SMR that 89.7% of the acidifying emissions are

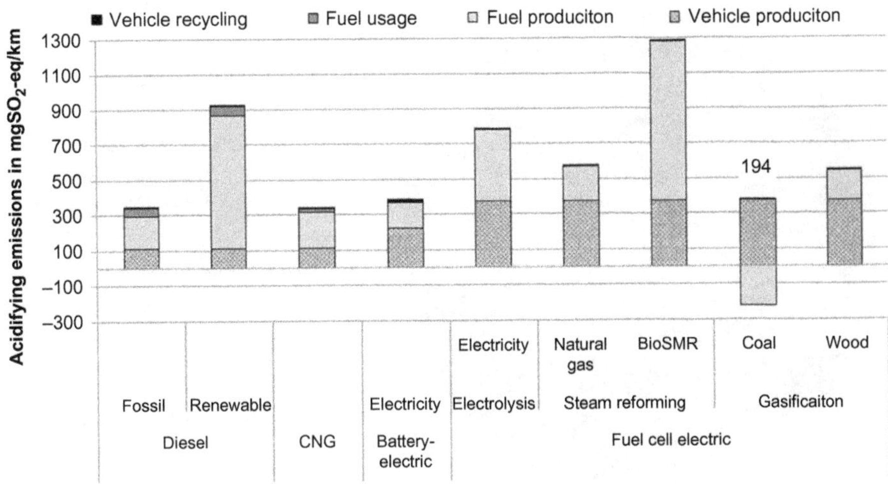

Figure 15.8 Acidifying emission results of the discussed mobility options.

released into the atmosphere during the cultivation of the biomass. The coal gasification discussed here (Section 15.3.2) is equipped with a very good recovery process for sulphur. This means, on the one hand, that only very little direct sulphur dioxide emissions occur. On the other hand, a considerable credit is given for the recovered sulphur. As a result the gasification of coal saves acidifying emissions. Only through the emissions caused by the production of the vehicle are acidifying emissions released. In summary, this accounts for 194 mg SO_2-eq/km. If such a high share of sulphur removal is not assumed, the result might be significantly different.

Another result for the FCEV is the high emission coming from vehicle production. These emissions originate from the mining of platinum needed for the production of fuel cells. As car manufacturers try to reduce the amount of platinum in fuel cells due to the high costs, it is most likely that these emissions will decrease over the next few years. In the last several years, diesel with a low sulphur content (CNG also contains only very little sulphur) was introduced into the market and catalytic converters in the vehicles are improving due to high limit values for oxides of nitrogen. For these reasons the "Fuel usage" emissions for the internal combustion engines are very low. Also, electricity generation, due to the high emission reduction measures at big thermal power plants, causes manageable amounts of acidifying emission.

As hydrogen production from electrolysis is the only option for producing hydrogen onsite directly at the HRS, certain emissions can be spared. Figure 15.9 shows the breakdown of the acidifying emissions into the different categories. As expected, most of the emissions originate from electricity production. Also, the emissions during "distribution" are dominated by the use of electricity. However, this is the only mobility option in which, for the acidifying emissions, the production of the conversion plant (i.e. electrolyzer construction) is noteworthy. These emissions originate from the usage of nickel in the electrolyzer cells.

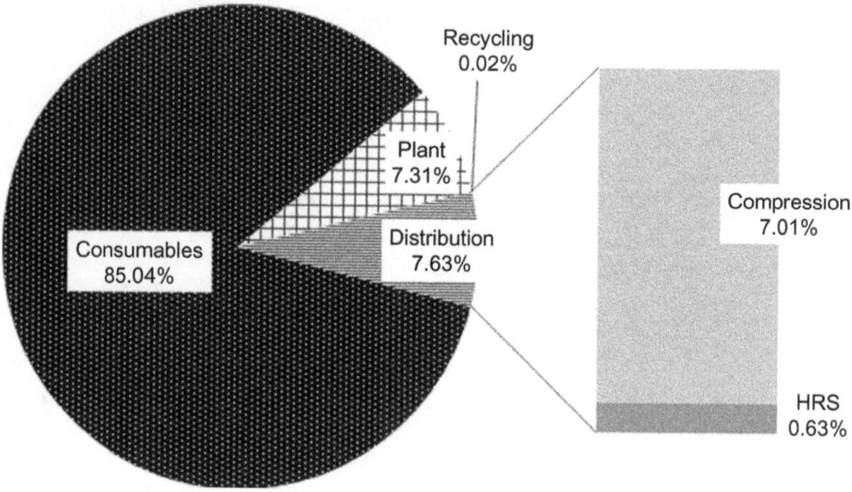

Figure 15.9 Detailed look at the acidifying emissions of the hydrogen provision electrolysis.

15.4.2 Discussion

There are some parameters that influence the results immensely. The **electricity mix** used is such a parameter. The EU-mix causes 460 g CO_2-eq/kWh. This includes Poland, which generates electricity mainly by coal (1010 g CO_2-eq/kWh), as well as Sweden, which uses hydropower as well as nuclear power (50 g CO_2-eq/kWh). However, in the long term, renewable energies will gain most likely a bigger share in the European electricity mix. To illustrate the influence of this parameter for the GHG emissions for driving one kilometre – again including the vehicle – in Figure 15.10 the GHG emissions of the electricity production are varied from an increase of 30% to a decrease to 10% of the start value. This is done for the three hydrogen production pathways,

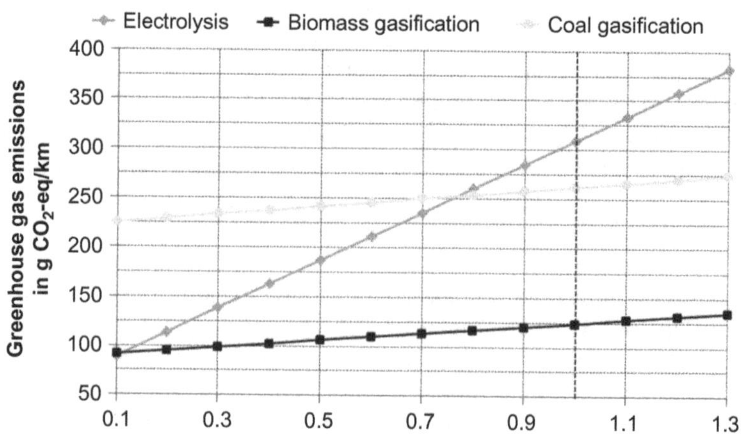

Figure 15.10 Sensitivity analysis of greenhouse gas emissions of the electricity mix on one driven kilometre using fuel cells (100% ≡ 460 gCO_2-eq/kWh electricity).

including the distribution most sensitive to the electricity used. It is demonstrated that a decrease to at least 48% (*ca.* 220 gCO_2-eq/kWh electricity) of the original GHG emissions has to be realised before hydrogen produced by an electrolyzer has the same GHG emissions as a vehicle operated by fossil diesel. To fulfil the requirements of the European Commission (EC) for 2020 with regard to the reduction of GHGs (well-to-wheel), a decrease at least to 38% (*ca.* 170 g CO_2-eq/kWh electricity) is necessary. At the moment only Sweden and France reach such low values for their electricity mix. As a consequence, the share of renewable energy in the European electricity mix has to be increased greatly or renewable energies have to be dedicated to the production of hydrogen from electrolysis.

Another possibility for greatly reducing the emissions of some hydrogen production pathways is **carbon capture and storage** (CCS). There are different technologies for capturing carbon dioxide from gasification or reformation processes and storing it in the underground (e.g. within depleted oil or gas fields). However, such technological solutions are still in a pilot stage. They are also quite expensive and the efficiency of the conversion plants decreases significantly (around 10% points). Due to the resulting high costs it is most likely that this technology will only be applied with the gasification of coal and not the gasification of biomass. Although, it would be possible to generate a carbon dioxide sink for hydrogen from biomass gasification. If carbon dioxide emissions from coal gasification could be decreased by 70% (UBA, 2006), the GHG emissions from the pathway with gasification of coal would be in the same range as the pathway based on the reforming of biomethane. However, in some countries this technology is looked at very skeptically (e.g. Germany and Austria) due to many unknown risks of the carbon storage.

15.5 Final considerations

This chapter assessed how hydrogen mobility can help to reduce GHG and other emissions against the background of minimising the increase of the global mean temperature due to anthropogenic global warming.

The assessment showed the areas where hydrogen for mobility is already environmentally benign regarding GHG and acidifying emissions, and where things have to be improved in order to reach this goal:

- Gasification of wood residues is, under the described conditions, the most environmentally benign hydrogen option. However, wood residues are already used in several energy-providing systems (e.g. small-scale heating). Therefore, it is a limited and valuable resource that cannot be the only solution for hydrogen mobility.
- Conventional ICEs have a large technological lead against FCEVs just because ICE vehicles have been used at a large scale for over 100 years. This is not only expressed by the higher costs (Greimel, 2014) but also by higher acidifying and GHG emissions due to the usage of expensive and materials that are costly to produce. Technical development will help to decrease these emissions of vehicle production.
- Another challenge includes the high GHG emissions due to distribution. With shorter transport distances or the installation of pipelines, a part of these emissions can be avoided.

Also alternative compression technologies, e.g., electrochemical compression, are in development to reduce the energy demand. At the moment, compression consumes around 14% of the energy content of hydrogen.

- A higher share of renewable electricity will reduce the acidifying and GHG emissions for the compression of hydrogen and production by electrolysis.

There are some further remarks regarding this or other LCAs:

- The two environmental impact categories assessed here show very different results regarding which technology is more environmentally benign. That makes it difficult to identify the most promising option. Introducing more impact categories might make it even more difficult.
- Other LCA studies might come to different quantitative results. The reasons for that can be the usage of a different database for LCI data for resources or a different electricity mix. Also other assumptions (e.g. fuel consumption of the vehicles or conversion efficiencies) can lead to altered results. However, the qualitative results stay the same: for example, regarding GHG emissions coal will always produce the highest results and needs CCS to be a discussable option.

References

Belau, T., 2012. Energiepflanzen. Daten für die Planung des Energiepflanzenanbaus, second ed. Kuratorium für Technik und Bauwesen in der Landwirtschaft (KTBL-Datensammlung mit Internetangebot), Darmstadt.

BMU German Federal Ministry for the Environment, Nature Conservation, Building and Nuclear Safety; BDI The Voice of German Industry e.V., 2010. Produktbezogene Klimaschutzstrategien – Product Carbon Footprint verstehen und nutzen. Berlin.

Boyano, A., Blanco-Marigorta, A., Morosuk, T., Tsatsaronis, G., 2011. Exergoenvironmental analysis of a steam methane reforming process for hydrogen production. Energy 36 (4), 2202–2214. http://dx.doi.org/10.1016/j.energy.2010.05.020.

Brinkmann, N., Wang, M., Weber, T., Darlington, T., 2005. Well-to-wheels analysis of advanced fuel/vehicle systems – A North American Study of Energy Use, Greenhouse Gas Emissions, and Criteria Pollutant Emissions.

Daisho, Y., et al., 2011. Final Report of Joint Meeting between the Automobile Evaluation Standards Subcommittee, Energy Efficiency Standards Subcommittee of the Advisory Committee for Natural Resources and Energy and the Automobile Fuel Efficiency Standards Subcommittee, Automobile Section, Land Transport Division of the Council for Transport Policy.

Deutsche Automobil Treuhand GmbH DAT, 2014. Leitfaden über den Kraftstoffverbrauch, die CO_2-Emissionen und den Stromverbrauch aller neuen Personenkraftwagenmodelle, die in Deutschland zum Verkauf angeboten werden. 2. Quartal 2014, Ostfildern.

Doctor, R., Molburg, J., Brockmeier, N., Manfredo, L., Gorokhov, V., Ramezan, M., Stiegel, G., 2001. Life-Cycle Analysis of a Shell Gasification-Based Multi-Product System with CO_2 Recovery. In: The First National Conference on Carbon Sequestration, Washington, 15–17 May, 2001.

Dubbers, E.; Forin, S.; Glensor, K.; Ichiyanagi, Emi and Jörgensen-Ullmann, K., 2012. Reducing automobile CO_2 emissions – can the EU draw lessons from Japan? Berlin.

Edwards, R., Larivé, J.-F., Rickeard, D., Weindorf, W., 2014. Well-to-wheels analysis of future automotive fuels and powertrains in the European context – Well-to-tank report version 4.a. Joint Research Centre (JRC), European Union.

EN ISO 14040:2006, 2006. Environmental management – life cycle assessmentprinciples and framework, German and English version. Beuth Verlag, Berlin.

EN ISO 14044:2006, 2006. Environmental management – life cycle assessment – requirements and guidelines, German and English version. Beuth Verlag, Berlin.

European Parliament and the Council of the European Union, 2014. Amending Regulation (EC) No 443/2009 to define the modalities for reaching the 2020 target to reduce CO_2 emissions from new passenger cars. Regulation (EU) No. 333/2014.

European Commission, 2013. EU Energy in Figures. Publications Office of the European Union (Statistical pocketbook), Luxembourg.

Gellert, S., 2013. Thermochemische Herstellung von Wasserstoff aus Biomasse unter besonderer Berücksichtigung der Rohgasreformierung, Dissertation, Hamburg University of Technology, Institute of Energy systems.

Global Footprint Network, 2012. CO_2-Footprint available from: http://www.footprintnetwork.org/de/index.php/GFN/page/carbon_footprint [Accessed 6 August 2014].

Greimel, H., 2014. Toyota fuel cell car to cost $69,000 in Japan, debut in U.S., Europe in 2015 available from http://www.autonews.com/article/20140625/OEM05/140629935/toyota-fuel-cell-car-to-cost-$69000-in-japan-debut-in-u.s.-europe-in [Accessed 2 September 2014].

Guinée, J., Gorrée, M., Heijungs, R., Huppes, G., Kleijn, R., Koning, A.de, et al., 2001. LCA – An operational guide to the ISO-standards. Part 2b: Operational annex, Universiteit Leiden; Institute of Environmental Sciences, Leiden.

Hofbauer, H.; Fürnsinn, S., Aichernig, C., 2005. Energiezentrale Güssing. Energiezentrale zur Umwandlung von biogenen Roh- und Reststoffen einer Region in Wärme, Strom, BioSNG und flüssige Kraftstoffe, Technische Universität Wien; Repotec, Wien (Energie Systeme der Zukunft).

Hydrogenics (Ed.), 2011. HySTAT® Hydrogen Generators, Oevel.

IPCC, 2007. Climate Change 2007: The Physical Science Basis. In: Contribution of Working Group I to the Fourth Assessment Report of the Intergovernmental Panel on Climate Change. Cambridge Press, New York.

ISO/TS 14067:2013-05, 2013. Greenhouse gases – carbon footprint of products – requirements and guidelines for quantification and communication. Tentative standard. Beuth Verlag, Berlin.

Kaltschmitt, M., Weinberg, J., Stegelmeier, M., Bohnenschäfer, W., Gansler, J., Ebert, M., Eilbericht, B., 2012. Analyse der Erneuerbaren Energien am Wärmemarkt. In: Schultz, R., Hochi, J., Personn, H. (Eds.), Ökologische und ökonomische Optimierung des Wärmemarktes. Unter besonderer Berücksichtigung des Endenergiebedarfs und von Biogas/Bioerdgas. Ibidem-Verlag (Schriftenreihe des Biogasrat e.V., 1), Stuttgart, pp. 109–176.

PAS 2050:2011, 2011. Specification for the assessment of the life cycle greenhouse gas emissions of goods and services. BSI, London.

Pehnt, M., 2002. Ganzheitliche Bilanzierung von Brennstoffzellen in der Energie- und Verkehrstechnik. VDI-Verlag (Forschungsbericht), Düsseldorf.

Roddy, J.D., 2012. Biomass and biofuel production. In: Sayigh, A.A.M. (Eds.), Comprehensive Renewable Energy. Elsevier Science, Amsterdam.

Scheftelowitz, M., Schicketanz, S., Reinicke, F., Beil, M., 2013. Stromerzeugung aus Biomasse. Zwischenbericht, Deutsches Biomasseforschungszentrum (DBFZ).

Seifert, M., 2010. Methodische Vorgehensweise zur Standortidentifikation und Planung der Biomassebereitstellung für Konversionsanlagen am Beispiel von Bio-SNG-Produktionsanlagen, Dissertation, Hamburg University of Technology, Institute of Environmental Technology and Energy Economics.

Spath, P.L., Mann, M.K., 2001. Life Cycle Assessment of Hydrogen Production via Natural Gas Steam Reforming, National Renewable Energy Laboratory NREL.

SWISS CENTRE FOR LIFE CYCLE INVENTORIES, 2013. ecoinvent Version 3.

U.S. Government, 2014. Electronic Code of Federal Regulations Title 49: Transportation §523.2 Definitions.

UBA German Federal Environment Agency, 2006. Technische Abscheidung und Speicherung von CO2 – nur eine Übergangslösung – Mögliche Auswirkungen, Potenziale und Anforderungen.

United Nations, 2009. Copenhagen Accord. Decision 2/CP.15. Part Two: Action taken by the Conference of the Parties at Its Fifteenth Session, 12/7/2009.

United States Environmental Protection Agency, 2012. EPA and NHTSA Set Standards to Reduce Greenhouse Gases and Improve Fuel Economy for Model Years 2017–2025 Cars and Light Trucks.

Weinberg, J., Wulf, C., Kaltschmitt, M., 2013. Ökobilanz der Elektromobile. Mobility 2.0 (3), 42–45.

Wiedmann, T., Minx, J., 2008. A definition of 'carbon footprint'. In: Pertsova, C.C. (Ed.), Ecological Economics Research Trends. Nova Science Publishers, Hauppauge, pp. 1–11.

Wulf, C., Kaltschmitt, M., 2013. Wasserstoff als Kraftstoff im Deutschen Verkehrssektor. Z. Energiewirtsch 37 (2), 127–141. http://dx.doi.org/10.1007/s12398-013-0105-9.

Index

Note: Page numbers followed by *f* indicate figures and *t* indicate tables.

R

S

T

CPSIA information can be obtained
at www.ICGtesting.com
Printed in the USA
LVHW021451191021
700863LV00001B/22

9 781782 423645